Sergei M. Kopeikin (Ed.)

Frontiers in Relativistic Celestial Mechanics

De Gruyter Studies in Mathematical Physics

Volume 22

This Festschrift is dedicated to Professor *Victor A. Brumberg*,
for his enthusiasm and devotion to the science of relativistic celestial mechanics,
and to celebrate his 80th birthday.

List of contributors

Prof. Dr. Ignazio Ciufolini
University of Salento
Department of Engineering for Innovation
Complesso Ecotekne, edificio "Corpo O"
Via per Monteroni
73100 Lecce
Italy
E-mail: ignazio.ciufolini@gmail.com

Prof. Dr. Toshio Fukushima
National Astronomical Observatory of Japan
NAOJ Public Relation Center
2-21-1 Osawa, Mitaka-shi
Tokyo 181-8588
Japan
E-mail: Toshio.Fukushima@nao.ac.jp

Prof. Dr. Vahe G. Gurzadyan
Center for Cosmology and Astrophysics
A. I. Alikhanian National Laboratory
2 Alikhanian Brothers Str.
0036 Yerevan
Armenia
E-mail: gurzadyan@mail.yerphi.am

Prof. Dr. Richard A. Matzner
The University of Texas at Austin
Center for Relativity
Department of Physics
Austin, TX 78712-1081
USA
E-mail: matzner2@physics.utexas.edu

Prof. Dr.-Ing. habil. Jürgen Müller
Leibniz University Hannover
Institute of Geodesy (IfE)
Schneiderberg 50
30167 Hannover
Germany
E-mail: mueller@mbox.ife.uni-hannover.de

Prof. Dr. Antonio Paolozzi
Sapienza University Rome
School of Aerospace Engineering
Dept. of Astronautica, Electrical and Energy Engineering (DIAEE)
Centro Fermi
Rome
Italy
E-mail: antonio.paolozzi@uniroma1.it

Prof. Dr. Erricos C. Pavlis
University of Maryland
Goddard Earth Science and Technology Center
Greenbelt, MD 20771
USA
E-mail: epavlis@umbc.edu

Prof. Sir Roger Penrose
University of Oxford
Mathematical Institute
Andrew Wiles Building
Radcliffe Observatory Quarter
Woodstock Road
Oxford OX2 6GG
United Kingdom
E-mail: Roger.Penrose@maths.ox.ac.uk

Prof. Dr. Pierre Teyssandier
Paris Observatory
SYRTE CNRS/UMR 8630, UPMC
61 avenue de l'Observatoire
F-75014 Paris
France
E-mail: Pierre.Teyssandier@obspm.fr

Prof. Dr. Norbert Wex
Max Planck Institute for Radio Astronomy
Auf dem Hügel 69
D-53121 Bonn
Germany
E-mail: wex@mpifr-bonn.mpg.de

Dr. Pacôme Delva
Paris Observatory
SYRTE CNRS, LNE, UPMC
61 av. de l'Observatoire
F-75014 Paris
France
E-mail: Pacome.Delva@obspm.fr

Dr. Agnes Fienga
Côte d'Azur Observatory
Géoazur-CNRS UMR 7329
250 avenue A. Einstein
06250 Valbonne
France
E-mail: fienga@geoazur.unice.fr

Dr. Rolf König
Helmholtz Centre Potsdam
GFZ German Research Centre for Geosciences
Muenchner Str. 20
82234 Oberpfaffenhofen-Wessling
Germany
E-mail: rolf.koenig@gfz-potsdam.de

Dr.-Ing. habil. Enrico Mai
Leibniz University Hannover
Institute of Geodesy (IfE)
Schneiderberg 50
30167 Hannover
Germany
E-mail: mai@ife.uni-hannover.de

Dr. Claudio Paris
Sapienza University Rome
School of Aerospace Engineering
Dept. of Astronautica, Electrical and Energy
Engineering (DIAEE)
Centro Fermi
Rome
Italy
E-mail: claudio.paris@uniroma1.it

Dr. Gérard Petit
International Bureau of Weights and Measures
(BIPM)
Pavillon de Breteuil
F-92312 Sèvres Cedex
France
E-mail: gpetit@bipm.org

Dr. John C. Ries
The University of Texas at Austin
Center for Space Research
3925 W Braker Ln 200
Austin, TX 78759
USA
E-mail: ries@csr.utexas.edu

Dr. Jean-Louis Simon
Paris Observatory
IMCCE-CNRS UMR 8028
77 Av. Denfert-Rochereau
F-75014 Paris
France
E-mail: simon@imcce.fr

Dr. Giampiero Sindoni
Sapienza University Rome
School of Aerospace Engineering
Dept. of Astronautica, Electrical and Energy
Engineering (DIAEE)
Centro Fermi
Rome
Italy
E-mail: giampiero.sindoni@uniroma1.it

Dr. Peter Wolf
Paris Observatory
SYRTE CNRS, LNE, UPMC
61 av. de l'Observatoire
F-75014 Paris
France
E-mail: Peter.Wolf@obpsm.fr

Dipl.-Ing. Liliane Biskupek
Leibniz University Hannover
Institute of Geodesy (IfE)
Schneiderberg 50
30167 Hannover
Germany
E-mail: biskupek@mbox.ife.uni-hannover.de

Dipl.-Ing. Franz Hofmann
Leibniz University Hannover
Institute of Geodesy (IfE)
Schneiderberg 50
30167 Hannover
Germany
E-mail: hofmann@ife.uni-hannover.de

Contents

List of figures —— xiv

List of tables —— xxiii

Preface —— xxv

Pierre Teyssandier

New tools for determining the light travel time in static, spherically symmetric spacetimes beyond the order G^2 —— 1

1 Introduction —— 1
2 Notations and conventions —— 3
3 Generalities —— 4
4 Specific assumptions on the metric and the light rays —— 6
4.1 Post-Minkowskian expansion of the metric —— 6
4.2 Time transfer function for a quasi-Minkowskian light ray —— 7
5 Fundamental properties of functions $\mathcal{T}^{(n)}$ —— 9
5.1 Recurrence relation satisfied by functions $\mathcal{T}^{(n)}$ —— 9
5.2 Analyticity of the functions $\mathcal{T}^{(n)}$ —— 11
6 First procedure: determination of the $\mathcal{T}^{(n)}$'s from the recurrence relation for $n = 1, 2, 3$ —— 12
7 Second procedure: determination of the $\mathcal{T}^{(n)}$'s from the geodesic equations —— 18
7.1 Null geodesic equations —— 18
7.2 Post-Minkowskian expansion of the impact parameter —— 19
7.3 Implementation of the method —— 20
8 Simplification of the second procedure —— 22
8.1 Use of a constraint equation —— 23
8.2 Explicit calculation of $\mathcal{T}^{(1)}$, $\mathcal{T}^{(2)}$, and $\mathcal{T}^{(3)}$ —— 26
9 Direction of light propagation up to order G^3 —— 27
10 Light ray emitted at infinity —— 30
11 Enhanced terms in $\mathcal{T}^{(1)}$, $\mathcal{T}^{(2)}$, and $\mathcal{T}^{(3)}$ —— 32
12 Application to some Solar System experiments —— 34
13 Concluding remarks —— 35
References —— 36

Norbert Wex

Testing relativistic gravity with radio pulsars —— 39

1 Introduction —— 39
1.1 Radio pulsars and pulsar timing —— 41

1.2 Binary pulsar motion in gravity theories — 43
1.3 Gravitational spin effects in binary pulsars — 45
1.4 Phenomenological approach to relativistic effects in binary pulsar observations — 47
2 Gravitational wave damping — 51
2.1 The Hulse–Taylor pulsar — 51
2.2 The Double Pulsar – The best test for Einstein's quadrupole formula, and more — 55
2.3 PSR J1738+0333 – The best test for scalar–tensor gravity — 59
2.4 PSR J0348+0432 – A massive pulsar in a relativistic orbit — 63
2.5 Implications for gravitational wave astronomy — 67
3 Geodetic precession — 69
3.1 PSR B1534+12 — 70
3.2 The Double Pulsar — 71
4 The strong equivalence principle — 73
4.1 The Damour–Schäfer test — 75
4.2 Direct tests — 77
5 Local Lorentz invariance of gravity — 78
5.1 Constraints on $\hat{\alpha}_1$ from binary pulsars — 79
5.2 Constraints on $\hat{\alpha}_2$ from binary and solitary pulsars — 80
5.3 Constraints on $\hat{\alpha}_3$ from binary pulsars — 83
6 Local position invariance of gravity — 84
7 A varying gravitational constant — 85
8 Summary and outlook — 89
References — 92

Jürgen Müller, Liliane Biskupek, Franz Hofmann, and Enrico Mai
Lunar laser ranging and relativity — 103
1 Introduction — 103
2 Model — 106
2.1 Overview — 107
2.2 Ephemerides — 109
3 Analysis — 137
3.1 Software package LUNAR — 137
3.2 Newtonian parameters — 139
4 Results for relativistic parameters — 140
4.1 Gravitational constant — 140
4.2 Equivalence principle — 141
4.3 Yukawa term — 144
4.4 Geodetic precession — 146
4.5 Metric parameter β — 147

4.6 Preferred-frame parameters α_1, α_2 — 149
5 Summary and outlook — 150
References — 152

Ignazio Ciufolini, Antonio Paolozzi, Vahe Gurzadyan, Erricos C. Pavlis, Rolf König, John Ries, Richard Matzner, Roger Penrose, Giampiero Sindoni, and Claudio Paris
Dragging of inertial frames, fundamental physics, and satellite laser ranging — 157
1 Introduction — 157
2 Dragging of inertial frames — 159
3 Tests of string theory and the LAGEOS and LARES space experiments — 162
4 LAGEOS and Gravity Probe B: two independent space experiments measuring frame dragging — 164
5 The LARES mission — 169
5.1 The LARES satellite — 170
5.2 The LARES satellite and geodesic motion — 174
5.3 Preliminary orbital analyses — 176
5.4 Error analyses — 178
5.5 Monte Carlo simulations — 180
6 Conclusions — 181
References — 182

Toshio Fukushima
Elliptic functions and elliptic integrals for celestial mechanics and dynamical astronomy — 187
1 Introduction — 187
2 Notations — 188
2.1 Glossary — 188
2.2 First input argument: φ, u, and x — 190
2.3 Second input argument: k, m, and α — 190
2.4 Sign of n — 191
2.5 Third input argument: n and α — 191
2.6 Ordering of arguments — 191
2.7 Omission of parameters — 192
3 Elliptic functions — 192
3.1 General elliptic function — 192
3.2 Weierstrass elliptic function — 193
3.3 Jacobian elliptic functions — 194
3.4 Jacobi's amplitude function — 195
3.5 Differential equations of Jacobian elliptic functions — 197
3.6 Addition theorem of Jacobian elliptic functions — 197

3.7 Jacobi's form of incomplete elliptic integrals — 199
3.8 Addition theorem of incomplete elliptic integrals — 201
3.9 Jacobi's original form of incomplete elliptic integral of the third kind — 202
4 Elliptic integrals — 202
4.1 General elliptic integral — 202
4.2 Legendre's form of incomplete elliptic integrals — 203
4.3 Associate incomplete elliptic integrals — 205
4.4 Complete elliptic integrals — 207
4.5 Generalized elliptic integrals — 209
4.6 Symmetric elliptic integrals — 211
5 Numerical computation of elliptic functions and elliptic integrals — 212
5.1 Overview — 212
5.2 Transformation method — 213
5.3 Example of transformation method — 214
5.4 Simultaneous computation of Jacobian elliptic functions — 214
5.5 Better computation of Jacobian elliptic functions — 215
5.6 Computation of Jacobi's form of incomplete elliptic integrals — 216
5.7 Computation of Legendre's form of incomplete elliptic integral of the first kind — 217
5.8 Computation of other incomplete elliptic integrals — 218
5.9 Computation of complete elliptic integrals other than $\Pi(n|m)$ and $J(n|m)$ — 218
5.10 Computation of complete elliptic integrals of the third kind — 220
5.11 CPU time comparison — 221
5.12 Software — 223
6 Conclusion — 223
References — 223

Jean-Louis Simon and Agnes Fienga
Victor Brumberg and the French school of analytical celestial mechanics — 227
1 Introduction — 227
2 Analytical formulism for planetary perturbations — 227
2.1 Development of the perturbative function — 228
2.2 Calculation of the Hansen coefficients — 230
3 General planetary theory (GPT) — 231
3.1 Introduction — 231
3.2 General theory by V. Brumberg — 231
4 Planetary theories with the aid of the expansions of elliptic functions — 234
4.1 Notations — 234

4.2 Expansion of the right-hand members of the equations: A change of the time variable — 235
4.3 Application to planetary problems — 237
5 Reference frames, time scales, and units for planetary ephemerides — 241
5.1 Victor Brumberg's contributions — 241
5.2 Planetary ephemerides — 244
5.3 Conclusions — 246
References — 247

Gérard Petit, Peter Wolf, and Pacôme Delva
Atomic time, clocks, and clock comparisons in relativistic spacetime: a review — 249
1 Introduction — 249
2 Atomic time and atomic clocks — 251
2.1 Atomic time scales — 252
2.2 Atomic clocks — 253
3 Relativistic framework for time and frequency comparisons — 254
3.1 Introduction — 254
3.2 Simultaneity and synchronization — 255
3.3 Relativistic definitions of spacetime coordinate systems — 256
3.4 Time scales in the barycentric and geocentric systems — 257
3.5 Relativistic theory for time transformations in the Solar System (BCRS) — 258
4 Relativistic treatment for time and frequency comparisons in the vicinity of the Earth (GCRS) — 260
4.1 One-way time transfer — 262
4.2 Two-way time transfer using artificial satellites — 263
4.3 Frequency comparisons — 264
5 Time and frequency transfer techniques — 267
5.1 Established time and frequency transfer techniques: GNSS and two-way time transfer — 267
5.2 Some novel two-way techniques — 269
6 Clocks in relativistic geodesy — 275
6.1 Review of chronometric geodesy — 275
6.2 The chronometric geoid — 277
7 Conclusions and prospects — 278
References — 279

Index — 285

List of figures

Contribution 2 (Norbert Wex):
Fig. 1 Illustration of the different gravity regimes used in this review. —— **41**
Fig. 2 The P–$\dot{P}$ diagram for radio pulsars. Binary pulsars are indicated by a red circle. Pulsars that play a particular role in this review are marked with a green dot and have their name as a label. The data are taken from the *ATNF Pulsar Catalogue* [22]. —— **42**
Fig. 3 Spacetime diagram illustration of pulsar timing. Pulsar timing connects the proper time of emission $\tau_{\rm psr}$, defined by the pulsar's intrinsic rotation, and the proper time of the observer on Earth $\tau_{\rm obs}$, measured by the atomic clock at the location of the radio telescope. The timing model, which expresses $\tau_{\rm obs}$ as a function of $\tau_{\rm psr}$, accounts for various "relativistic effects" associated with the metric properties of the spacetime, i.e. the world line of the pulsar and the null-geodesic of the radio signal. In addition, it contains a number of terms related to the Earth motion and relativistic corrections in the Solar System, like time dilation and signal propagation delays (see [27] for details). —— **42**
Fig. 4 Mass–mass diagram for PSR B1913+16 based on GR and the three observed post-Keplerian parameters $\dot{\omega}$ (black), γ (red), and $\dot{P}_b$ (blue). The dashed $\dot{P}_b$ curve is based on the observed $\dot{P}_b$, without corrections for Galactic and Shklovskii effects. The solid $\dot{P}_b$ curve is based on the corrected (intrinsic) $\dot{P}_b$, where the thin lines indicate the one-sigma boundaries. Values are taken from Table 2. —— **53**
Fig. 5 Shift in the time of periastron passage of PSR B1913+16 due to gravitational wave damping. The parabola represents the GR prediction and the data points the timing measurements, with (vertical) error bars mostly too small to be resolved. The observed shift in periastron time is a direct measurement of the change in the world line of the pulsar due to the back reaction of the emitted gravitational waves (cf. Figure 3). The corresponding spatial shift amounts to about 20 000 km. Figure is taken from [31]. —— **54**
Fig. 6 Short-orbital-period (P_b < 1 day) binary pulsars used for gravity tests. The velocity V_b (divided by the speed of light c) is a direct measure for the strength of post-Newtonian effects in the orbital dynamics. The gravitational wave luminosity $L_{\rm GW}$ is an indicator for the strength of radiative effects that cause secular changes to the orbital elements due to gravitational wave damping. —— **56**

Fig. 7 GR mass–mass diagram based on timing observations of the Double Pulsar. The orange areas are excluded simply by the fact that $\sin i \leq 1$. The figure is taken from [75] (Ω_{SO} lines removed) and based on the timing solution published in [76]. — **58**

Fig. 8 Optical finding chart for the PSR J1738+0333 companion. Indicated are the white dwarf companion (WD), the slit orientation used during the observation and the comparison star (C) that was included in the slit. The white dwarf is sufficiently bright to allow for high signal-to-noise spectroscopy (see [84] for details, where this figure is taken from). — **60**

Fig. 9 GR mass–mass diagram based on the timing observations of PSR J1738+0333 and the optical observations of its white-dwarf companion, respectively. The thin lines indicate the one-sigma errors of the measured parameters. The gray area is excluded by the condition $\sin i \leq 1$. — **61**

Fig. 10 Constraints on the class of $T_1(\alpha_0, \beta_0)$ scalar–tensor theories of [57, 58], from different binary pulsar and Solar System (Cassini and Lunar Laser Ranging) experiments. The gray area indicates the still allowed T_1 theories, and includes GR ($\alpha_0 = \beta_0 = 0$). It is obvious that PSR J1738+0333 is the most constraining experiment for most of the β_0 range, and is even competitive with Cassini in testing the Jordan–Fierz–Brans–Dicke theory ($\beta_0 = 0$). As can be clearly seen, the double neutron-star systems PSR B1534+12 [89], PSR B1913+16 (Hulse–Taylor pulsar) and PSR J0737-3039A/B (Double Pulsar) are considerably less constraining, as explained in the text. PSR J1141-6545 is also well suited for a dipolar radiation test [90], since it also has a white dwarf companion [91]. Figure is taken from [30]. — **62**

Fig. 11 Spectroscopically measured radial velocities for the white-dwarf companion of PSR J0348+0432. For illustration purposes the data are plotted twice. The fitted sinusoidal curve (blue) has an amplitude of $351 \pm 4\,\mathrm{km/s}$. As a comparison, the sinusoidal green line shows the radial velocity of the pulsar as derived from the timing solution. The amplitude of the green line is known with very high precision: $30.008235 \pm 0.000016\,\mathrm{km/s}$. The ratio of the amplitudes gives the mass ratio R. Figure is taken from [94]. — **63**

Fig. 12 GR mass–mass diagram based on timing and optical observations of the PSR J0348+0432 system. The thin lines indicate the one-sigma errors of the measured parameters. The gray area is excluded by the condition $\sin i \leq 1$. — **64**

Fig. 13 Fractional gravitational binding energy of a neutron star as a function of its (inertial) mass, based on equation of state MPA1 [85]. The plot clearly shows the prominent position of PSR J0348+0432. The other dots indicate the neutron star masses of the individual test systems in Figure 10. — **65**

Fig. 14 Effective scalar coupling as a function of the neutron-star mass, in the $T_1(\alpha_0, \beta_0)$ mono-scalar–tensor gravity theory of [57, 58]. For the linear coupling of matter to the scalar field we have chosen $\alpha_0 = 10^{-4}$, a value well below the sensitivity of any near-future Solar System experiment, like GAIA [95]. The blue curves correspond to stable neutron-star configurations for different values of the quadratic coupling β_0: -5 to -4 (top to bottom) in steps of 0.1. The yellow area indicates the parameter space still allowed by the limit (2.7) [label "J1738"], whereas only the green area is in agreement with the limit (2.9) [label "J0348"]. The plot shows clearly how the massive pulsar PSR J0348+0432 probes deep into a new gravity regime. Neutron-star calculations are based on equation of state MPA1 [85] (see [94] for a different equation of state). — **66**

Fig. 15 Mass–mass diagram based on timing and optical observations of the PSR J0348+0432 system, for the mono-scalar–tensor gravity $T_1(10^{-4}, -4.5)$. The thin lines indicate the one-sigma errors of the measured parameters. The vertical gray line is at the maximum mass of a neutron star for the given theory and equation of state (MPA1). The gray area is excluded by the condition $\sin i \leq 1$. Obviously $T_1(10^{-4}, -4.5)$ is clearly falsified by this test, as there is no common region for the curves of the three parameters m_c, R and $\dot{P}_b$. — **67**

Fig. 16 Average eclipse profile of pulsar A observed at 820 MHz over a 5-day period around 11 April 2007 (black line). The model based on a co-rotating magnetosphere gives a good explanation of the eclipse profile (red dashed line). Figure is taken from [112]. — **71**

Fig. 17 GR mass–mass diagram for the Double Pulsar. Same as in Section 2.2 (Figure 7), plus the inclusion of the constraints from the geodetic precession of pulsar B (Ω_{SO}). Figure is taken from [75]. — **72**

Fig. 18 Time evolution of the orbital eccentricity vector $\mathbf{e} = \mathbf{e}_{\mathrm{PN}} + \mathbf{e}_{\Delta}$ for a small-eccentricity binary, in the presence of a SEP violation. The vector $\mathbf{g}_{\perp}$ represents the projection of the external acceleration in the orbital plane. — **76**

Fig. 19 Comparison of two pulse profiles of PSR B1937+21 obtained at two different epochs. The blue one was obtained on September 2, 1997, while the red one was obtained on June 6, 2009. The main peak is aligned and scaled to have the same intensity. There exists no visible difference within the noise level. Profiles were taken from [134]. — **82**

Fig. 20 Sensitivity s_A for Jordan–Fierz–Brans–Dicke theory, i.e. $T_1(\alpha_0, 0)$, and for two different equations of state (red: MPA1 [85], blue: AP4 [144]). For each equation of state four lines have been calculated, corresponding to $|\alpha_0| = 0.5, 0.2, 0.1, 0.01$ as the maximum mass decreases. For $|\alpha_0| < 0.01$, s_A is practically independent of α_0. — **86**

Fig. 21 A joint $\dot{G}$–κ_D test based on PSRs J1738+0333 and PSR J0437-4715. The inner contour includes 68.3% and the outer contour 95.4% of all probability. GR ($\dot{G} = \kappa_D = 0$) is well within the inner contour and close to the peak of probability density. The vertical gray band includes regions consistent with the one-sigma constraints for $\dot{G}/G$ from LLR (equation (7.3)). Generally only the upper half of the diagram has physical meaning, as the radiation of dipolar gravitational waves is expected to make the system lose orbital energy. Figure is taken from [30]. — **88**

Fig. 22 Enhancement of $\dot{G}$ in a pulsar-white dwarf system as a function of the pulsar mass m_p. Figure shows the ratio $(\dot{\mathcal{G}}/\mathcal{G})$ / $(\dot{G}/G)$ as given in equations (7.9) and (7.10) for $T_1(10^{-4}, -4.3)$ gravity, a theory which still passes the PSR J0348+0432 test (see Figure 14). The gray vertical lines indicate the mass range for PSR J0437-4715 (mean and one-sigma uncertainties) [29]. PSR J1614-2230, with its mass of $1.97 \pm 0.04\, M_\odot$ and orbital period of 8.7 days [77], seems to be a promising future test for $\dot{G}$, once tight constraints on the intrinsic $\dot{P}_b$ can be derived from the timing observations. — **89**

Contribution 3 (Jürgen Müller, Liliane Biskupek, Franz Hofmann, and Enrico Mai):

Fig. 1 Distribution of LLR observatories on the Moon (upper panel) and reflector arrays on Earth (lower panel). — **104**

Fig. 2 Distribution of 20 061 LLR normal points taken by the major observatories over the years. — **105**

Fig. 3 Measurement statistics of observatories (left) and reflectors (right). — **106**

Fig. 4 LLR measurement setup with basis vectors. — **108**

Fig. 5 Construction steps for an ephemeris. — **110**

Fig. 6 Weighted annual residuals of the LLR analysis of 20 061 NP from 1970 to 2013. — **138**

Fig. 7 Power spectrum of the difference in the Earth–Moon distance with and without a $\dot{G}$ perturbation with $\dot{G}/G = 1.5 \times 10^{-13}\ \mathrm{yr}^{-1}$. Prominent periods and their frequencies are labeled. Here, “syn” denotes the synodic frequency and “anomal” the anomalistic frequency. — **141**

Fig. 8 Power spectrum of the difference in the Earth–Moon distance with and without a perturbation due to a violation of the equivalence principle with $\Delta(m_g/m_i)_{\mathrm{EM}} = 2 \times 10^{-13}$. See Figure 7 for labeling. — 143

Fig. 9 Power spectrum of the difference in the Earth–Moon distance with and without a perturbation due to an additional Yukawa term with $\alpha = 1.5 \times 10^{-11}$. See Figure 7 for labeling. — 145

Fig. 10 Power spectrum of the difference in the Earth–Moon distance with and without a perturbation due to an additional geodetic precession with $h = 0.003$. See Figure 7 for labeling. — 147

Fig. 11 Power spectrum of the difference in the Earth–Moon distance with and without a perturbation due to a nonunity value for the β PPN parameter with $\beta = 1 + 3 \times 10^{-4}$. See Figure 7 for labeling. — 148

Fig. 12 Power spectrum of the difference in the Earth–Moon distance with and without a perturbation due to a non-zero value for the α_1 preferred-frame parameter with $\alpha_1 = 4 \times 10^{-5}$. "Ann" denotes the annual frequency, for further labels see Figure 7. — 149

Fig. 13 Power spectrum of the difference in the Earth–Moon distance with and without a perturbation due to a non-zero value for the α_2 preferred-frame parameter with $\alpha_2 = 3 \times 10^{-5}$. See Figure 7 for labeling. — 150

Contribution 4 (Ignazio Ciufolini, Antonio Paolozzi, Vahe Gurzadyan, Erricos C. Pavlis, Rolf König, John Ries, Richard Matzner, Roger Penrose, Giampiero Sindoni, and Claudio Paris):

Fig. 1 The gravitomagnetic field [27] $\boldsymbol{H}$, in weak field and slow motion, generated in general relativity by the angular momentum $\boldsymbol{J}$ of a central body looks similar to the magnetic field $\boldsymbol{B}$ generated by a magnetic moment $\boldsymbol{m}$ in electrodynamics. Frame dragging, $\dot{\boldsymbol{\Omega}}$, is represented on a test gyroscope $\boldsymbol{S}$. — 161

Fig. 2 The dependence of the Chern–Simons constant m_{CS} on the frame-dragging measurement parameter n as expected by the LARES mission [48]. — 164

Fig. 3 The GFZ EIGEN-6C Geoid Model from 2011. EIGEN-6C (European Improved Gravity model of the Earth by New techniques – model 6C) is a high-resolution global gravity field model. It is the first combination model that includes GOCE data. In this figure the geoid is plotted with respect to the Earth's reference ellipsoid to remove the dominant Earth's quadrupole component. Its role is fundamental in geodesy and Earth sciences and ranges from practical purposes, like orbit determination, to scientific applications, like the investigation of the density structure of the Earth's interior. The recent EIGEN released in 2012 is called EIGEN-6C2 and has been created from a combination of a multitude of data (courtesy of GFZ-Potsdam). — **166**

Fig. 4 Measurement of frame dragging obtained in 2004 analyzing the LAGEOS and LAGEOS 2 orbits with GEODYN, using the 2004 GRACE determination of the Earth's gravitational field [71, 77]. The top figure (in black) shows the cumulative residual of the nodal longitudes, $\delta\Omega$, of the LAGEOS satellites combined to eliminate the uncertainty in the quadrupole moment of Earth. The best fit line (not reported here for sake of clarity) through this observed residual has a slope of 47.9 milliarcsec per year. The bottom figure (in red) shows the general relativistic theoretical predictions of the Lense–Thirring effect for the combination of the nodal longitudes of the LAGEOS satellites, i.e. a line with a slope of 48.2 milliarcsec per year (figure taken from [71]). — **167**

Fig. 5 Measurement of frame dragging with the LAGEOS and LAGEOS 2 satellites obtained independently in 2008 by the CSR team of the University of Texas at Austin using UTOPIA and the GRACE models: EIGEN-GRACE02S, GGM02S, EIGEN-CG03C, GIF22a, JEM04G, EIGEN-GL04C, JEM01-RL03B, GGM03S, ITG-GRACE03S and EIGEN-GL05C. The average value of frame dragging measured by Ries et al. using these models is 0.99, where 1 is the value predicted by general relativity. The total error budget of the CSR-UT team is about 12% of the frame dragging effect. Adapted from [74, 75]. — **168**

Fig. 6 Measurement of frame dragging with the LAGEOS and LAGEOS 2 satellites obtained independently in 2012 by the German GFZ team of Potsdam–Munich using EPOS-OC and the GRACE models: EIGEN-6C, EIGEN-6C (without considering the annual and semiannual variations in the Earth's gravitational field), EIGEN-6S, EIGEN-51C, and EIGEN-GRACE03S. The average value of frame dragging measured by Koenig et al. using these models is 0.95, where 1 is the value predicted by general relativity. Adapted from [76]. — **168**

Fig. 7 Concept of the LARES space experiment, also shown in the figure are the LAGEOS, LAGEOS 2 and GRACE satellites that play a key role in the experiment. The spacetime distortion due to frame-dragging induced by the Earth's rotation is represented by the distortion of the radial curves. The Earth's colors represent the EIGEN-GRACE02S anomalies of the Earth's gravitational field obtained with GRACE. — **170**
Fig. 8 LARES semifinished sphere mounted on the lathe machine. — **171**
Fig. 9 Micrograph of LARES material. The material structure consists of tungsten particles surrounded by a Ni–Cu phase (picture taken from [85]). — **172**
Fig. 10 Cross section of the LAGEOS satellites, showing the internal structure and in particular the tension stud to assemble and constrains the three pieces together. — **172**
Fig. 11 LARES demonstration model on the integration stand. — **173**
Fig. 12 (left) LARES satellite mounted on the separation and support systems. (center) LARES system on the upper stage, (right) VEGA payload compartment (fairing). — **174**
Fig. 13 Deviations in meters, from a spacetime geodesic followed by an ideal "test particle," of the orbit of a satellite perturbed by an average along-track acceleration with magnitude respectively equal to the unmodeled along-track accelerations observed on the LARES (blue curve), LAGEOS (green curve) and STARLETTE (red curve) satellites. The vertical axis may be thought of to represent the spacetime geodesic in a reference frame co-moving with the test particle. The blue, green, and red curves were obtained using an average along-track acceleration with magnitude, respectively, equal to: $0.4 \times 10^{-12}\ \mathrm{m/s^2}$, $1 \times 10^{-12}\ \mathrm{m/s^2}$, and $40 \times 10^{-12}\ \mathrm{m/s^2}$, i.e. the unmodeled residual along-track acceleration observed on each of the three satellites (adapted from [78]). — **177**
Fig. 14 Percentage error in the measurement of frame dragging by the three satellites LARES, LAGEOS, and LAGEOS 2 corresponding to each even zonal harmonic using the published uncertainties of the EIGEN-GRACE02S model (left panel) and using, as uncertainties, the difference between the coefficients of the models EIGEN-GRACE02S and GGM02S (right panel). From [50] — **179**

Contribution 5 (Toshio Fukushima):
Fig. 1 Sketch of Jacobian elliptic functions: local. Illustrated are the graphs of three principal Jacobian elliptic functions, $\mathrm{sn}(u|m)$, $\mathrm{cn}(u|m)$, and $\mathrm{dn}(u|m)$ for u in its standard domain $0 \le u \le K(m)$ for various values of m as $m = 0.0, 0.1, \ldots, 0.9$. — **194**
Fig. 2 Sketch of Jacobian elliptic functions: global. Same as Figure 1 but for a longer period. — **195**

Fig. 3 Sketch of Jacobi's amplitude function: local. Same as Figure 1 but for $\mathrm{am}(u|m)$. — 196
Fig. 4 Sketch of Jacobi's Epsilon function: local. Same as Figure 1 but for $E_u(u|m)$. — 200
Fig. 5 Sketch of Jacobi's form of incomplete elliptic integral of the third kind: local. Same as Figure 1 but for $\Pi_u(u, n|m)$ when $n = 0.5$. Added is the straight line u plotted by a broken line for reference. — 200
Fig. 6 Sketch of Legendre's form of incomplete elliptic integrals: local. Plotted are the graphs of $F(\varphi|m)$ and $E(\varphi|m)$ as functions of φ in its standard domain, $0 \leq \varphi \leq \pi/2$ for various values of m as $m = 0.1, 0.2, \dots, 0.9$. Notice inequalities $E(\varphi|0.9) < E(\varphi|0.8) < \cdots < F(\varphi|0.8) < F(\varphi|0.9)$. The results when $m = 0$ become the same straight line, $F(\varphi|0) = E(\varphi|0) = \varphi$, which is shown by a broken line. — 204
Fig. 7 Sketch of Legendre's form of incomplete elliptic integrals: global. Same as Figure 6 but for a longer period. — 204
Fig. 8 Sketch of associate incomplete elliptic integral of the third kind: local. Same as Figure 6 but for $J(\varphi, n|m)$ when $n = 0.5$. For reference, a straight line φ is shown by a broken line again. — 206
Fig. 9 Sketch of associate incomplete elliptic integral of the third kind: global. Same as Figure 8 but for a longer period. — 206
Fig. 10 Sketch of five complete elliptic integrals. Illustrated are the graphs of five complete elliptic integrals, $K(m), E(m), B(m), D(m)$, and $S(m)$ for m in its standard domain $0 \leq m < 1$. — 208
Fig. 11 Sketch of $J(n|m)$. Same as Figure 10 but of $J(n|m)$ for various values of n as $n = 0.0, 0.1, \dots, 0.9$. Notice that $J(n|m)$ is monotonically increasing with respect to n when m is fixed. — 208

Contribution 6 (Jean-Louis Simon and Agnes Fienga):
Fig. 1 Planets P and P' orbiting the Sun. — 228
Fig. 2 Barycentric and geocentric reference systems with B stands for barycentric, G for geocentric, V for VLBI, C ecliptical, Q equatorial, D dynamical, K kinematical, and $^+$ rotating. Extracted from [11]. $\vec{X}$, $\vec{X}_C$, and $\vec{X}_Q$ are the barycentric coordinate systems — 241

Contribution 7 (Gérard Petit, Peter Wolf, and Pacôme Delva):
Fig. 1 A photon of frequency ν_A is emitted at point A toward point B, where the measured frequency is ν_B. (a) A and B are two points at rest in an accelerated frame, with acceleration $\vec{a}$ in the same direction as the emitted photon. (b) A and B are at rest in a nonaccelerated (locally inertial) frame in the presence of a gravitational field such that $\vec{g} = -\vec{a}$. — 265

Fig. 2 Two clocks A and B are measuring proper time along their trajectory. One signal with phase S is emitted by A at proper time τ_A, and another one with phase $S + dS$ at time $\tau_A + d\tau_A$. They are received by clock B, respectively, at times τ_B and $\tau + d\tau_B$. — **266**
Fig. 3 Schematic of the T2L2 principle of operation (see the text for details). — **270**
Fig. 4 Time stability of T2L2 ground-space clock comparisons for different ground stations, as a function of averaging time. — **271**
Fig. 5 Principle of operation of the ACES MWL. See the text for details. — **272**
Fig. 6 Performance objective of the ACES clocks (SHM = Space Hydrogen Maser, PHARAO = Cold Atom Cs clock) and the MWL (MicroWave Link) in terms of time stability as a function of averaging time. — **273**

List of tables

Contribution 1 (Pierre Teyssandier):

Table 1 Numerical values in ps of the main stationary contributions to the light travel time in the Solar System for various values of $r_c/R_\odot$. In each case, $r_A = 50$ au and $r_B = 1$ au. The parameters γ and κ are taken as $\gamma = 1$ and $\kappa = 15/4$, respectively. For the numerical estimates of $|\mathcal{T}_S^{(1)}|$ and $\mathcal{T}_{J_2}^{(1)}$, the light ray is assumed to propagate in the equatorial plane of the Sun. The dynamical effects due to the planetary perturbations are not taken into account. —— **34**

Contribution 2 (Norbert Wex):

Table 1 Examples of precision measurements using pulsar timing. A number in bracket indicates the (one-sigma) uncertainty in the last digit of each value. The symbol $M_\odot$ stands for the Solar mass. (cf. Table 1 in [28]). —— **43**

Table 2 Observed orbital timing parameters of PSR B1913+16, based on the Damour–Deruelle timing model (taken from [31]). Figures in parentheses represent estimated uncertainties in the last quoted digit. —— **52**

Table 3 A selection of observed orbital timing parameters of the Double Pulsar, based on the Damour–Deruelle timing model (taken from [76]). All post-Keplerian parameters below are obtained from the timing of pulsar A. The timing precision for pulsar B is considerably lower, and allows only for a, in comparison, low precision measurement ($\sim 0.3\%$) of $\dot{\omega}$ [76]. Figures in parentheses represent estimated uncertainties in the last quoted digit. —— **57**

Table 4 Contributions to the accumulated number of gravitational wave cycles in the frequency band of 20 Hz to 1350 Hz for a $2M_\odot/1.25M_\odot$ neutron-star pair (cf. equations (235) and (236) in [36]). The frequency of 1350 Hz corresponds to the innermost circular orbit of the merging binary system [36]. —— **68**

Contribution 3 (Jürgen Müller, Liliane Biskupek, Franz Hofmann, and Enrico Mai):

Table 1 If E results for relativistic parameters. See Section 4 for a detailed explanation and comparison with results from other groups. —— **151**

Contribution 4 (Ignazio Ciufolini, Antonio Paolozzi, Vahe Gurzadyan, Erricos C. Pavlis, Rolf König, John Ries, Richard Matzner, Roger Penrose, Giampiero Sindoni, and Claudio Paris):

Table 1 The LAGEOS satellites [56, 57]. —— **165**

Table 2 LARES orbit data. —— **170**

Table 3 Comparison of mechanical characteristics of LARES tungsten alloy, steel used on the brackets of the separation system and the material used for the shells of LAGEOS. — 171

Contribution 5 (Toshio Fukushima):
Table 1 Averaged CPU times to compute various elliptic functions and integrals. The unit of CPU time is ns at a PC with an Intel Core i7-2675QM processor running at 2.20 GHz where one machine clock cycle is 0.455 ns. — 222

Contribution 6 (Jean-Louis Simon and Agnes Fienga):
Table 1 First-order perturbations of the pair Jupiter–Saturn ($\alpha = 0.544$) and of the pair Venus–Earth ($\alpha = 0.723$): number of the terms with amplitude larger than $0.001''$ (or $0''.001 \times \mathrm{AU}$ for $\Delta^1 a$) for each variable, in the approximation in $\bar{\lambda}, \bar{\lambda}'$ (N) and in the approximation in w, w' $(\tilde{N})$. (Results taken from [15]). — 238
Table 2 Integration constants J2000 for the pair Jupiter–Saturn for the solution $H_{\overline{\lambda},\overline{\lambda'}}$ [23] and for the solution $H_{w,w'}$ [15]. Units are $''/1000$ yrs for $\bar{n}$, ua for a_0, ° for ε_0, ϖ_0, i_0, and Ω_0. e_0 is without dimension. — 239
Table 3 Mean longitudes of Jupiter (λ) and Saturn (λ'): number of terms of magnitude up to $0''.001$ in the solution $H_{w,w'}$ of [15] (series $\widetilde{S}_i$) and in the solution $H_{\overline{\lambda},\overline{\lambda'}}$ of [25] (series S_i). — 240
Table 4 Planetary ephemerides data sets given by data types (column 1), by objects (column 2) and mean accuracy (columns 3, 4, and 5) in right ascension α, declination δ and geocentric distances ρ. The numbers of each type of observations used in the INPOP10a adjustment are given in column 6. — 246
Table 5 Maximum differences between INPOP13a, INPOP10a and DE421 over 1 century in spherical geocentric coordinates and distances and barycentric Earth spherical coordinates. For the Moon, INPOP10e is used for the comparisons to INPOP10a. — 246

Preface

A key figure of relativistic celestial mechanics of the second half of twentieth century has been Victor A. Brumberg, a scholar who presently lives in Boston (USA) and who is still active in research. Victor A. Brumberg has made a significant contribution to general relativity and the science of relativistic planetary ephemerides of the Solar System. He mentored and inspired many researchers around the globe (including the editor of this book) to start working in the field of relativistic celestial mechanics. The very term "relativistic celestial mechanics" was introduced by Victor. A. Brumberg in his famous monograph "Relativistic Celestial Mechanics" published in 1972 by Nauka (Science) – the main scientific publisher of the USSR – in Moscow. For next two decades this monograph remained the most authoritative reference and the source of invaluable information for researchers working on relativistic equations of motion and experimental testing of general relativity. Brumberg received the 2008 Brower Award from the Division of Dynamic Astronomy of the American Astronomical Society. The Brouwer Award was established to recognize outstanding contributions in the field of dynamical astronomy, including celestial mechanics, astrometry, geophysics, stellar systems, galactic, and extragalactic dynamics.

This book is a second volume of Festschrift aimed to honour the scientific achievements of V. A. Brumberg and to celebrate his 80th birthday on February 12, 2013. The book was launched on the eve of another remarkable date – 100 years of Einstein's general relativity – the theory which dramatically changed the world of theoretical physics by opening new fascinating opportunities in the scientific study of the fundamental laws of Nature.

The volume consists of seven chapters discussing the recent experimental advances in relativistic celestial mechanics including pulsar timing, lunar laser ranging, ephemeris astronomy, and time-keeping metrology. Relativistic celestial mechanics is an essential branch of the modern gravitational physics, a branch exploring the fundamental structure of spacetime by studying motion of massive bodies like black holes, stars, planets, and elementary particles, including photons, in gravitational field. It establishes basic theoretical principles for calculation and interpretation of various relativistic effects observed in astrophysical systems and in the Solar System. Relativistic celestial mechanics of photons is more known among astronomers as relativistic astrometry. For an indefeasible branch of gravitational physics, it is required to map the coordinate description of motion of celestial bodies into parameter space of observables. Theoretical progress in understanding the orbital motion of celestial bodies would be inconceivable without a corresponding improvement in mathematical description of propagation of light rays in stationary and time-dependent gravitational field. High-precision astrometric catalogs of stars and radio sources are of paramount importance for astrophysical studies of galactic dynamics and conducting relativistic experiments in the Solar System and beyond. The catalogues materialize the inertial coordinate system in the sky, in which orthogonal grid is a primary reference for astrometric measurement of positions and proper motions of celestial bodies. General relativity predicts that Sun and major planets of the Solar System curve spacetime by their own gravitational field which warps and twists the reference grid of the inertial coordinate system in the sky. Fortunately, for astronomers, gravitational field of the Solar System is weak and the spacetime deformation is small. It allows us to subtract the spacetime distortions induced by gravity and restore the inertial system in the sky. Various theories predict slightly different magnitude and direction of warping and twisting of the inertial coordinates. Making astrometric observations of stars and comparing them with theory give access to the experimental testing of general relativity.

Chapter 1, written by P. Teyssandier, describes advances in mathematical modeling of the relativistic effects produced by static, spherically symmetric gravitational field on high-precision stellar catalogues in a large class of different metric theories of gravity. It focuses on determination of the travel time of a photon as a function – often called *time transfer function* – of the positions of emitter and receiver. This function is of crucial interest for gravity tests in the Solar System. Until very recently, it was known only up to the second order in the Newtonian gravitational constant G for a 3-parameter family of static, spherically symmetric metrics generalizing the Schwarzschild metric. The chapter presents two procedures enabling to derive the time transfer function at any order of approximation with respect to the small parameter m/r, with m being half of the Schwarzschild radius of the central massive body and r a radial coordinate. The first procedure is based on solving the Hamilton–Jacobi (eikonal) equation satisfied by the time transfer function. The second procedure sets up the iterative solution of an integro-differential equation derived from the null geodesic equations. It is shown that the two methods lead to the same expression for the time transfer function up to the order of G^3 but the second procedure seems to be more advantageous as it relies upon simple elementary integrations which may be performed with any symbolic computer program. The vector functions characterizing the direction of light propagation at the points of emission and reception are derived up to the third order in G. The relevance of the third-order terms in the time transfer function is briefly discussed with regard to the Solar System experiments.

Chapter 2, written by N. Wex, gives an account of the progress having been reached in experimental testing of general relativity with pulsar timing over the last decade. Before the 1970s, precision tests for gravity theories were constrained to the weak gravitational fields of the Solar System. Hence, only the weak-field, slow-motion aspects of relativistic celestial mechanics could be investigated. Testing gravity beyond the first post-Newtonian contributions was for a long time out of reach. The discovery of the first binary pulsar by Russell Hulse and Joseph Taylor in the summer of 1974 initiated a completely new field for testing the relativistic dynamics of gravitationally interacting bodies. For the first time the back reaction of gravitational wave emission on the binary orbital motion could be studied. Furthermore, the Hulse–Taylor pulsar provided the first test bed for the orbital dynamics of strongly self-gravitating bodies. To date there are a number of binary pulsars known, which can be utilized for precision test of gravity. Depending on their orbital properties and their companion, these pulsars provide access to testing different aspects of relativistic dynamics. Besides tests of specific gravity theories, like general relativity or scalar-tensor gravity, there are binary pulsars that allow for generic constraints on potential deviations of gravity from general relativity in the quasi-stationary strong-field and the radiative regime. The chapter presents a brief overview of this modern field of relativistic celestial mechanics, reviews some of the highlights of gravity tests with radio pulsars, and discusses their implications for gravitational physics and astronomy, including the upcoming gravitational-wave astronomy.

Chapter 3, authored by the team of four authors – J. Müller, L. Biskupek, F. Hofmann, and E. Mai, outlines the achievements and the future prospects of Lunar Laser Ranging (LLR) experiment that provides an ongoing time series of highly accurate Earth–Moon distance measurements since 1969. Over the years, the range precision has improved from the meter level of accuracy to the few millimeter level, the latter, thanks to Apache Point Observatory Lunar Laser-ranging Operation (APOLLO) facility located in the Sacramento Mountains in Sunspot, New Mexico (USA). The authors describe in full detail the current LLR analysis software consisting of a collection of sophisticated software modules yielding ranging precision at the centimeter level of accuracy. The whole measurement process is modeled at appropriate post-Newtonian approximation and includes fitting parameters of the orbits of the major bodies of the Solar System, the rotation of Earth and Moon, signal propagation, the reference and time systems as well as the time-variable terrestrial positions of the LLR observatories and reflectors. The chapter provides the reader with an overview of the current LLR experimental constrains of various relativistic effects within the framework of PPN formalism which includes testing of:

- time variability of gravitational constant G,
- geodetic precession of the lunar orbit,
- violation of strong equivalence principle,
- dark matter acceleration,
- Yukawa term of modified Newton's law of gravity,
- preferred-frame effects and PPN parameters β and γ,
- gravitomagnetic and spin–orbit coupling effects.

The chapter comments on future progress in relativistic LLR experiment which is intimately related to improving the model of LLR analysis down to the mm level.

Chapter 4, is written by I. Ciufolini with collaborators including Sir R. Penrose. It introduces the reader to the theory of gravitomagnetic effect known as the dragging of inertial frames predicted in general relativity by Lense and Thirring in 1919, and describes its spin-orbital test performed with laser ranging of the Laser Geodynamic Satellites, LAGEOS and LAGEOS II, and spin–spin test done with a gyroscopes on-board of Gravity Probe B space mission. The major part of the chapter is devoted to the discussion of the Laser Relativity Satellite (LARES) that was successfully launched in February 2012 to improve the accuracy in measurement of the frame-dragging effect and for testing other theories of gravity including the string theories. The chapter describes the results of the orbital data analyzes and shows that LARES motion approximates the geodesic motion of a point-like particle in general relativity much better than any other geodetic satellite. Finally, the chapter provides a forecast in measuring the frame-dragging effect with LARES based on the detailed study of its orbital motion in the next decade performed with the extensive Monte Carlo simulations. It yields a weighty argument in support of the mission target – the measurement of the gravitomagnetic effect with an accuracy of a few percent.

Chapter 5, authored by T. Fukushima, describes the basic mathematical toolkit of modern ephemeris astronomy – elliptic functions and elliptic integrals – which frequently appear in the analytical expressions of the variables of dynamical systems in celestial mechanics. The elliptic functions and the corresponding elliptic integrals are difficult to approach, understand, and utilize in practice. In order to enhance their accessibility for astronomers the chapter overviews numerous notations of the elliptic functions and integrals, which have been introduced by different researchers, and, then, explains their properties, key formulas, and internal relationships. The chapter presents the key ingredients of the new procedures of numerical computation technique for fast and precise calculation of the elliptic functions and integrals having been recently developed by the author and his research group for doing orbital dynamics of planetary systems and satellites.

Chapter 6, written by J.-L. Simon and A. Fienga, discusses the modern planetary ephemerides created in the Institut de Mécanique Céleste et de Calcul des Éphémérides (IMCCE) and the scientific impact on their development having been made over many years by Prof. Victor A. Brumberg who has done a fundamental work in the construction of analytical relativistic theory of the lunar and planetary motions. The chapter introduces the reader to the basic elements of general planetary theory and explains the iterative process of its construction by making use of the Brumberg expansions of elliptic functions in the hypergeometric polynomials. Specific examples of the application of this procedure are given for pairs of planets Jupiter–Saturn and Venus–Earth. The chapter also describes the contribution of Prof. Victor A. Brumberg to the relativistic theory of reference frames, time scales and units for planetary ephemerides which led to the development of the Brumberg–Kopeikin (BK) and Damour–Soffel-Xu (DSX) formalism of reference frames adopted in the form of resolutions by the International Astronomical Union (IAU) in 2000. Finally, the authors of this chapter give an account of the planetary ephemerides worked out by a team of IMCCE researchers, namely, the semi-analytical theory VSOP and numerical theory INPOP, and compare them with respect to NASA JPL planetary ephemerides DE.

Chapter 7, written by G. Petit, P. Wolf, and P. Delva, is a review of the state of the art achieved in our understanding of atomic time, clocks and clock comparison in relativistic spacetime. In particular, it explains why and how the relativistic BK-DSX formalism is applied in the procedure of time and frequency comparisons. The chapter starts from a brief introduction to the physical realization of atomic time scales and the actual performances of atomic clocks, and recalls the IAU relativistic framework that resulted from the historical development of BK-DSX formalism. Then, the authors tell how the relativistic framework is applied for time and frequency comparisons in the geocentric celestial reference system and show that the current relativistic modeling is adequate for the present and most upcoming needs of clocks and time transfer techniques covering one-way and two-way transmissions. The authors discuss future development of the time transfer techniques with emphasis on two new techniques: the Time Transfer by Laser Link (T2L2) currently deployed on board Jason-2 satellite, and the microwave link developed for the forthcoming Atomic Clock Ensemble in Space (ACES) mission on board of the International Space Station (ISS). The chapter describes novel implications of clocks in relativistic geodesy, notably the use of frequency standards to measure the gravitational potential and to establish the reference geoid with an unprecedented accuracy that will revolutionize the field of geophysics.

Over the past 30 years, relativistic celestial mechanics has experienced radical progress in developing new methods and techniques of experimental testing of general relativity. This volume cannot embrace it in its entirety. For further reading on the experimental status of relativistic celestial mechanics and astrometry, we recommend the following review articles and textbooks:

Brumberg, V. A., "Celestial mechanics: past, present, future," *Solar System Research*, Vol. 47, Issue 5, pp. 347–358 (2013)

Brumberg, V. A., "Essential Relativistic Celestial Mechanics," Adam Hilger: Bristol, 1991

Brumberg, V. A., "Analytical Techniques of Celestial Mechanics," Springer: Heidelberg, 1995

Kopeikin, S., Efroimsky, M. and G. Kaplan, "Relativistic Celestial Mechanics of the Solar System," Wiley-VCH: Berlin, 2011

Takahashi, F., Kondo, T., Takahashi, Y. and Koyama, Y., "Very Long Baseline Interferometry," Wave Summit Course, Ohmsha: Tokyo, 2000

C. M. Will, "Theory and Experiment in Gravitational Physics," Cambridge University Press: Cambridge, 1993

Ashby, N., "Relativity in the Global Positioning System," Living Reviews in Relativity 6, 1 (2003). URL (cited on January 14, 2014): www.livingreviews.org/lrr-2003-1

C. M. Will, "The Confrontation between General Relativity and Experiment," Living Reviews in Relativity 9, 3 (2006). URL (cited on January 14, 2014) www.livingreviews.org/lrr-2006-3

February 12, 2014

Editor: Sergei Kopeikin
University of Missouri, USA

Pierre Teyssandier

New tools for determining the light travel time in static, spherically symmetric spacetimes beyond the order G^2

1 Introduction

Many experiments designed to test relativistic gravity involve photons traveling between an emitter and a receiver both located at a finite distance. Some of these experiments are based on the measurement of a time delay or a comparison of distant clocks, while the other ones aim to measure the gravitational deflection of light. In spite of their differences, however, all these tests can be modeled by a single mathematical tool, namely the expression of the light travel time as a function of the positions of the emitter and the receiver for a given time of reception (or emission). Indeed, it has been shown that knowing such an expression, that we call a time transfer function, makes possible to determine not only the frequency shift and the Doppler-tracking between the emitter and the observer, but also the direction of light propagation [1–7].

The aim of this chapter is to give an overview of the two procedures which are currently at our disposal for calculating the time transfer function in static, spherically symmetric spacetimes at least up to the order G^3, with G being the Newtonian gravitational constant. The necessity of tackling the calculation of terms of order G^3 and beyond may be questioned since it is generally believed that the most accurate projects for testing the metric theories of gravity in the Solar System, like SAGAS [8], ODYSSEY [9], LATOR [10], or ASTROD [11] require the knowledge of the time transfer function only up to the order G^2, (see, e.g. [12] and references therein). This reasoning neglects the fact that some so-called enhanced term of order G^3 in the time transfer function may become comparable to the 'regular' term of order G^2, that is the term which can be estimated as const $\cdot \frac{m^2}{cr_c}$, with m being half the Schwarzschild radius of the central body and r_c the zeroth-order distance of closest approach of the light ray [13]. The enhancement occurs in a close superior conjunction, i.e. in the case where the emitter and the receiver are almost on the opposite sides of the central mass – a configuration of crucial importance in experimental gravitation. This effect is recovered from the full expression of the time transfer function up to order G^3 obtained in a recent paper [14]. In the same work, it is shown that this term must be taken into account in Solar System experiments aiming to determine the post-Newtonian param-

Pierre Teyssandier: SYRTE–CNRS/UMR 8630, UPMC, Observatoire de Paris, 61 avenue de l'Observatoire, F-75014 Paris, France

eter γ with an accuracy of 10^{-8}. Consequently, performing the calculations beyond the second order is fully relevant. We confine our exploration of the higher orders to the static, spherically symmetric spacetimes. In the present state of the art, indeed, it appears justified to neglect the relativistic contributions due to the nonsphericity or to dynamic interactions between the Sun and the planets beyond the linear regime (see, e.g. [1, 15–20] and refs. therein).

We focus our attention on the theories of gravity in which it is possible to suppose that the components of the metric are analytic expansions in powers of m/r. The cosmological constant is neglected. The metric is thus regarded as a generalization of the Schwarzschild metric characterized by an infinity of dimensionless constants including the well-known post-Newtonian parameters β and γ. We restrict our attention to the case where the paths followed by light are what we called quasi-Minkowskian light rays in [5], namely null geodesics described as perturbations in powers of G of a Minkowskian null segment passing through the spatial positions of the emitter and of the receiver. The corresponding time transfer function is then represented by a series in powers of G. For the sake of brevity, a term of order G^n is said to be of order n.

Upon these assumptions, the first-order term in the time transfer function reduces to the well-known Shapiro time delay [21], which can be obtained by different reasonings, some of them involving only elementary calculations (see, e.g. [22] or [23]). Until recently, the contributions beyond the linear regime were calculated only up to the second order. Two kinds of methods were available.

(a) Integration of the null geodesic equations. After the pioneering work [24], the post-post-Newtonian expression of the time transfer function in the Schwarzschild metric has been obtained by Brumberg for a class of quasi-Galilean coordinate systems of interest in celestial mechanics [25, 26]. The analytic integration of the null geodesic equations in a three-parameter family of static, spherically symmetric spacetimes has been recently performed in [27] for discussing the astrometric Gaia mission. These approaches work well, but present the drawback to be indirect, since the expression of the time transfer function is deduced from a solution which corresponds to a light ray emitted at infinity in a given direction.
(b) Methods natively adapted to the generic case where both the emitter and the receiver of the light rays are located at a finite distance from the origin of the spatial coordinates. These methods are based either on an iterative determination of the Synge world function (see [28] for the Schwarzschild metric and [2] for a three-parameter family of static, spherically symmetric spacetimes), or on an iterative integration of the Hamilton–Jacobi (or eikonal) equation satisfied by the time transfer function (see [29], and [13] for a more recent analysis). The two variants have been successfully employed. The results obtained by the different above-mentioned procedures are equivalent, up to a coordinate transformation.

However, the matter is currently making substantial progress with the new procedure developed in [14]. This procedure allows us to determine the time transfer function by an iterative solution of an integrodifferential equation derived from the null geodesic equations. The calculations only involve elementary integrations which can be performed with any symbolic computer program whatever the order of approximation. It must be emphasized that the expression of the time transfer function up to the third order obtained in [14] is not to be confused with the formulas found in [30] and [31] (these formulas involve the radial coordinate of the pericenter of the ray without calculating this quantity as a function of the positions of the emitter and the receiver). Moreover, this expression markedly improves the result previously given in [13]. In [13], indeed, only the asymptotic form of the time transfer function when the emitter and the receiver tend to be in conjunction is found.

Faced with such a success, it is legitimate to ask whether the iterative procedure elaborated in [29] allows us to determine the time transfer function up to the third order. We prove here that this is effectively the case and that the result found in [14] is recovered. So we have now two procedures at our disposal for a large class of static, spherically symmetric spacetimes.

The paper is organized as follows. Section 2 lists the notations and conventions we use. In Section 3, the fundamental relations that link the light propagation direction and the frequency shift to the time transfer function are reminded for a general static, spherically symmetric metric. In Section 4, the specific assumptions made on the metric and on the light rays are stated. Section 5 yields a recurrence relation satisfied by the perturbations terms involved in the expansion of the time transfer function. A fundamental property of analyticity is established for these terms. Section 6 is devoted to the first procedure presented in this paper. It is shown how the recurrence relation established in Section 5 enables to determine an explicit expression of the time transfer up to the third order. Section 7 gives an overview of the procedure recently proposed in [14]. Section 8 reminds how this procedure can be noticeably simplified and leads to streamlined calculations for the time transfer function at any order. In Section 9, the vector functions giving the light propagation direction of a quasi-Minkowskian light ray are determined up to the third order. The results of Section 9 are applied in Section 10 to a ray emitted at infinity in an arbitrary direction and observed at a given point. In Section 11, the appearance of enhanced terms is rigorously proved up to the third order. The relevance of these terms for some Solar System experiments is discussed in Section 12. Concluding remarks are given in Section 13.

2 Notations and conventions

We use notations and conventions as follow.
- The signature of the metric is $(+, -, -, -)$.
- Greek indices run from 0 to 3, and Latin indices run from 1 to 3.

- Any bold italic letter refers to an ordered triple: $(a^1, a^2, a^3) = (a^i) = \boldsymbol{a}$ and $(b_1, b_2, b_3) = (b_i) = \underline{\boldsymbol{b}}$. All the triples are regarded as 3-vectors of the ordinary Euclidean space.
- Given triples $\boldsymbol{a}$, $\boldsymbol{b}$, $\underline{\boldsymbol{c}}$, $\underline{\boldsymbol{d}}$, we put $\boldsymbol{a}.\boldsymbol{b} = a^i b^i$, $\boldsymbol{a}.\underline{\boldsymbol{c}} = a^i c_i$ and $\underline{\boldsymbol{c}}.\underline{\boldsymbol{d}} = c_i d_i$, with Einstein's convention on repeated indices being used.
- $|\boldsymbol{a}|$ denotes the formal Euclidean norm of the triple $\boldsymbol{a}$: $|\boldsymbol{a}| = (\boldsymbol{a}.\boldsymbol{a})^{1/2}$. If $|\boldsymbol{a}| = 1$, $\boldsymbol{a}$ is conventionally called a unit (Euclidean) 3-vector.
- $\boldsymbol{a} \times \boldsymbol{b}$ is the triple obtained by the usual rule giving the exterior product of two vectors of the Euclidean space.
- Given a bi-scalar function $F(\boldsymbol{x}, \boldsymbol{y})$, $\nabla_{\boldsymbol{x}} F(\boldsymbol{x}, \boldsymbol{y})$ and $\nabla_{\boldsymbol{y}} F(\boldsymbol{x}, \boldsymbol{y})$ denote the gradients of F with respect to $\boldsymbol{x}$ and $\boldsymbol{y}$, respectively.

3 Generalities

Before entering into the main subject of this paper, it may be useful to remind the most relevant results obtained in [2] and [29] concerning the relations between the light travel time and the quantities involved in the time/frequency transfers experiments or in astrometry.

Throughout this chapter, spacetime is assumed to be a four-dimensional manifold endowed with a static, spherically symmetric metric g. We suppose that there exists a domain $\mathcal{D}_h$ in which the metric is regular, asymptotically flat and may be interpreted as the gravitational field of a central body having a mass M. We put $m = GM/c^2$. The domain of regularity $\mathcal{D}_h$ is assumed to be covered by a single quasi-Cartesian coordinate system $x^\mu = (x^0, x^i)$ adapted to the symmetries of the metric. We use the time coordinate t defined by $x^0 = ct$ and we put $\boldsymbol{x} = (x^i)$, $i = 1, 2, 3$. For convenience, the coordinates $(x^0, \boldsymbol{x})$ are chosen so that the metric takes an isotropic form

$$ds^2 = \mathcal{A}(r)(dx^0)^2 - \frac{1}{\mathcal{B}(r)}\delta_{ij}dx^i dx^j \,, \tag{3.1}$$

where $r = |\boldsymbol{x}|$. Using the corresponding spherical coordinates (r, ϑ, φ), one has

$$\delta_{ij}dx^i dx^j = dr^2 + r^2 d\vartheta^2 + r^2 \sin^2\vartheta d\varphi^2 \,.$$

We generically consider a photon emitted at a point-event x_A and received at a point-event x_B, with x_A and x_B being located in the domain of regularity $\mathcal{D}_h$. We put $x_A = (ct_A, \boldsymbol{x}_A)$ and $x_B = (ct_B, \boldsymbol{x}_B)$. It is assumed that the photon propagates along a null geodesic path of the metric g. This geodesic is denoted by $\Gamma(\boldsymbol{x}_A, \boldsymbol{x}_B)$[1], or simply Γ in the absence of ambiguity. We suppose that x_A and x_B cannot be linked by two distinct null

1 In a static spacetime, the mention of the initial time t_A may be omitted.

geodesic paths (configurations like the Einstein ring are not taken into account). Then the light travel time $t_B - t_A$ can be considered as a function of $\boldsymbol{x}_A$ and $\boldsymbol{x}_B$, so that one can write

$$t_B - t_A = \mathcal{T}(\boldsymbol{x}_A, \boldsymbol{x}_B; \Gamma) . \tag{3.2}$$

We call $\mathcal{T}(\boldsymbol{x}_A, \boldsymbol{x}_B; \Gamma)$ the time transfer function associated with Γ.

The importance of the notion of time transfer function for the astrometry and the frequency transfers rests on the fact that the light direction at x_A and x_B can be fully determined when $\mathcal{T}(\boldsymbol{x}_A, \boldsymbol{x}_B; \Gamma)$ is explicitly known. The argument may be summarized as follows. Since the light rays are null geodesic paths, the propagation direction of a photon traveling along $\Gamma(x_A, x_B)$ is completely characterized by the light direction triple defined as

$$\widehat{\underline{l}}_x = \left(\frac{l_i}{l_0}\right)_x , \tag{3.3}$$

where x denotes a point of $\Gamma(x_A, x_B)$ and the quantities l_α are the covariant components of a 4-vector tangent to $\Gamma(x_A, x_B)$ at x. The value of $\widehat{\underline{l}}_x$ is independent of the parameter describing $\Gamma(x_A, x_B)$. Denote by $\widehat{\underline{l}}_A$ and $\widehat{\underline{l}}_B$ the values of $\widehat{\underline{l}}_x$ at points x_A and x_B, respectively. It is shown in [2] that $\widehat{\underline{l}}_A$ and $\widehat{\underline{l}}_B$ can be inferred from the time transfer function $\mathcal{T}$ by using the relations

$$\widehat{\underline{l}}_A = \widehat{\underline{l}}_e(\boldsymbol{x}_A, \boldsymbol{x}_B; \Gamma) , \tag{3.4}$$

$$\widehat{\underline{l}}_B = \widehat{\underline{l}}_r(\boldsymbol{x}_A, \boldsymbol{x}_B; \Gamma) , \tag{3.5}$$

where the functions $\widehat{\underline{l}}_e$ and $\widehat{\underline{l}}_r$ are defined as

$$\widehat{\underline{l}}_e(\boldsymbol{x}_A, \boldsymbol{x}_B; \Gamma) = c\nabla_{x_A} \mathcal{T}(\boldsymbol{x}_A, \boldsymbol{x}_B; \Gamma) , \tag{3.6}$$

$$\widehat{\underline{l}}_r(\boldsymbol{x}_A, \boldsymbol{x}_B; \Gamma) = -c\nabla_{x_B} \mathcal{T}(\boldsymbol{x}_A, \boldsymbol{x}_B; \Gamma) . \tag{3.7}$$

We can conclude from (3.6) and (3.7) that knowing the time transfer function associated to a given null geodesic is extremely useful in astrometry. Let us briefly examine the problem of modeling frequency shifts. Let u_A^α and u_B^α be the unit 4-velocity vectors of the emitter at x_A and of the receiver at x_B, respectively. Denote by ν_A the frequency of the signal emitted at x_A as measured by a standard clock comoving with the emitter and by ν_B the frequency of the signal received at x_B as measured by a standard clock comoving with the receiver. The ratio ν_B/ν_A is given by the well-known formula [32]

$$\frac{\nu_B}{\nu_A} = \frac{u_B^\beta (l_\beta)_B}{u_A^\alpha (l_\alpha)_A} . \tag{3.8}$$

Denote by $\boldsymbol{\beta}_A$ the coordinate velocity divided by c of the emitter at the instant of emission and by $\boldsymbol{\beta}_B$ the coordinate velocity divided by c of the receiver at the instant of reception, namely the triples defined as

$$\boldsymbol{\beta}_A = \left(\frac{d\boldsymbol{x}_A(t)}{cdt}\right)_{x_A} , \quad \boldsymbol{\beta}_B = \left(\frac{d\boldsymbol{x}_B(t)}{cdt}\right)_{x_B} .$$

Noting that l_0 is conserved along a geodesic of (3.1), it is immediately seen that (3.8) may be written in the form

$$\frac{\nu_B}{\nu_A} = \frac{\sqrt{\mathcal{A}(r_A) - \mathcal{B}^{-1}(r_A)\boldsymbol{\beta}_A^2}}{\sqrt{\mathcal{A}(r_B) - \mathcal{B}^{-1}(r_B)\boldsymbol{\beta}_B^2}} \frac{1 + \boldsymbol{\beta}_B.\widehat{\underline{\boldsymbol{l}}}_r(\boldsymbol{x}_A, \boldsymbol{x}_B; \Gamma)}{1 + \boldsymbol{\beta}_A.\widehat{\underline{\boldsymbol{l}}}_e(\boldsymbol{x}_A, \boldsymbol{x}_B; \Gamma)}, \tag{3.9}$$

where $\widehat{\underline{\boldsymbol{l}}}_e$ and $\widehat{\underline{\boldsymbol{l}}}_r$ are given by the right-hand side of (3.6) and (3.7), respectively. Formula (3.9) completes the proof of the relevance of the time transfer functions for experimental gravitation.

From a theoretical point of view, the problem of determining the time transfer functions in a given spacetime is inextricably complicated. Indeed, given two spatial positions $\boldsymbol{x}_A$ and $\boldsymbol{x}_B$, and an instant t_A, there exists in general an infinity of light rays emitted at the point-event $(ct_A, \boldsymbol{x}_A)$ and passing through point-events located at $\boldsymbol{x}_B$. Rigorously established a long time ago for the exact Schwarzschild metric (see, e.g. [33, 34] and refs. therein), this feature occurs in a very large class of spacetimes [35]. Moreover, the full expressions of the different functions $\mathcal{T}(\boldsymbol{x}_A, \boldsymbol{x}_B; \Gamma)$ are unknown, even for the Schwarzschild spacetime. Fortunately, the gravitational field in the Solar System may be regarded as weak, so that it may be assumed that the photons involved in experiments propagate along what we call quasi-Minkowskian light rays (see Section 4.2). We shall see below that the corresponding time transfer function is then unique and may be determined by iterative procedures whatever the required order of approximation.

4 Specific assumptions on the metric and the light rays

4.1 Post-Minkowskian expansion of the metric

Metric (3.1) is considered as a generalization of the exterior Schwarzschild metric, which may be written in the form

$$ds^2_{\mathrm{Sch}} = \frac{\left(1 - \frac{m}{2r}\right)^2}{\left(1 + \frac{m}{2r}\right)^2}(dx^0)^2 - \left(1 + \frac{m}{2r}\right)^4 \delta_{ij}dx^i dx^j \tag{4.1}$$

in the region outside the event horizon located at $r = m/2$. So we henceforth assume that there exists a value $r_h > 0$ of the radial coordinate such that the domain of regularity $\mathcal{D}_h$ is the region outside the sphere of radius r_h. If there exists at least one event horizon, we must take for r_h the value of r on the outer horizon. By analogy with general relativity we consider that $r_h \sim m$ and we suppose that whatever $r > r_h$, $\mathcal{A}(r)$

and $\mathcal{B}^{-1}(r)$ are positive functions represented by analytical expansions as follows:

$$\mathcal{A}(r) = 1 - \frac{2m}{r} + 2\beta\frac{m^2}{r^2} - \frac{3}{2}\beta_3\frac{m^3}{r^3} + \beta_4\frac{m^4}{r^4} + \sum_{n=5}^{\infty}\frac{(-1)^n n}{2^{n-2}}\beta_n\frac{m^n}{r^n}\,, \tag{4.2}$$

$$\frac{1}{\mathcal{B}(r)} = 1 + 2\gamma\frac{m}{r} + \frac{3}{2}\epsilon\frac{m^2}{r^2} + \frac{1}{2}\gamma_3\frac{m^3}{r^3} + \frac{1}{16}\gamma_4\frac{m^4}{r^4} + \sum_{n=5}^{\infty}(\gamma_n - 1)\frac{m^n}{r^n}\,, \tag{4.3}$$

where the coefficients $\beta, \beta_3, \dots, \beta_n, \gamma, \epsilon, \gamma_3, \dots, \gamma_n, \dots$ are generalized post-Newtonian parameters chosen so that

$$\beta = \gamma = \epsilon = 1\,, \quad \beta_n = \gamma_n = 1 \quad \text{for} \quad n \geq 3 \tag{4.4}$$

in general relativity.

The light rays of the metric (3.1) are also the light rays of any metric $d\tilde{s}^2$ conformal to (3.1). This feature enables us to carry out our calculations for a metric containing only one potential. We choose $d\tilde{s}^2 = \mathcal{A}^{-1}(r)ds^2$, that is

$$d\tilde{s}^2 = (dx^0)^2 - \mathcal{U}(r)\delta_{ij}dx^i dx^j\,, \tag{4.5}$$

where $\mathcal{U}$ is defined by

$$\mathcal{U}(r) = \frac{1}{\mathcal{A}(r)\mathcal{B}(r)}\,. \tag{4.6}$$

It results from (4.2) and (4.3) that the potential $\mathcal{U}(r)$ occurring in (4.5) may be written as

$$\mathcal{U}(r) = 1 + 2(1+\gamma)\frac{m}{r} + \sum_{n=2}^{\infty} 2\kappa_n\frac{m^n}{r^n} \tag{4.7}$$

for $r > r_h$, with the coefficients κ_n being constants which can be expressed in terms of the generalized post-Newtonian parameters involved in the expansions of $\mathcal{A}(r)$ and $\mathcal{B}(r)$. Taking into account a notation already introduced in [5], namely

$$\kappa = 2(1+\gamma) - \beta + \tfrac{3}{4}\epsilon\,, \tag{4.8}$$

κ_2 and κ_3 are given by

$$\kappa_2 = \kappa\,, \quad \kappa_3 = 2\kappa - 2\beta(1+\gamma) + \tfrac{3}{4}\beta_3 + \tfrac{1}{4}\gamma_3\,. \tag{4.9}$$

4.2 Time transfer function for a quasi-Minkowskian light ray

In this chapter, we restrict our attention to the special class of null geodesic paths we have called the quasi-Minkowskian light rays in [5]. This means that in what follows, the path covered by the photon is assumed to be entirely confined in $\mathcal{D}_h$ and to be described by parametric equations of the form

$$x^0 = ct_A + \xi|\boldsymbol{x}_B - \boldsymbol{x}_A| + \sum_{n=1}^{\infty} X^0_{(n)}(\boldsymbol{x}_A, \boldsymbol{x}_B, \xi)\,, \tag{4.10}$$

$$\boldsymbol{x} = \boldsymbol{z}(\xi) + \sum_{n=1}^{\infty} \boldsymbol{X}_{(n)}(\boldsymbol{x}_A, \boldsymbol{x}_B, \xi)\,, \tag{4.11}$$

where ξ is the affine parameter varying on the range $0 \leq \xi \leq 1$, $\boldsymbol{z}(\xi)$ is defined by

$$\boldsymbol{z}(\xi) = \boldsymbol{x}_A + \xi(\boldsymbol{x}_B - \boldsymbol{x}_A) \tag{4.12}$$

and the functions $X^0_{(n)}$ and $\boldsymbol{X}_{(n)}$ are terms of order n obeying the boundary conditions

$$X^0_{(n)}(\boldsymbol{x}_A, \boldsymbol{x}_B, 0) = 0\,, \tag{4.13}$$

$$\boldsymbol{X}_{(n)}(\boldsymbol{x}_A, \boldsymbol{x}_B, 0) = \boldsymbol{X}_{(n)}(\boldsymbol{x}_A, \boldsymbol{x}_B, 1) = 0\,. \tag{4.14}$$

According to a notation already introduced in [5], such a null geodesic path will be denoted by $\Gamma_s(\boldsymbol{x}_A, \boldsymbol{x}_B)$. For the sake of brevity, the time transfer function associated with $\Gamma_s(\boldsymbol{x}_A, \boldsymbol{x}_B)$ will be henceforth denoted by $\mathcal{T}(\boldsymbol{x}_A, \boldsymbol{x}_B)$ or simply by $\mathcal{T}$. Setting $\xi = 1$ in (4.10), it may be seen that this function can be expanded in power series of G as follows:

$$\mathcal{T}(\boldsymbol{x}_A, \boldsymbol{x}_B) = \frac{|\boldsymbol{x}_B - \boldsymbol{x}_A|}{c} + \sum_{n=1}^{\infty} \mathcal{T}^{(n)}(\boldsymbol{x}_A, \boldsymbol{x}_B)\,, \tag{4.15}$$

where $\mathcal{T}^{(n)}$ stands for the term of order n.

Expansion (4.15) is easy to determine when $\boldsymbol{x}_A$ and $\boldsymbol{x}_B$ are linked by a radial null geodesic entirely lying in $\mathcal{D}_h$. In this case, indeed, it is immediately deduced from (4.5) that the expression of $\mathcal{T}$ is given by the exact formula

$$\mathcal{T}(r_A, r_B) = \operatorname{sgn}(r_B - r_A)\frac{1}{c}\int_{r_A}^{r_B} \sqrt{\mathcal{U}(r)}dr\,, \tag{4.16}$$

where $r_A = |\boldsymbol{x}_A|$ and $r_B = |\boldsymbol{x}_B|$. Substituting for $\mathcal{U}(r)$ from (4.7) into (4.16) shows that $\mathcal{T}$ may be expanded as follows:

$$\mathcal{T}(r_A, r_B) = \frac{|r_B - r_A|}{c} + \sum_{n=1}^{\infty} \mathcal{T}^{(n)}(r_A, r_B)\,, \tag{4.17}$$

where the first three perturbation terms are given by

$$\mathcal{T}^{(1)}(r_A, r_B) = \frac{(1+\gamma)m}{c}\left|\ln\frac{r_B}{r_A}\right|\,, \tag{4.18}$$

$$\mathcal{T}^{(2)}(r_A, r_B) = \left[\kappa - \tfrac{1}{2}(1+\gamma)^2\right]\frac{m^2}{r_A r_B}\frac{|r_B - r_A|}{c}\,, \tag{4.19}$$

$$\mathcal{T}^{(3)}(r_A, r_B) = \tfrac{1}{2}\left[\kappa_3 - (1+\gamma)\kappa + \tfrac{1}{2}(1+\gamma)^3\right]\frac{m^3}{r_A r_B}\left(\frac{1}{r_A} + \frac{1}{r_B}\right)\frac{|r_B - r_A|}{c}\,. \tag{4.20}$$

Determining the right-hand side of (4.15) is much more complicated when $\Gamma_s(\boldsymbol{x}_A, \boldsymbol{x}_B)$ is not a radial geodesic. As it has been recalled in introduction, the perturbations terms $\mathcal{T}^{(n)}$ might be obtained by an iterative integration of the null geodesic equations. Indeed, taking into account that $d\tilde{s}^2 = 0$ along a null geodesic, it results from (4.5) and (4.11) that the time transfer function is given by

$$\mathcal{T}(\boldsymbol{x}_A, \boldsymbol{x}_B) = \frac{1}{c}\int_0^1 \sqrt{\mathcal{U}(r(\xi))}\left|\boldsymbol{x}_B - \boldsymbol{x}_A + \sum_{n=1}^{\infty}\frac{d\boldsymbol{X}_{(n)}(\boldsymbol{x}_A, \boldsymbol{x}_B, \xi)}{d\xi}\right| d\xi\,, \tag{4.21}$$

where the integral is taken along $\Gamma_s(\boldsymbol{x}_A, \boldsymbol{x}_B)$. Taking into account the boundary conditions (4.14), it may be inferred from (4.21) that each function $\mathcal{T}^{(n)}$ is theoretically calculable if the perturbations terms $\boldsymbol{X}_{(1)}, \ldots, \boldsymbol{X}_{(n-1)}$ involved in (4.11) are determined by solving the null geodesic equations. This procedure is cumbersome, however. Fortunately, more workable methods can be developed, as we shall see in the next sections.

5 Fundamental properties of functions $\mathcal{T}^{(n)}$

5.1 Recurrence relation satisfied by functions $\mathcal{T}^{(n)}$

Let $\boldsymbol{x}$ be an arbitrary spatial position such that $\boldsymbol{x} \neq \boldsymbol{x}_A$. Consider a quasi-Minkowskian light ray $\Gamma_s(\boldsymbol{x}_A, \boldsymbol{x})$ joining $\boldsymbol{x}_A$ and $\boldsymbol{x}$. The covariant components of a vector tangent to $\Gamma_s(\boldsymbol{x}_A, \boldsymbol{x})$ at $\boldsymbol{x}$ satisfy the equation

$$(l_0)_x^2 - \mathcal{U}^{-1}(r)\delta^{ij}(l_i)_x(l_j)_x = 0 \tag{5.1}$$

since $\Gamma_s(\boldsymbol{x}_A, \boldsymbol{x})$ is a null geodesic of metric (4.5). Dividing (5.1) side by side by $[(l_0)_x]^2$, and then taking into account (3.7), it is easily seen that $\mathcal{T}(\boldsymbol{x}_A, \boldsymbol{x})$ satisfies an eikonal equation as follows:

$$c^2|\nabla_x\mathcal{T}(\boldsymbol{x}_A, \boldsymbol{x})|^2 = \mathcal{U}(r)\,. \tag{5.2}$$

This equation could be solved by applying the iterative procedure developed in [29] for a general metric. Nevertheless, this procedure is so simple for an eikonal equation like (5.2) that a specific proof deserves to be explicited as follows.

Replacing $\mathcal{T}(\boldsymbol{x}_A, \boldsymbol{x})$ by its expansion in powers of G and $\mathcal{U}(r)$ by (4.7), it is immediately seen that equation (5.2) is equivalent to the infinite system of equations

$$c\frac{\boldsymbol{x} - \boldsymbol{x}_A}{|\boldsymbol{x} - \boldsymbol{x}_A|}.\nabla_x\mathcal{T}^{(1)}(\boldsymbol{x}_A, \boldsymbol{x}) = (1+\gamma)\frac{m}{r}\,, \tag{5.3}$$

$$c\frac{\boldsymbol{x} - \boldsymbol{x}_A}{|\boldsymbol{x} - \boldsymbol{x}_A|}.\nabla_x\mathcal{T}^{(n)}(\boldsymbol{x}_A, \boldsymbol{x}) = \kappa_n\frac{m^n}{r^n} - \frac{c^2}{2}\sum_{p=1}^{n-1}\nabla_x\mathcal{T}^{(p)}(\boldsymbol{x}_A, \boldsymbol{x}).\nabla_x\mathcal{T}^{(n-p)}(\boldsymbol{x}_A, \boldsymbol{x}) \tag{5.4}$$

for $n \geq 2$. This system is valid for any point $\boldsymbol{x}$. Consequently, we may suppose that $\boldsymbol{x} = z(\xi)$, with $z(\xi)$ being defined by (4.12), which means that $\boldsymbol{x}$ is varying along the straight segment joining $\boldsymbol{x}_A$ and $\boldsymbol{x}_B$. Then we have for any $n \geq 1$

$$\left[\frac{\boldsymbol{x}-\boldsymbol{x}_A}{|\boldsymbol{x}-\boldsymbol{x}_A|}.\nabla_{\boldsymbol{x}}\mathcal{T}^{(n)}(\boldsymbol{x}_A,\boldsymbol{x})\right]_{\boldsymbol{x}=z(\xi)} = \boldsymbol{N}_{AB}.\left[\nabla_{\boldsymbol{x}}\mathcal{T}^{(n)}(\boldsymbol{x}_A,\boldsymbol{x})\right]_{\boldsymbol{x}=z(\xi)}, \tag{5.5}$$

where $\boldsymbol{N}_{AB}$ is defined by

$$\boldsymbol{N}_{AB} = \frac{\boldsymbol{x}_B-\boldsymbol{x}_A}{|\boldsymbol{x}_B-\boldsymbol{x}_A|}. \tag{5.6}$$

But a straightforward calculation shows that

$$\boldsymbol{N}_{AB}.\left[\nabla_{\boldsymbol{x}}\mathcal{T}^{(n)}(\boldsymbol{x}_A,\boldsymbol{x})\right]_{\boldsymbol{x}=z(\xi)} = \frac{1}{|\boldsymbol{x}_B-\boldsymbol{x}_A|}\frac{d}{d\xi}\mathcal{T}^{(n)}(\boldsymbol{x}_A,z(\xi)), \tag{5.7}$$

where $d\mathcal{T}^{(n)}(\boldsymbol{x}_A,z(\xi))/d\xi$ denotes the total derivative of $\mathcal{T}^{(n)}(\boldsymbol{x}_A,z(\xi))$ with respect to ξ along the segment joining $\boldsymbol{x}_A$ and $\boldsymbol{x}_B$. Consequently, the system of equations (5.3)–(5.4) may be written in the form

$$\frac{d}{d\xi}\mathcal{T}^{(1)}(\boldsymbol{x}_A,z(\xi)) = \frac{1}{c}|\boldsymbol{x}_B-\boldsymbol{x}_A|\frac{(1+\gamma)m}{|z(\xi)|}, \tag{5.8}$$

$$\begin{aligned}\frac{d}{d\xi}\mathcal{T}^{(n)}(\boldsymbol{x}_A,z(\xi)) = \frac{1}{c}|\boldsymbol{x}_B-\boldsymbol{x}_A|\Bigg\{&\kappa_n\frac{m^n}{|z(\xi)|^n}\\ &-\frac{c^2}{2}\sum_{p=1}^{n-1}\left[\nabla_{\boldsymbol{x}}\mathcal{T}^{(p)}(\boldsymbol{x}_A,\boldsymbol{x}).\nabla_{\boldsymbol{x}}\mathcal{T}^{(n-p)}(\boldsymbol{x}_A,\boldsymbol{x})\right]_{\boldsymbol{x}=z(\xi)}\Bigg\}\end{aligned} \tag{5.9}$$

for $n \geq 2$. Integrating equations (5.8) and (5.9) on the range $0 \leq \xi \leq 1$ and noting that $\mathcal{T}^{(n)}(\boldsymbol{x}_A,\boldsymbol{x}_A) = 0$, we get the fundamental proposition which follows.

Proposition 1. *The perturbation terms $\mathcal{T}^{(n)}$ involved in expansion (4.15) may be written in the form*

$$\mathcal{T}^{(n)}(\boldsymbol{x}_A,\boldsymbol{x}_B) = \frac{1}{c}|\boldsymbol{x}_B-\boldsymbol{x}_A|F^{(n)}(\boldsymbol{x}_A,\boldsymbol{x}_B), \tag{5.10}$$

where the functions $F^{(n)}$ are determined by the recurrence relation

$$F^{(1)}(\boldsymbol{x}_A,\boldsymbol{x}_B) = (1+\gamma)m\int_0^1\frac{d\xi}{|z(\xi)|}, \tag{5.11}$$

$$\begin{aligned}F^{(n)}(\boldsymbol{x}_A,\boldsymbol{x}_B) = {}&\kappa_n m^n\int_0^1\frac{d\xi}{|z(\xi)|^n}\\ &-\frac{c^2}{2}\int_0^1\sum_{p=1}^{n-1}\left[\nabla_{\boldsymbol{x}}\mathcal{T}^{(p)}(\boldsymbol{x}_A,\boldsymbol{x}).\nabla_{\boldsymbol{x}}\mathcal{T}^{(n-p)}(\boldsymbol{x}_A,\boldsymbol{x})\right]_{\boldsymbol{x}=z(\xi)}d\xi\end{aligned} \tag{5.12}$$

for $n \geq 2$, with the integrals being taken along the segment defined by the parametric equation $\boldsymbol{x} = \boldsymbol{z}(\xi)$, $0 \leq \xi \leq 1$.

The recurrence relation explicited in Proposition 1 shows that $\Gamma_s(\boldsymbol{x}_A, \boldsymbol{x}_B)$ is unique provided that expansion (4.15) is an admissible representation of the time transfer function. However, determining the most general conditions under which our construction is valid remains an open problem. According to (5.11) and (5.12), the functions $F^{(n)}$ are given by integrals involving the analytic expansion of the metric along the straight segment joining $\boldsymbol{x}_A$ and $\boldsymbol{x}_B$ (see also [29]). Consequently, we shall henceforth assume that the expression $|\boldsymbol{x}_B - \boldsymbol{x}_A|[1 + \sum_{p=1}^{n} F^{(p)}(\boldsymbol{x}_A, \boldsymbol{x}_B)]/c$ constitutes a reliable approximation of the time transfer function as long as the straight segment joining $\boldsymbol{x}_A$ and $\boldsymbol{x}_B$ does not intersect the hypersurface $r = r_h$, a condition expressed by the inequality

$$|\boldsymbol{z}(\xi)| > r_h \quad \text{for} \quad 0 \leq \xi \leq 1\,. \tag{5.13}$$

We shall see in Section 11 that this condition is largely satisfied by a light ray emitted in the Solar System (or coming from a star) and observed in the Solar System after having grazed the Sun.

5.2 Analyticity of the functions $\mathcal{T}^{(n)}$

A property of analyticity which is indispensable for justifying the procedure developed in Sections 7 and 8 can easily be inferred from Proposition 1. Let us begin with proving the following lemma.

Lemma 1. *The functions $F^{(n)}$ recursively determined by (5.11) et (5.12) are analytic in $\boldsymbol{x}_A$ and $\boldsymbol{x}_B$, except when $\boldsymbol{x}_A$ and $\boldsymbol{x}_B$ are such that $\boldsymbol{n}_B = -\boldsymbol{n}_A$, with $\boldsymbol{n}_A$ and $\boldsymbol{n}_B$ being defined as*

$$\boldsymbol{n}_A = \frac{\boldsymbol{x}_A}{r_A}, \quad \boldsymbol{n}_B = \frac{\boldsymbol{x}_B}{r_B}\,. \tag{5.14}$$

Proof of Lemma 1. The proposition is obviously true for $n = 1$, since the integrand $1/|\boldsymbol{z}(\xi)|$ in (5.11) is analytic in $\boldsymbol{x}_A$ and $\boldsymbol{x}_B$ for any ξ such that $0 \leq \xi \leq 1$ provided that $\boldsymbol{n}_B \neq -\boldsymbol{n}_A$. Suppose now the validity of Lemma 1 for $F^{(1)}, \dots, F^{(n)}$. Assuming p to be such that $1 \leq p \leq n$, and then substituting $\boldsymbol{z}(\xi)$ for $\boldsymbol{x}$ into $\nabla_{\boldsymbol{x}}\mathcal{T}^{(p)}(\boldsymbol{x}_A, \boldsymbol{x})$, it is immediately inferred from (5.10) that

$$\begin{aligned} c\left[\nabla_{\boldsymbol{x}}\mathcal{T}^{(p)}(\boldsymbol{x}_A, \boldsymbol{x})\right]_{\boldsymbol{x}=\boldsymbol{z}(\xi)} &= \boldsymbol{N}_{AB}F^{(p)}(\boldsymbol{x}_A, \boldsymbol{z}(\xi)) \\ &\quad + \xi|\boldsymbol{x}_B - \boldsymbol{x}_A|\left[\nabla_{\boldsymbol{x}}F^{(p)}(\boldsymbol{x}_A, \boldsymbol{x})\right]_{\boldsymbol{x}=\boldsymbol{z}(\xi)}\,. \end{aligned} \tag{5.15}$$

Using (5.15) leads to

$$c^2 \left[\nabla_x \mathcal{T}^{(p)}(x_A, x).\nabla_x \mathcal{T}^{(n+1-p)}(x_A, x)\right]_{x=z(\xi)}$$
$$= F^{(p)}(x_A, z(\xi))F^{(n+1-p)}(x_A, z(\xi))$$
$$+ \xi(x_B - x_A).\left[F^{(p)}(x_A, x)\nabla_x F^{(n+1-p)}(x_A, x)\right.$$
$$\left. + F^{(n+1-p)}(x_A, x)\nabla_x F^{(p)}(x_A, x)\right]_{x=z(\xi)}$$
$$+ \xi^2 |x_B - x_A|^2 \left[\nabla_x F^{(p)}(x_A, x).\nabla_x F^{(n+1-p)}(x_A, x)\right]_{x=z(\xi)} . \tag{5.16}$$

It follows from our assumption that the right-hand side of (5.16) is a sum of functions which are analytic in x_A and x_B for any ξ such that $0 \leq \xi \leq 1$, except if $n_B = -n_A$. Each integral

$$\int_0^1 \left[\nabla_x \mathcal{T}^{(p)}(x_A, x).\nabla_x \mathcal{T}^{(n+1-p)}(x_A, x)\right]_{x=z(\xi)} d\xi$$

is therefore analytic if $n_B \neq -n_A$. The same property is obviously possessed by the integral $\int_0^1 d\xi/|z(\xi)|^{n+1}$. Lemma 1 is thus proved by recurrence. □

Since $|x_B - x_A|$ is analytic except if $x_B \neq x_A$, we can state the proposition below.

Proposition 2. *The functions $\mathcal{T}^{(n)}$ involved in expansion (4.15) are analytic in x_A and x_B when both the following conditions are met:*
(a) $x_B \neq x_A$;
(b) $n_B \neq -n_A$.

The importance of this property will clearly appear in Section 7.3. It is worth of noting that the second condition in Proposition 2 is automatically fulfilled when inequality (5.13) is satisfied. This fact explains why the condition *b)* is never explicitly involved in the assumptions of the propositions enunciated below.

6 First procedure: determination of the $\mathcal{T}^{(n)}$'s from the recurrence relation for $n = 1, 2, 3$

The recurrence relation yielded by Proposition 1 enables us to determine explicitly the functions $\mathcal{T}^{(n)}$ at least up to the third order. The calculations are made easier by using the expressions of the light direction triples up to the order G^2 performed in [5]. Indeed, substituting $\mathcal{T}$ from (4.15) into (3.6) and (3.7), it may be seen that the vector functions $\widehat{\underline{l}}_e$ and $\widehat{\underline{l}}_r$ can be expanded in power series of G as follows:

$$\widehat{\underline{l}}_e(\boldsymbol{x}_A, \boldsymbol{x}_B) = -\boldsymbol{N}_{AB} + \sum_{n=1}^{\infty} \widehat{\underline{l}}_e^{(n)}(\boldsymbol{x}_A, \boldsymbol{x}_B)\,, \tag{6.1}$$

$$\widehat{\underline{l}}_r(\boldsymbol{x}_A, \boldsymbol{x}_B) = -\boldsymbol{N}_{AB} + \sum_{n=1}^{\infty} \widehat{\underline{l}}_r^{(n)}(\boldsymbol{x}_A, \boldsymbol{x}_B)\,, \tag{6.2}$$

where the contributions of order n are determined by

$$\widehat{\underline{l}}_e^{(n)}(\boldsymbol{x}_A, \boldsymbol{x}_B) = c\nabla_{\boldsymbol{x}_A}\mathcal{T}^{(n)}(\boldsymbol{x}_A, \boldsymbol{x}_B)\,, \tag{6.3}$$

$$\widehat{\underline{l}}_r^{(n)}(\boldsymbol{x}_A, \boldsymbol{x}_B) = -c\nabla_{\boldsymbol{x}_B}\mathcal{T}^{(n)}(\boldsymbol{x}_A, \boldsymbol{x}_B)\,. \tag{6.4}$$

As a consequence, the recurrence relation (5.12) may be written in the form

$$F^{(n)}(\boldsymbol{x}_A, \boldsymbol{x}_B) = \kappa_n m^n \int_0^1 \frac{d\xi}{|z(\xi)|^n} - \frac{1}{2}\int_0^1 \sum_{p=1}^{n-1}\left[\widehat{\underline{l}}_r^{(p)}(\boldsymbol{x}_A, z(\xi)).\widehat{\underline{l}}_r^{(n-p)}(\boldsymbol{x}_A, z(\xi))\right] d\xi\,. \tag{6.5}$$

whatever $n \geq 2$. The results inferred from (5.10), (5.11), and (6.5) for $n = 1, 2, 3$ may be enunciated as follows.

Proposition 3. *Let $\boldsymbol{x}_A$ and $\boldsymbol{x}_B$ be two points in $\mathcal{D}_h$ such that both the conditions $\boldsymbol{n}_A \neq \boldsymbol{n}_B$ and (5.13) are met. For $n = 1, 2, 3$, the functions $\mathcal{T}^{(n)}$ are yielded by*

$$\mathcal{T}^{(1)}(\boldsymbol{x}_A, \boldsymbol{x}_B) = \frac{(1+\gamma)m}{c}\ln\left(\frac{r_A + r_B + |\boldsymbol{x}_B - \boldsymbol{x}_A|}{r_A + r_B - |\boldsymbol{x}_B - \boldsymbol{x}_A|}\right), \tag{6.6}$$

$$\mathcal{T}^{(2)}(\boldsymbol{x}_A, \boldsymbol{x}_B) = \frac{m^2}{r_A r_B}\frac{|\boldsymbol{x}_B - \boldsymbol{x}_A|}{c}\left[\kappa\frac{\arccos \boldsymbol{n}_A.\boldsymbol{n}_B}{|\boldsymbol{n}_A \times \boldsymbol{n}_B|} - \frac{(1+\gamma)^2}{1 + \boldsymbol{n}_A.\boldsymbol{n}_B}\right], \tag{6.7}$$

$$\mathcal{T}^{(3)}(\boldsymbol{x}_A, \boldsymbol{x}_B) = \frac{m^3}{r_A r_B}\left(\frac{1}{r_A} + \frac{1}{r_B}\right)\frac{|\boldsymbol{x}_B - \boldsymbol{x}_A|}{c(1 + \boldsymbol{n}_A.\boldsymbol{n}_B)}\left[\kappa_3 - (1+\gamma)\kappa\frac{\arccos \boldsymbol{n}_A.\boldsymbol{n}_B}{|\boldsymbol{n}_A \times \boldsymbol{n}_B|} + \frac{(1+\gamma)^3}{1 + \boldsymbol{n}_A.\boldsymbol{n}_B}\right], \tag{6.8}$$

where the coefficients κ and κ_3 are determined by (4.8) and (4.9), respectively. In general relativity, γ, κ, and κ_3 are given by

$$\gamma = 1\,, \quad \kappa = \tfrac{15}{4}, \quad \kappa_3 = \tfrac{9}{2}. \tag{6.9}$$

Before entering the proof of this proposition, it is worthy of note that equations (4.18)–(4.20) corresponding to a radial light ray are recovered by taking the limit of equations (6.6)–(6.8) when $\boldsymbol{n}_B \to \boldsymbol{n}_A$. One has indeed

$$\lim_{n_B \to n_A}\frac{\arccos \boldsymbol{n}_A.\boldsymbol{n}_B}{|\boldsymbol{n}_A \times \boldsymbol{n}_B|} = 1\,. \tag{6.10}$$

Expression (6.6) is straightforwardly obtained from (5.11) under the form

$$\mathcal{T}^{(1)}(\boldsymbol{x}_A, \boldsymbol{x}_B) = \frac{(1+\gamma)m}{c} \ln\left(\frac{r_B + \boldsymbol{N}_{AB}.\boldsymbol{x}_B}{r_A + \boldsymbol{N}_{AB}.\boldsymbol{x}_A}\right), \tag{6.11}$$

which coincides with the well-known Shapiro time delay expressed in a standard post-Newtonian gauge (see, e.g. [22]). The equivalent formula given by (6.6) is more convenient for deriving the first-order light direction triples (see, e.g. [1] and [23]).

Equation (6.7) has been obtained in [2] and [29] (see also [27] for an equivalent expression in an harmonic gauge). Nevertheless, we give a detailed proof of Proposition 3 also for $n = 2$ because our calculation is based on a new procedure which in principle can be efficient at any order.

Proof of Proposition 3 *for $n = 2$ and $n = 3$.* For calculating $F^{(2)}$, we just need the triple $\hat{\underline{l}}_r^{(1)}(\boldsymbol{x}_A, \boldsymbol{x}_B)$, which is easily deduced from (6.4) and (6.6). One has (see, e.g. [5]):

$$\hat{\underline{l}}_r^{(1)}(\boldsymbol{x}_A, \boldsymbol{x}_B) = -\frac{(1+\gamma)m|\boldsymbol{N}_{AB} \times \boldsymbol{n}_B|}{r_c}\left[\boldsymbol{N}_{AB} - \frac{|\boldsymbol{n}_A \times \boldsymbol{n}_B|}{1 + \boldsymbol{n}_A.\boldsymbol{n}_B}\boldsymbol{P}_{AB}\right], \tag{6.12}$$

where $\boldsymbol{P}_{AB}$ is defined as

$$\boldsymbol{P}_{AB} = \boldsymbol{N}_{AB} \times \left(\frac{\boldsymbol{n}_A \times \boldsymbol{n}_B}{|\boldsymbol{n}_A \times \boldsymbol{n}_B|}\right). \tag{6.13}$$

Substituting $\boldsymbol{z}(\xi)$ for $\boldsymbol{x}_B$ in (6.12), defining $\boldsymbol{n}(\xi)$ as

$$\boldsymbol{n}(\xi) = \frac{\boldsymbol{z}(\xi)}{|\boldsymbol{z}(\xi)|}, \tag{6.14}$$

and then pointing out that

$$\frac{\boldsymbol{z}(\xi) - \boldsymbol{x}_A}{|\boldsymbol{z}(\xi) - \boldsymbol{x}_A|} = \boldsymbol{N}_{AB}, \tag{6.15}$$

it may be seen that relation (6.5) reduces to

$$F^{(2)}(\boldsymbol{x}_A, \boldsymbol{x}_B) = \kappa m^2 \int_0^1 \frac{d\xi}{|\boldsymbol{z}(\xi)|^2} - (1+\gamma)^2 m^2 \int_0^1 \frac{1}{1 + \boldsymbol{n}_A.\boldsymbol{n}(\xi)} \frac{1}{|\boldsymbol{z}(\xi)|^2} d\xi \tag{6.16}$$

when $n = 2$. The first integral in (6.16) is elementary. The second one is easily calculated using a relation already exploited in [29], namely

$$\frac{1}{1 + \boldsymbol{n}_A.\boldsymbol{n}(\xi)} \frac{1}{|\boldsymbol{z}(\xi)|^2} = \frac{d}{d\xi}\left[\frac{\xi}{r_A|\boldsymbol{z}(\xi)| + \boldsymbol{x}_A.\boldsymbol{z}(\xi)}\right].$$

However, finding such a procedure for the higher order terms cannot be reasonably expected. So in this section, we propose another method based on a change of variable which appreciably simplifies the calculations and may be successfully applied at least up to the third order.

Since the metric is spherically symmetric, a nonradial light ray $\Gamma_s(\boldsymbol{x}_A, \boldsymbol{x}_B)$ is confined to the plane defined by $\boldsymbol{x}_A$ and $\boldsymbol{x}_B$. We assume that this plane is the equatorial plane defined by $\vartheta = \pi/2$. The parameter ξ is a monotonic function of the angular coordinate φ. As a consequence, φ can be used as a variable of integration in formulas (5.11) and (5.12). For the sake of brevity, we assume that the direction of the light propagation is such that $\varphi - \varphi_A > 0$ during the motion of the photon. Let us denote by H the foot of the perpendicular drawn from the origin O of the spatial coordinates to the straight line passing through $\boldsymbol{x}_A$ and $\boldsymbol{x}_B$. If φ_c is the value of φ for H, an elementary geometric reasoning shows that

$$|z(\xi)| = \frac{r_c}{\cos(\varphi - \varphi_c)} \tag{6.17}$$

and

$$d\xi = \frac{|z(\xi)|^2}{r_c|\boldsymbol{x}_B - \boldsymbol{x}_A|}\, d\varphi\,, \tag{6.18}$$

where r_c is the zeroth-order distance of closest approach of the light ray to O, namely

$$r_c = OH = \frac{r_A r_B}{|\boldsymbol{x}_B - \boldsymbol{x}_A|}|\boldsymbol{n}_A \times \boldsymbol{n}_B|\,. \tag{6.19}$$

Taking into account the relation

$$\boldsymbol{n}_A.\boldsymbol{n}(\xi) = \cos(\varphi - \varphi_A)\,, \tag{6.20}$$

it appears that $F^{(2)}$ may be written in the form

$$F^{(2)}(\boldsymbol{x}_A, \boldsymbol{x}_B) = \frac{m^2}{r_c|\boldsymbol{x}_B - \boldsymbol{x}_A|}[\kappa\Phi_{12} - (1+\gamma)^2\Phi_{22}]\,, \tag{6.21}$$

where

$$\Phi_{12} = \int_{\varphi_A}^{\varphi_B} d\varphi\,, \quad \Phi_{22} = \int_{\varphi_A}^{\varphi_B} \frac{d\varphi}{1 + \cos(\varphi - \varphi_A)}\,. \tag{6.22}$$

Integrals Φ_{12} and Φ_{22} are elementary. Finding expressions of these integrals in terms of r_A, r_B, $\boldsymbol{n}_A$ and $\boldsymbol{n}_B$ is easily performed by supplementing (6.20) with a relation as follows:

$$|\boldsymbol{n}_A \times \boldsymbol{n}(\xi)| = \sin(\varphi - \varphi_A)\,. \tag{6.23}$$

We get

$$\Phi_{12} = \arccos \boldsymbol{n}_A.\boldsymbol{n}_B, \quad \Phi_{22} = \frac{|\boldsymbol{n}_A \times \boldsymbol{n}_B|}{1 + \boldsymbol{n}_A.\boldsymbol{n}_B}\,. \tag{6.24}$$

Hence expression (6.7) for $\mathcal{T}^{(2)}$.

The same procedure may be applied for determining $F^{(3)}$. Substituting $\mathcal{T}^{(2)}$ from (6.7) in (6.4) yields (see [5])

$$\widehat{\underline{l}}_r^{(2)}(\boldsymbol{x}_A, \boldsymbol{x}_B) = -\frac{m^2|\boldsymbol{N}_{AB} \times \boldsymbol{n}_B|}{r_c^2}\Bigg\{|\boldsymbol{N}_{AB} \times \boldsymbol{n}_B|\left[\kappa - \frac{(1+\gamma)^2}{1+\boldsymbol{n}_A.\boldsymbol{n}_B}\right]\boldsymbol{N}_{AB} + \Bigg\{\kappa\left[\frac{\arccos \boldsymbol{n}_A.\boldsymbol{n}_B}{|\boldsymbol{n}_A \times \boldsymbol{n}_B|}\boldsymbol{N}_{AB}.\boldsymbol{n}_A - \boldsymbol{N}_{AB}.\boldsymbol{n}_B\right] + (1+\gamma)^2\frac{\boldsymbol{N}_{AB}.\boldsymbol{n}_B - \boldsymbol{N}_{AB}.\boldsymbol{n}_A}{1+\boldsymbol{n}_A.\boldsymbol{n}_B}\Bigg\}\boldsymbol{P}_{AB}\Bigg\} . \quad (6.25)$$

Then, calculating $\widehat{\underline{l}}_r^{(1)}.\widehat{\underline{l}}_r^{(2)}$ from (6.12) and (6.25), it is easily deduced from (6.5) that $F^{(3)}$ may be written in the form

$$F^{(3)}(\boldsymbol{x}_A, \boldsymbol{x}_B) = \frac{m^3}{r_c^2|\boldsymbol{x}_B - \boldsymbol{x}_A|}\left[\kappa_3\Phi_{13} - (1+\gamma)\kappa\,\Phi_{23} + (1+\gamma)^3\Phi_{33}\right] , \quad (6.26)$$

where Φ_{13}, Φ_{23} and Φ_{33} are defined by

$$\Phi_{13} = r_c^2|\boldsymbol{x}_B - \boldsymbol{x}_A|\int_0^1 \frac{d\xi}{|z(\xi)|^3} , \quad (6.27)$$

$$\Phi_{23} = \frac{|\boldsymbol{x}_B - \boldsymbol{x}_A|}{r_c}\int_0^1 \Bigg[|\boldsymbol{N}_{AB} \times \boldsymbol{n}(\xi)| + \frac{|\boldsymbol{n}_A \times \boldsymbol{n}(\xi)|(\boldsymbol{N}_{AB}.\boldsymbol{n}(\xi))}{1+\boldsymbol{n}_A.\boldsymbol{n}(\xi)} - (\boldsymbol{N}_{AB}.\boldsymbol{n}_A)\frac{\arccos \boldsymbol{n}_A.\boldsymbol{n}(\xi)}{1+\boldsymbol{n}_A.\boldsymbol{n}(\xi)}\Bigg]|\boldsymbol{N}_{AB} \times \boldsymbol{n}(\xi)|^2 d\xi , \quad (6.28)$$

$$\Phi_{33} = \frac{|\boldsymbol{x}_B - \boldsymbol{x}_A|}{r_c}\int_0^1 \Bigg\{\frac{|\boldsymbol{N}_{AB} \times \boldsymbol{n}(\xi)|}{1+\boldsymbol{n}_A.\boldsymbol{n}(\xi)} + |\boldsymbol{n}_A \times \boldsymbol{n}(\xi)|\frac{\boldsymbol{N}_{AB}.\boldsymbol{n}(\xi) - \boldsymbol{N}_{AB}.\boldsymbol{n}_A}{[1+\boldsymbol{n}_A.\boldsymbol{n}(\xi)]^2}\Bigg\}|\boldsymbol{N}_{AB} \times \boldsymbol{n}(\xi)|^2 d\xi . \quad (6.29)$$

Using (6.17), (6.18), (6.20), and (6.23) supplemented with

$$|\boldsymbol{N}_{AB} \times \boldsymbol{n}(\xi)| = \cos(\varphi - \varphi_c), \quad \boldsymbol{N}_{AB}.\boldsymbol{n}(\xi) = \sin(\varphi - \varphi_c) , \quad (6.30)$$

Φ_{13}, Φ_{23}, and Φ_{33} may be rewritten in the form

$$\Phi_{13} = \int_{\varphi_A}^{\varphi_B} \cos(\varphi - \varphi_c)d\varphi , \quad (6.31)$$

$$\Phi_{23} = \int_{\varphi_A}^{\varphi_B} \Bigg[\cos(\varphi - \varphi_c) + \frac{\sin(\varphi - \varphi_A)\sin(\varphi - \varphi_c)}{1+\cos(\varphi - \varphi_A)} - \sin(\varphi_A - \varphi_c)\frac{\varphi - \varphi_A}{1+\cos(\varphi - \varphi_A)}\Bigg] d\varphi , \quad (6.32)$$

$$\Phi_{33} = \int_{\varphi_A}^{\varphi_B} \left\{ \frac{\cos(\varphi - \varphi_c)}{1 + \cos(\varphi - \varphi_A)} \right.$$

$$\left. + \sin(\varphi - \varphi_A) \frac{\sin(\varphi - \varphi_c) - \sin(\varphi_A - \varphi_c)}{[1 + \cos(\varphi - \varphi_A)]^2} \right\} d\varphi \,. \tag{6.33}$$

Noting that $\varphi - \varphi_c = \varphi - \varphi_A + \varphi_A - \varphi_c$, and then using the trigonometric formulas developing the sine and cosine of a sum of angles, it may be seen that (6.32) and (6.33) transform into

$$\Phi_{23} = -\sin(\varphi_A - \varphi_c) \int_{\varphi_A}^{\varphi_B} \frac{\varphi - \varphi_A + \sin(\varphi - \varphi_A)}{1 + \cos(\varphi - \varphi_A)} d\varphi$$

$$+ (\varphi_B - \varphi_A) \cos(\varphi_A - \varphi_c) \,, \tag{6.34}$$

$$\Phi_{33} = \int_{\varphi_A}^{\varphi_B} \left\{ \frac{\cos(\varphi_A - \varphi_c)}{1 + \cos(\varphi - \varphi_A)} - 2 \frac{\sin(\varphi_A - \varphi_c)\sin(\varphi - \varphi_A)}{[1 + \cos(\varphi - \varphi_A)]^2} \right\} d\varphi \,. \tag{6.35}$$

Integrating expressions (6.34) and (6.35) is straightforward. Taking into account (6.30) written for $\xi = 1$ and noting that

$$|\boldsymbol{N}_{AB} \times \boldsymbol{n}_A| = \frac{r_B |\boldsymbol{n}_A \times \boldsymbol{n}_B|}{|\boldsymbol{x}_B - \boldsymbol{x}_A|}, \quad |\boldsymbol{N}_{AB} \times \boldsymbol{n}_B| = \frac{r_A |\boldsymbol{n}_A \times \boldsymbol{n}_B|}{|\boldsymbol{x}_B - \boldsymbol{x}_A|} \,, \tag{6.36}$$

we get

$$\Phi_{13} = \frac{r_A + r_B}{|\boldsymbol{x}_B - \boldsymbol{x}_A|}(1 - \boldsymbol{n}_A.\boldsymbol{n}_B) \,,$$

$$\Phi_{23} = \Phi_{13} \frac{\arccos \boldsymbol{n}_A.\boldsymbol{n}_B}{|\boldsymbol{n}_A \times \boldsymbol{n}_B|} \,,$$

$$\Phi_{33} = \Phi_{13} \frac{1}{1 + \boldsymbol{n}_A.\boldsymbol{n}_B} \,.$$

Hence equation (6.8) for $\mathcal{T}^{(3)}$. □

The results of this section show that Proposition 1 enables us to perform the calculation of $\mathcal{T}^{(1)}$, $\mathcal{T}^{(2)}$ and $\mathcal{T}^{(3)}$. It is probable that the recurrence relation (5.12) allows explicit calculations for $n \geq 4$. However, it is to be feared that unwieldy calculations have to be performed. So we set out another procedure, recently proposed in [14]. As it has been emphasized in the introduction, this procedure only involves elementary integrations whatever the order of approximation.

7 Second procedure: determination of the $\mathcal{T}^{(n)}$'s from the geodesic equations

7.1 Null geodesic equations

Let Γ be an arbitrary nonradial null geodesic path of the metric $d\tilde{s}^2$. We suppose that Γ is confined in the region $\mathcal{D}_h$ and described by parametric equations $x^\alpha = x^\alpha(\zeta)$, where ζ is an arbitrarily chosen affine parameter. We choose again the spherical coordinates (r, ϑ, φ) so that $\vartheta = \pi/2$ for any point of this path. Denoting by $\tilde{l}_\alpha$ the covariant components of the vector tangent to $\Gamma_s(\boldsymbol{x}_A, \boldsymbol{x}_B)$, an equation as follows:

$$\tilde{l}_0 dx^0 + \tilde{l}_r dr + \tilde{l}_\varphi d\varphi = 0 \tag{7.1}$$

is satisfied along Γ since $\tilde{l}_\alpha$ is a null vector. Owing to the symmetries of the metric, we have

$$\tilde{l}_0 = E\,, \tag{7.2}$$

$$\tilde{l}_\varphi = -J\,, \tag{7.3}$$

with E and J being constants of the motion. For convenience, the affine parameter ζ is chosen in such a way that $E > 0$. Furthermore, it is always possible to suppose $J > 0$ without lack of generality when calculating the time transfer function in a static, spherically symmetric spacetime. Then the quantity defined as

$$b = \frac{J}{E} \tag{7.4}$$

is the impact parameter of the light ray (see, e.g. [36] and [5])[2]. It may be noted that $b = 0$ would correspond to a radial null geodesic.

Since $d\tilde{s}^2 = 0$ along Γ, it follows from (7.2), (7.3), and (7.4) that

$$\tilde{l}_r = -\varepsilon \frac{E}{r} \sqrt{r^2 \mathfrak{U}(r) - b^2}\,, \tag{7.5}$$

where $\varepsilon = 1$ when r is an increasing function of time and $\varepsilon = -1$ when r is a decreasing function of time[3]. Substituting $\tilde{l}_r$ from (7.5) into (7.1), and then dividing throughout by E, we get a relation enabling to determine the light travel time by an integration along Γ, namely

$$dx^0 = b\,d\varphi + \frac{\varepsilon}{r} \sqrt{r^2 \mathfrak{U}(r) - b^2}\,dr\,. \tag{7.6}$$

2 b is an intrinsic quantity attached to Γ since the constants of the motion E and J are themselves coordinate-independent quantities.

3 The sign of ε in equation (7.5) is changed if and only if the photon passes through a pericenter or an apocenter. The passage through an apocenter corresponds to an extreme relativistic case.

Let us now assume that Γ is a quasi-Minkowskian light ray $\Gamma_s(\boldsymbol{x}_A, \boldsymbol{x}_B)$. As we shall see in what follows, our procedure for calculating explicitly the corresponding perturbation functions $\mathcal{T}^{(n)}$ rests on the property shown in the next subsection that the impact parameter b can be determined as a function of $\boldsymbol{x}_A$ and $\boldsymbol{x}_B$ under the form of an expansion in a series in powers of G by taking the partial derivative of $\mathcal{T}$ with respect to the cosine of the angle formed by $\boldsymbol{x}_A$ and $\boldsymbol{x}_B$.

7.2 Post-Minkowskian expansion of the impact parameter

Let $[\varphi_A, \varphi_B]$ be the range of the angular function $\varphi(t)$ along a quasi-Minkowskian light ray $\Gamma_s(\boldsymbol{x}_A, \boldsymbol{x}_B)$. For the sake of brevity, we shall frequently use a notation as follows

$$\mu = \boldsymbol{n}_A.\boldsymbol{n}_B = \cos(\varphi_B - \varphi_A) . \tag{7.7}$$

Using this notation, the time transfer function may be considered as a function of r_A, r_B and μ:

$$\mathcal{T}(\boldsymbol{x}_A, \boldsymbol{x}_B) = \mathcal{T}(r_A, r_B, \mu) .$$

It is then possible to enunciate the following proposition.

Proposition 4. *Let $\boldsymbol{x}_A$ and $\boldsymbol{x}_B$ be two points in $\mathcal{D}_h$ such that both the conditions $\boldsymbol{n}_A \neq \boldsymbol{n}_B$ and (5.13) are fulfilled. The impact parameter b of a quasi-Minkowskian light ray joining $\boldsymbol{x}_A$ and $\boldsymbol{x}_B$ may be expanded in powers of G as follows:*

$$b = r_c\left[1 + \sum_{n=1}^{\infty}\left(\frac{m}{r_c}\right)^n q_n\right], \tag{7.8}$$

where r_c is defined by (6.19) and the quantities q_n are given by

$$q_n = -c\left(\frac{r_c}{m}\right)^n \frac{\sqrt{1-\mu^2}}{r_c}\,\frac{\partial \mathcal{T}^{(n)}(r_A, r_B, \mu)}{\partial \mu}. \tag{7.9}$$

Proof of Proposition 4. Noting that

$$|\boldsymbol{n}_A \times \boldsymbol{n}_B| = \sqrt{1-\mu^2}\,,$$

it is immediately inferred from equation (13) in [5] that the impact parameter of $\Gamma_s(\boldsymbol{x}_A, \boldsymbol{x}_B)$ may be rewritten in the form

$$b = -c\sqrt{1-\mu^2}\,\frac{\partial \mathcal{T}(r_A, r_B, \mu)}{\partial \mu} . \tag{7.10}$$

Substituting $\mathcal{T}$ from (4.15) into (7.10) directly leads to the expansion given by (7.8). The zeroth-order term is easily derived from the elementary formula

$$|\boldsymbol{x}_B - \boldsymbol{x}_A| = \sqrt{r_A^2 - 2r_A r_B \mu + r_B^2}\,. \tag{7.11}$$

Indeed, using (7.11) and taking (6.19) into account yield

$$\frac{\partial |\boldsymbol{x}_B - \boldsymbol{x}_A|}{\partial \mu} = -\frac{r_c}{\sqrt{1-\mu^2}} . \tag{7.12}$$

□

We shall see in the next section that the expression of the time transfer function corresponding to a quasi-Minkowskian light ray can be straightforwardly deduced from Proposition 4.

7.3 Implementation of the method

If $\Gamma_s(\boldsymbol{x}_A, \boldsymbol{x}_B)$ passes through a pericenter $\boldsymbol{x}_P$, the integration of (7.6) requires the determination of $|\boldsymbol{x}_P|$ as a function of $\boldsymbol{x}_A$ and $\boldsymbol{x}_B$. The calculation of the time transfer function is very complicated for such a configuration. Fortunately, owing to the analytic extension theorem, it follows from Proposition 2 that it is sufficient to determine the expression of each term $\mathcal{T}^{(n)}$ as a function of $\boldsymbol{x}_A$ and $\boldsymbol{x}_B$ in an arbitrarily chosen open subset of the domain of analyticity. For this reason, the calculation of the $\mathcal{T}^{(n)}$ are henceforth carried out under the assumption that $\boldsymbol{x}_A$ and $\boldsymbol{x}_B$ fulfill the following conditions:

(a) The radial variable r along a quasi-Minkowskian null geodesic joining $\boldsymbol{x}_A$ and $\boldsymbol{x}_B$ is an increasing function of t:

$$\frac{dr}{dt} > 0 , \quad t_A \leq t \leq t_B . \tag{7.13}$$

(b) An inequality as follows:

$$\boldsymbol{N}_{AB}.\boldsymbol{n}_A > 0 \tag{7.14}$$

is satisfied, with $\boldsymbol{N}_{AB}$ being defined by (5.6).

These conditions considerably simplify the calculations. Indeed, (7.13) eliminates the occurrence of any pericenter (or apocenter) between the emission and the reception of light and (7.14) implies that the projection H of the origin O on the straight line passing through $\boldsymbol{x}_A$ and $\boldsymbol{x}_B$ lies outside the straight segment linking $\boldsymbol{x}_A$ and $\boldsymbol{x}_B$. One has therefore

$$r_c < r_A \leq r \leq r_B \tag{7.15}$$

for any point of $\Gamma_s(\boldsymbol{x}_A, \boldsymbol{x}_B)$. These inequalities ensure that condition (5.13) is met, since $r_h < r_A$ for any point $\boldsymbol{x}_A$ located in $\mathcal{D}_h$.

Under these assumptions, integrating (7.6) along $\Gamma_s(\boldsymbol{x}_A, \boldsymbol{x}_B)$ is straightforward since the range of the angular function $\varphi(r)$ between the emission and the reception of the photon is given by

$$\varphi_B - \varphi_A = \arccos \mu . \tag{7.16}$$

Noting that in this case $\varepsilon = 1$, it may be seen that the time transfer function is then related to the impact parameter b by an equation as follows

$$\mathcal{T}(\boldsymbol{x}_A, \boldsymbol{x}_B) = \frac{1}{c}\left[b \arccos\mu + \int_{r_A}^{r_B} \frac{1}{r}\sqrt{r^2\mathcal{U}(r) - b^2 dr} \right]. \tag{7.17}$$

Since b is a function of $\boldsymbol{x}_A$ and $\boldsymbol{x}_B$ determined by (7.10), (7.17) has to be regarded as an integro-differential equation satisfied by $\mathcal{T}$. In order to solve this integro-differential equation by an iterative procedure, let us substitute (4.7) for $\mathcal{U}$ and (7.8) for b. Expanding $\sqrt{r^2\mathcal{U}(r) - b^2}/r$ in a power series in m/r_c, rearranging the terms and introducing the notation

$$s = \sqrt{r^2 - r_c^2}, \tag{7.18}$$

we get an expression as follows for $\mathcal{T}$:

$$\mathcal{T}(\boldsymbol{x}_A, \boldsymbol{x}_B) = \frac{1}{c}\left[r_c \arccos\mu + \int_{r_A}^{r_B} \frac{s}{r} dr \right] + \frac{1}{c}\sum_{n=1}^{\infty}\left(\frac{m}{r_c}\right)^n \left\{ r_c q_n \arccos\mu + \int_{r_A}^{r_B}\left[U_n - \frac{r_c^2 q_n}{rs}\right] dr \right\}, \tag{7.19}$$

where each U_n is a function of r which may be written in the form

$$U_1 = \frac{(1+\gamma) r_c}{s}, \tag{7.20}$$

$$U_n = \sum_{k=0}^{3n-4} U_{kn}(q_1, \dots, q_{n-1}) r_c^{3n-k-2} \frac{r^{k-n+1}}{s^{2n-1}} \tag{7.21}$$

for $n \geq 2$, with the quantities $U_{kn}(q_1, \dots, q_{n-1})$ being polynomials in $q_1, \dots, q_{n-1}$. Noting that

$$r_c \arccos\mu + \int_{r_A}^{r_B} \frac{s}{r} dr = |\boldsymbol{x}_B - \boldsymbol{x}_A| \tag{7.22}$$

and

$$\arccos\mu - r_c \int_{r_A}^{r_B} \frac{dr}{rs} = 0 \tag{7.23}$$

when conditions (7.13) and (7.14) are met[4], (4.15) is immediately recovered from (7.19), with each perturbation term being given by

$$\mathcal{T}^{(n)}(\boldsymbol{x}_A, \boldsymbol{x}_B) = \frac{1}{c}\left(\frac{m}{r_c}\right)^n \int_{r_A}^{r_B} U_n dr. \tag{7.24}$$

4 Note that (7.22) is just (7.17) written in the case where the gravitational field vanishes, i.e., $m = 0$.

As it has been explained in the beginning of this subsection, the expression of $\mathcal{T}^{(n)}$ as a function of $\boldsymbol{x}_A$ and $\boldsymbol{x}_B$ derived from (7.24) can be regarded as valid even when conditions (7.13) and (7.14) are not met. In this sense, (7.24) constitutes the main ingredient of the procedure developed in the present section.

The fact that the coefficient q_n is not involved in U_n and the property for each coefficient q_k to be proportional to a derivative of the function $\mathcal{T}^{(k)}$ imply that $\mathcal{T}^{(n)}$ can be determined when the sequence of functions $\mathcal{T}^{(1)}, \ldots, \mathcal{T}^{(n-1)}$ is known. To initiate the process, it is sufficient to infer the expression of $\mathcal{T}^{(1)}$ from (7.20) and (7.24). The integration is immediate. Noting that

$$\sqrt{r_A^2 - r_c^2} = r_A \boldsymbol{N}_{AB} . \boldsymbol{n}_A \,, \tag{7.25}$$

$$\sqrt{r_B^2 - r_c^2} = r_B \boldsymbol{N}_{AB} . \boldsymbol{n}_B \tag{7.26}$$

when conditions (7.13) and (7.14) are met, we get again (6.11), as it could be expected. Substituting $\mathcal{T}^{(1)}$ from (6.6) into (7.9) written for $n = 1$, and then using (7.12), it is easily seen that

$$q_1 = \frac{(1+\gamma) r_c}{1 + \boldsymbol{n}_A . \boldsymbol{n}_B} \left(\frac{1}{r_A} + \frac{1}{r_B} \right) . \tag{7.27}$$

Taking into account this determination of q_1, it becomes possible to carry out the calculation of $\mathcal{T}^{(2)}$ since U_2 only involves q_1. Then, q_2 can be derived from (7.9) taken for $n = 2$. Therefore, $\mathcal{T}^{(3)}$ can be calculated since U_3 only involves q_1 and q_2, and so on. It may be added that all the integrations involved on the right-hand side of (7.24) are elementary and can be carried out with any symbolic computer program. As a consequence, the procedure set up in this section allows the explicit calculation of $\mathcal{T}^{(n)}$ as a function of $\boldsymbol{x}_A$ and $\boldsymbol{x}_B$ whatever the order n.

8 Simplification of the second procedure

Even if it involves only elementary integrals, the procedure developed in the previous section is somewhat tedious. Nevertheless, the method can be notably simplified by making use of the differential equation governing the variation of the angular coordinate along the light ray.

8.1 Use of a constraint equation

Equations (7.3) and (7.5) are equivalent to the geodesic equations

$$\frac{d\varphi}{d\zeta} = \frac{J}{r^2\mathfrak{U}(r)}, \tag{8.1}$$

$$\frac{dr}{d\zeta} = \varepsilon\frac{E}{r\mathfrak{U}(r)}\sqrt{r^2\mathfrak{U}(r) - b^2}. \tag{8.2}$$

Eliminating the affine parameter ζ between (8.1) and (8.2) leads to

$$\frac{d\varphi}{dr} = \varepsilon\frac{b}{r}\frac{1}{\sqrt{r^2\mathfrak{U}(r) - b^2}}. \tag{8.3}$$

Since $\varepsilon = 1$ when conditions (7.13) and (7.14) are met, integrating (8.3) and taking into account (7.16) yield thereby

$$\arccos\mu = \int_{r_A}^{r_B} \frac{b}{r\sqrt{r^2\mathfrak{U}(r) - b^2}} dr. \tag{8.4}$$

Equation (8.4) may be regarded as a constraining equation which implicitly determines b as a function of $\boldsymbol{x}_A$ and $\boldsymbol{x}_B$. So it may be expected that this equation implies some conditions on the coefficients q_n which may be used to simplify the calculations.

Replacing $\mathfrak{U}$ by (4.7) and b by (7.8) into (8.4), it may be seen that

$$\arccos\mu = r_c\left[\int_{r_A}^{r_B} \frac{dr}{rs} + \frac{1}{r_c}\sum_{n=1}^{\infty}\left(\frac{m}{r_c}\right)^n \int_{r_A}^{r_B} W_n dr\right], \tag{8.5}$$

where the W_n's are functions of r which may be written in the form

$$W_1 = -(1+\gamma)\frac{r_c^3}{s^3} + q_1\frac{r_c^2 r}{s^3}, \tag{8.6}$$

$$W_n = \sum_{k=0}^{3(n-1)} W_{kn}(q_1,\dots,q_{n-1}) r_c^{3n-k}\frac{r^{k-n+1}}{s^{2n+1}} + q_n\frac{r_c^2 r}{s^3} \tag{8.7}$$

for $n \geq 2$, with the terms $W_{kn}(q_1,\dots,q_{n-1})$ being polynomials in $q_1,\dots,q_{n-1}$. Taking into account (7.23), it is immediately seen that (8.5) reduces to

$$\sum_{n=1}^{\infty}\left(\frac{m}{r_c}\right)^n \int_{r_A}^{r_B} W_n dr = 0 \tag{8.8}$$

for $n \geq 1$. Since (8.8) holds whatever m, it is clear that (8.5) is equivalent to the infinite set of equations

$$\int_{r_A}^{r_B} W_n dr = 0, \quad n = 1, 2, \dots \tag{8.9}$$

The set of constraint equations (8.9) may be systematically used for simplifying our problem. Let us consider the functions U_n^* defined as

$$U_1^* = U_1\,, \tag{8.10}$$

$$U_n^* = U_n + \sum_{p=1}^{n-1} k_{pn} W_p \tag{8.11}$$

for $n \geq 2$, where the k_{pn}'s are arbitrary quantities which do not depend on r. Taking into account (8.9), it is immediately seen that

$$\int_{r_A}^{r_B} U_n dr = \int_{r_A}^{r_B} U_n^* dr\,. \tag{8.12}$$

Hence $\mathcal{T}^{(n)}$ may be rewritten in the form

$$\mathcal{T}^{(n)}(\boldsymbol{x}_A, \boldsymbol{x}_B) = \frac{1}{c}\left(\frac{m}{r_c}\right)^n \int_{r_A}^{r_B} U_n^* dr\,. \tag{8.13}$$

Of course, the remark formulated just after (7.24) might be reproduced here.

It is easily seen that a judicious choice of the quantities k_{pn} enables us to shorten the expressions involved in (8.13) when $n \geq 2$. Until $n = 3$, only the expression of W_1 is needed. Indeed, it is easily inferred from the expansion of (7.17) that U_2 and U_3 are given by

$$U_2 = -\frac{\kappa r_c^4}{rs^3} + \frac{(1+\gamma)q_1 r_c^3}{s^3} + \frac{[2\kappa - (1+\gamma)^2 - q_1^2]r_c^2 r}{2s^3}\,, \tag{8.14}$$

$$\begin{aligned} U_3 = {} & \frac{\kappa_3 r_c^7}{r^2 s^5} - \frac{\kappa q_1 r_c^6}{rs^5} - \frac{[2\kappa_3 - (1+\gamma)(\kappa + q_1^2 - q_2)]r_c^5}{s^5} \\ & + \frac{[2\kappa - 3(1+\gamma)^2 - q_1^2 + 2q_2]q_1 r_c^4 r}{2s^5} \\ & + \frac{[2\kappa_3 - (1+\gamma)(2\kappa - q_1^2 - 2q_2) + (1+\gamma)^3]r_c^3 r^2}{2s^5} - \frac{q_1 q_2 r_c^2 r^3}{s^5}\,. \end{aligned} \tag{8.15}$$

Setting $k_{12} = \frac{1}{2}q_1$ removes the term in q_1^2 in U_2 and leads to

$$U_2^* = -\frac{\kappa r_c^4}{rs^3} + \frac{(1+\gamma)q_1 r_c^3}{2s^3} + \frac{[2\kappa - (1+\gamma)^2]r_c^2 r}{2s^3}\,. \tag{8.16}$$

Choosing $k_{13} = q_2$ and $k_{23} = 0$ remove the terms involving q_2 in U_3. Then U_3^* reduces to

$$\begin{aligned} U_3^* &= \frac{\kappa_3 r_c^7}{r^2 s^5} - \frac{\kappa q_1 r_c^6}{r s^5} - \frac{[2\kappa_3 - (1+\gamma)(\kappa + q_1^2)] r_c^5}{s^5} \\ &+ \frac{[2\kappa - 3(1+\gamma)^2 - q_1^2] q_1 r_c^4 r}{2s^5} \\ &+ \frac{[2\kappa_3 - (1+\gamma)(2\kappa - q_1^2) + (1+\gamma)^3] r_c^3 r^2}{2s^5} . \end{aligned} \tag{8.17}$$

It is thus proved that owing to the constraint equation (8.4), only the determination of q_1 is required for calculating the functions $\mathcal{T}^{(2)}$ and $\mathcal{T}^{(3)}$.

Remark. It may be pointed out that the coefficients q_n could be directly inferred from the constraint equation without differentiating the functions $\mathcal{T}^{(n)}$ with respect to μ. Indeed, it follows from (8.6), (8.7), and (8.9) that

$$q_1 = \frac{1+\gamma}{r_c} \frac{r_A \sqrt{r_B^2 - r_c^2} - r_B \sqrt{r_A^2 - r_c^2}}{\sqrt{r_B^2 - r_c^2} - \sqrt{r_A^2 - r_c^2}} , \tag{8.18}$$

$$\begin{aligned} q_n &= -\frac{1}{r_c} \frac{\sqrt{r_A^2 - r_c^2}\sqrt{r_B^2 - r_c^2}}{\sqrt{r_B^2 - r_c^2} - \sqrt{r_A^2 - r_c^2}} \\ &\times \sum_{k=0}^{3(n-1)} W_{kn}(q_1, \dots, q_{n-1}) r_c^{3n-k-1} \int_{r_A}^{r_B} \frac{r^{k-n+1}}{s^{2n+1}} dr \end{aligned} \tag{8.19}$$

for $n \geq 2$. Equation (8.19) shows that q_n can be determined once $q_1, \dots, q_{n-1}$ are known.

It is easily checked that (8.18) is equivalent to (7.27). Indeed, noting that

$$\sqrt{r_B^2 - r_c^2} - \sqrt{r_A^2 - r_c^2} = |\boldsymbol{x}_B - \boldsymbol{x}_A| \tag{8.20}$$

when conditions (7.13) and (7.14) are met, and then taking into account (6.19), (7.25) and (7.26), it may be seen that (8.18) transforms into

$$q_1 = (1+\gamma) \frac{\boldsymbol{N}_{AB}.\boldsymbol{n}_B - \boldsymbol{N}_{AB}.\boldsymbol{n}_A}{|\boldsymbol{n}_A \times \boldsymbol{n}_B|} . \tag{8.21}$$

Substituting $r_A \boldsymbol{n}_A$ for $\boldsymbol{x}_A$ and $r_B \boldsymbol{n}_B$ for $\boldsymbol{x}_B$ into the numerator of the right-handside of (5.6) yields

$$\boldsymbol{N}_{AB}.\boldsymbol{n}_B - \boldsymbol{N}_{AB}.\boldsymbol{n}_A = \frac{(r_A + r_B)(1 - \boldsymbol{n}_A.\boldsymbol{n}_B)}{|\boldsymbol{x}_B - \boldsymbol{x}_A|} . \tag{8.22}$$

Finally, substituting for $\boldsymbol{N}_{AB}.\boldsymbol{n}_B - \boldsymbol{N}_{AB}.\boldsymbol{n}_A$ from (8.22) into (8.21), and then noting that (6.19) is equivalent to

$$\frac{1}{|\boldsymbol{x}_B - \boldsymbol{x}_A|} = \frac{r_c}{r_A r_B} \frac{1}{|\boldsymbol{n}_A \times \boldsymbol{n}_B|} ,$$

it is immediately seen that (7.27) is recovered.

8.2 Explicit calculation of $\mathcal{T}^{(1)}$, $\mathcal{T}^{(2)}$, and $\mathcal{T}^{(3)}$

We are now in a position to determine the perturbation terms involved in the expansion of the time transfer function up to the order G^3. The term $\mathcal{T}^{(1)}$ has been already treated in Section 7.3. For $n = 2$ and $n = 3$, it follows from (8.16) and (8.17) that $\mathcal{T}^{(n)}$ may be written in the form

$$\mathcal{T}^{(n)}(\boldsymbol{x}_A, \boldsymbol{x}_B) = \frac{1}{c}\left(\frac{m}{r_c}\right)^n \sum_{k=0}^{\sigma(n)} U^*_{kn}(q_1) r_c^{3n-k-2} \int_{r_A}^{r_B} \frac{r^{k-n+1}}{s^{2n-1}}\, dr\,, \tag{8.23}$$

where $\sigma(2) = 2$ and $\sigma(3) = 4$, with the coefficients U^*_{kn} being polynomials in q_1. The integrals occurring into the right-hand side of (8.23) are elementary and can be expressed in terms of $r_A, r_B, r_c, \sqrt{r_A^2 - r_c^2}$ and $\sqrt{r_B^2 - r_c^2}$. For the explicit calculations, it is convenient to write (7.25) and (7.26) in the form

$$\sqrt{r_A^2 - r_c^2} = \frac{r_A(r_B\mu - r_A)}{|\boldsymbol{x}_B - \boldsymbol{x}_A|}\,, \tag{8.24}$$

$$\sqrt{r_B^2 - r_c^2} = \frac{r_B(r_B - r_A\mu)}{|\boldsymbol{x}_B - \boldsymbol{x}_A|}\,. \tag{8.25}$$

Using (6.19), (7.27), (8.24), and (8.25), it may be seen that $\mathcal{T}^{(2)}$ and $\mathcal{T}^{(3)}$ can be expressed in terms of $r_A r_B$, $1/r_A + 1/r_B$, $|\boldsymbol{x}_B - \boldsymbol{x}_A|$ and μ. It has been already emphasized that the explicit calculations can be performed with any symbolic computer program. Of course, a simple hand calculation is also possible. For $n = 2$ and $n = 3$, the calculations are greatly facilitated by noting that (8.16) and (8.17) are equivalent to

$$\frac{1}{r_c^2}U_2^* = \frac{\kappa}{rs} + \frac{(1+\gamma)q_1 r_c}{2s^3} - \frac{(1+\gamma)^2 r}{2s^3} \tag{8.26}$$

and

$$\frac{1}{r_c^3}U_3^* = \frac{\kappa_3}{r^2 s} - \frac{(1+\gamma)\kappa}{s^3} + \frac{\kappa q_1 r_c}{rs^3} + \frac{(1+\gamma)[(1+\gamma)^2 + q_1^2]r^2}{2s^5} - \frac{[3(1+\gamma)^2 + q_1^2]q_1 r_c r}{2s^5} + \frac{(1+\gamma)q_1^2 r_c^2}{s^5}\,, \tag{8.27}$$

respectively. Calculating $\mathcal{T}^{(2)}$ from (8.13) and (8.26) is elementary and straightforwardly yields (6.7). Calculating $\mathcal{T}^{(3)}$ from (8.13) and (8.27) requires somewhat tedious calculations, which are detailed in an appendix of [14]. The result coincides with (6.8). We have seen in Section 7.3 that the expressions thus obtained can be considered as valid even when conditions (7.13) and (7.14) are not fulfilled. So we can state that at least up to the third order, the procedures developed in Sections 6 and 7 lead to identical expressions for the first three perturbation terms involved in the expansion of the time transfer function. This concordance confirms the reliability of the second procedure presented in this chapter.

9 Direction of light propagation up to order G^3

We are now in a position to obtain explicit expressions for the triples giving the direction of light propagation at points $\boldsymbol{x}_A$ and $\boldsymbol{x}_B$ up to the third order in G. The vector functions $\widehat{\underline{\boldsymbol{l}}}_e^{(n)}$ and $\widehat{\underline{\boldsymbol{l}}}_r^{(n)}$ could be straightforwardly derived for $n = 1, 2, 3$ by substituting for $\mathcal{T}^{(n)}$ from (6.6)–(6.8) into (6.3) and (6.4). Nevertheless, the calculation is greatly facilitated by making use of formulas (17a) and (17b) given in [5]. Indeed, taking into account (7.8), these formulas lead to

$$\widehat{\underline{\boldsymbol{l}}}_e^{(n)}(\boldsymbol{x}_A, \boldsymbol{x}_B) = \left[c\frac{\partial \mathcal{T}^{(n)}}{\partial r_A}\boldsymbol{N}_{AB}.\boldsymbol{n}_A - \frac{m^n}{r_A^n}\frac{r_A^{n-1}}{r_c^{n-1}}q_n|\boldsymbol{N}_{AB}\times\boldsymbol{n}_A|\right]\boldsymbol{N}_{AB} + \left[c\frac{\partial \mathcal{T}^{(n)}}{\partial r_A}|\boldsymbol{N}_{AB}\times\boldsymbol{n}_A| + \frac{m^n}{r_A^n}\frac{r_A^{n-1}}{r_c^{n-1}}q_n\boldsymbol{N}_{AB}.\boldsymbol{n}_A\right]\boldsymbol{P}_{AB}, \tag{9.1}$$

$$\widehat{\underline{\boldsymbol{l}}}_r(\boldsymbol{x}_A, \boldsymbol{x}_B) = -\left[c\frac{\partial \mathcal{T}^{(n)}}{\partial r_B}\boldsymbol{N}_{AB}.\boldsymbol{n}_B + \frac{m^n}{r_B^n}\frac{r_B^{n-1}}{r_c^{n-1}}q_n|\boldsymbol{N}_{AB}\times\boldsymbol{n}_B|\right]\boldsymbol{N}_{AB} - \left[c\frac{\partial \mathcal{T}^{(n)}}{\partial r_B}|\boldsymbol{N}_{AB}\times\boldsymbol{n}_B| - \frac{m^n}{r_B^n}\frac{r_B^{n-1}}{r_c^{n-1}}q_n\boldsymbol{N}_{AB}.\boldsymbol{n}_B\right]\boldsymbol{P}_{AB}, \tag{9.2}$$

where $\boldsymbol{P}_{AB}$ is defined by (6.13).

Equations (9.1) and (9.2) show that knowing $c\partial\mathcal{T}^{(n)}/\partial r_A$, $c\partial\mathcal{T}^{(n)}/\partial r_B$ and q_n is sufficient to determine the triples $\widehat{\underline{\boldsymbol{l}}}_e^{(n)}$ and $\widehat{\underline{\boldsymbol{l}}}_r^{(n)}$. The coefficient q_1 is given by (7.27). Replacing $\boldsymbol{n}_A.\boldsymbol{n}_B$ by μ in (6.7) and (6.8), and then taking into account (7.11), q_2 and q_3 are straightforwardly derived from (7.9). Noting that

$$\frac{r_A r_B(1-\mu^2)}{|\boldsymbol{x}_B - \boldsymbol{x}_A|^2} = \frac{r_c^2}{r_A r_B}, \tag{9.3}$$

we can formulate the proposition below.

Proposition 5. *The coefficients q_1, q_2 and q_3 involved in the expansion of the impact parameter b of a quasi-Minkowskian light ray joining $\boldsymbol{x}_A$ and $\boldsymbol{x}_B$ are given by*

$$q_1(\boldsymbol{x}_A, \boldsymbol{x}_B) = (1+\gamma)\left(\frac{r_c}{r_A} + \frac{r_c}{r_B}\right)\frac{1}{1+\boldsymbol{n}_A.\boldsymbol{n}_B}, \tag{9.4}$$

$$q_2(\boldsymbol{x}_A, \boldsymbol{x}_B) = \kappa\left[1 - \left(\boldsymbol{n}_A.\boldsymbol{n}_B - \frac{r_c^2}{r_A r_B}\right)\frac{\arccos \boldsymbol{n}_A.\boldsymbol{n}_B}{|\boldsymbol{n}_A\times\boldsymbol{n}_B|}\right] - \frac{(1+\gamma)^2}{1+\boldsymbol{n}_A.\boldsymbol{n}_B}\left(1 - \boldsymbol{n}_A.\boldsymbol{n}_B + \frac{r_c^2}{r_A r_B}\right), \tag{9.5}$$

$$q_3(\boldsymbol{x}_A,\boldsymbol{x}_B) = \left(\frac{r_c}{r_A}+\frac{r_c}{r_B}\right)\frac{1}{1+\boldsymbol{n}_A.\boldsymbol{n}_B}\left\{\kappa_3\left(1-\boldsymbol{n}_A.\boldsymbol{n}_B+\frac{r_c^2}{r_Ar_B}\right)\right.$$
$$-(1+\gamma)\kappa\left[1+\left(1-2\boldsymbol{n}_A.\boldsymbol{n}_B+\frac{r_c^2}{r_Ar_B}\right)\frac{\arccos \boldsymbol{n}_A.\boldsymbol{n}_B}{|\boldsymbol{n}_A\times\boldsymbol{n}_B|}\right]$$
$$\left.+\frac{(1+\gamma)^3}{1+\boldsymbol{n}_A.\boldsymbol{n}_B}\left(2-2\boldsymbol{n}_A.\boldsymbol{n}_B+\frac{r_c^2}{r_Ar_B}\right)\right\}. \tag{9.6}$$

Noting that

$$\frac{1}{r_A}=\frac{|\boldsymbol{N}_{AB}\times\boldsymbol{n}_A|}{r_c}, \quad \frac{1}{r_B}=\frac{|\boldsymbol{N}_{AB}\times\boldsymbol{n}_B|}{r_c} \tag{9.7}$$

and

$$\frac{r_c^2}{r_Ar_B}=\boldsymbol{n}_A.\boldsymbol{n}_B-(\boldsymbol{N}_{AB}.\boldsymbol{n}_A)(\boldsymbol{N}_{AB}.\boldsymbol{n}_B), \tag{9.8}$$

it is easily seen that formulas (9.4)–(9.6) are equivalent to the expressions of q_1, q_2, q_3 obtained in [14].

Deriving now $c\partial\mathcal{T}^{(n)}/\partial r_A$ and $c\partial\mathcal{T}^{(n)}/\partial r_B$ for $n = 1, 2, 3$ from (6.6)–(6.8), and then substituting for the q_n from equations (9.4)–(9.6) into (9.1) and (9.2), straightforward calculations lead to the explicit expressions of the light direction triples up to order G^3. In fact, we may content ourselves with calculating $\widehat{\underline{l}}_e^{(n)}$ or $\widehat{\underline{l}}_r^{(n)}$ since a relation as follows

$$\widehat{\underline{l}}_r^{(n)}(\boldsymbol{x}_A,\boldsymbol{x}_B) = -\widehat{\underline{l}}_e^{(n)}(\boldsymbol{x}_B,\boldsymbol{x}_A) \tag{9.9}$$

results from (6.3)–(6.4) when the symmetry law $\mathcal{T}^{(n)}(\boldsymbol{x}_A,\boldsymbol{x}_B) = \mathcal{T}^{(n)}(\boldsymbol{x}_B,\boldsymbol{x}_A)$ is taken into account. We get the proposition which follows.

Proposition 6. *Under the assumption of Proposition 3, the triples* $\widehat{\underline{l}}_e^{(n)}$ *and* $\widehat{\underline{l}}_r^{(n)}$ *are given for* $n = 1, 2, 3$ *by*

$$\widehat{\underline{l}}_e^{(1)}(\boldsymbol{x}_A,\boldsymbol{x}_B) = -\frac{(1+\gamma)m}{r_A}\left[\boldsymbol{N}_{AB}+\frac{|\boldsymbol{n}_A\times\boldsymbol{n}_B|}{1+\boldsymbol{n}_A.\boldsymbol{n}_B}\boldsymbol{P}_{AB}\right], \tag{9.10}$$

$$\widehat{\underline{l}}_e^{(2)}(\boldsymbol{x}_A,\boldsymbol{x}_B) = -\frac{m^2}{r_A^2}\left[\kappa-\frac{(1+\gamma)^2}{1+\boldsymbol{n}_A.\boldsymbol{n}_B}\right]\boldsymbol{N}_{AB}$$
$$-\frac{m^2}{r_A^2}\frac{1}{|\boldsymbol{n}_A\times\boldsymbol{n}_B|}\left\{\kappa\left[\frac{r_A}{r_B}-\boldsymbol{n}_A.\boldsymbol{n}_B\right.\right.$$
$$\left.+\left(1-\frac{r_A}{r_B}\boldsymbol{n}_A.\boldsymbol{n}_B\right)\frac{\arccos \boldsymbol{n}_A.\boldsymbol{n}_B}{|\boldsymbol{n}_A\times\boldsymbol{n}_B|}\right]$$
$$\left.-(1+\gamma)^2\left(1+\frac{r_A}{r_B}\right)\frac{1-\boldsymbol{n}_A.\boldsymbol{n}_B}{1+\boldsymbol{n}_A.\boldsymbol{n}_B}\right\}\boldsymbol{P}_{AB}, \tag{9.11}$$

$$
\begin{aligned}
\widehat{\underline{l}}_e^{(3)}(\boldsymbol{x}_A,\boldsymbol{x}_B) = &-\frac{m^3}{r_A^3}\Bigg\{\kappa_3 - \frac{(1+\gamma)\kappa}{1+\boldsymbol{n}_A.\boldsymbol{n}_B}\Bigg[1+\frac{r_A}{r_B} \\
&+\left(1-\frac{r_A}{r_B}\boldsymbol{n}_A.\boldsymbol{n}_B\right)\frac{\arccos \boldsymbol{n}_A.\boldsymbol{n}_B}{|\boldsymbol{n}_A\times\boldsymbol{n}_B|}\Bigg] \\
&+\frac{(1+\gamma)^3}{(1+\boldsymbol{n}_A.\boldsymbol{n}_B)^2}\left[2+\frac{r_A}{r_B}(1-\boldsymbol{n}_A.\boldsymbol{n}_B)\right]\Bigg\}\boldsymbol{N}_{AB} \\
&-\frac{m^3}{r_A^3}\frac{1}{|\boldsymbol{n}_A\times\boldsymbol{n}_B|}\Bigg\{\kappa_3(1-\boldsymbol{n}_A.\boldsymbol{n}_B)\left[1+\left(1+\frac{r_A}{r_B}\right)^2\frac{1}{1+\boldsymbol{n}_A.\boldsymbol{n}_B}\right] \\
&-\frac{(1+\gamma)\kappa}{1+\boldsymbol{n}_A.\boldsymbol{n}_B}\Bigg\{\left(1+\frac{r_A}{r_B}\right)\left(\frac{r_A}{r_B}-\boldsymbol{n}_A.\boldsymbol{n}_B\right) \\
&+\Bigg[\left(1+\frac{r_A}{r_B}\right)^2(1-2\boldsymbol{n}_A.\boldsymbol{n}_B) \\
&+\left(1+\frac{r_A}{r_B}\boldsymbol{n}_A.\boldsymbol{n}_B\right)(1+\boldsymbol{n}_A.\boldsymbol{n}_B)\Bigg]\frac{\arccos \boldsymbol{n}_A.\boldsymbol{n}_B}{|\boldsymbol{n}_A\times\boldsymbol{n}_B|}\Bigg\} \\
&+(1+\gamma)^3\frac{1-\boldsymbol{n}_A.\boldsymbol{n}_B}{1+\boldsymbol{n}_A.\boldsymbol{n}_B}\left[\left(1+\frac{r_A}{r_B}\right)^2\frac{2}{1+\boldsymbol{n}_A.\boldsymbol{n}_B}-\frac{r_A}{r_B}\right]\Bigg\}\boldsymbol{P}_{AB}
\end{aligned}
\tag{9.12}
$$

and

$$
\widehat{\underline{l}}_r^{(1)}(\boldsymbol{x}_A,\boldsymbol{x}_B) = -\frac{(1+\gamma)m}{r_B}\left[\boldsymbol{N}_{AB}-\frac{|\boldsymbol{n}_A\times\boldsymbol{n}_B|}{1+\boldsymbol{n}_A.\boldsymbol{n}_B}\boldsymbol{P}_{AB}\right], \tag{9.13}
$$

$$
\begin{aligned}
\widehat{\underline{l}}_r^{(2)}(\boldsymbol{x}_A,\boldsymbol{x}_B) = &-\frac{m^2}{r_B^2}\left[\kappa-\frac{(1+\gamma)^2}{1+\boldsymbol{n}_A.\boldsymbol{n}_B}\right]\boldsymbol{N}_{AB} \\
&+\frac{m^2}{r_B^2}\frac{1}{|\boldsymbol{n}_A\times\boldsymbol{n}_B|}\Bigg\{\kappa\Bigg[\frac{r_B}{r_A}-\boldsymbol{n}_A.\boldsymbol{n}_B \\
&+\left(1-\frac{r_B}{r_A}\boldsymbol{n}_A.\boldsymbol{n}_B\right)\frac{\arccos \boldsymbol{n}_A.\boldsymbol{n}_B}{|\boldsymbol{n}_A\times\boldsymbol{n}_B|}\Bigg] \\
&-(1+\gamma)^2\left(1+\frac{r_B}{r_A}\right)\frac{1-\boldsymbol{n}_A.\boldsymbol{n}_B}{1+\boldsymbol{n}_A.\boldsymbol{n}_B}\Bigg\}\boldsymbol{P}_{AB},
\end{aligned}
\tag{9.14}
$$

$$
\begin{aligned}
\widehat{\underline{l}}_r^{(3)}(\boldsymbol{x}_A,\boldsymbol{x}_B) = &-\frac{m^3}{r_B^3}\Bigg\{\kappa_3-\frac{(1+\gamma)\kappa}{1+\boldsymbol{n}_A.\boldsymbol{n}_B}\Bigg[1+\frac{r_B}{r_A} \\
&+\left(1-\frac{r_B}{r_A}\boldsymbol{n}_A.\boldsymbol{n}_B\right)\frac{\arccos \boldsymbol{n}_A.\boldsymbol{n}_B}{|\boldsymbol{n}_A\times\boldsymbol{n}_B|}\Bigg] \\
&+\frac{(1+\gamma)^3}{(1+\boldsymbol{n}_A.\boldsymbol{n}_B)^2}\left[2+\frac{r_B}{r_A}(1-\boldsymbol{n}_A.\boldsymbol{n}_B)\right]\Bigg\}\boldsymbol{N}_{AB}
\end{aligned}
$$

$$+\frac{m^3}{r_B^3}\frac{1}{|\boldsymbol{n}_A\times\boldsymbol{n}_B|}\left\{\kappa_3(1-\boldsymbol{n}_A.\boldsymbol{n}_B)\left[1+\left(1+\frac{r_B}{r_A}\right)^2\frac{1}{1+\boldsymbol{n}_A.\boldsymbol{n}_B}\right]\right.$$
$$-\frac{(1+\gamma)\kappa}{1+\boldsymbol{n}_A.\boldsymbol{n}_B}\left\{\left(1+\frac{r_B}{r_A}\right)\left(\frac{r_B}{r_A}-\boldsymbol{n}_A.\boldsymbol{n}_B\right)\right.$$
$$+\left[\left(1+\frac{r_B}{r_A}\right)^2(1-2\boldsymbol{n}_A.\boldsymbol{n}_B)\right.$$
$$\left.\left.+\left(1+\frac{r_B}{r_A}\boldsymbol{n}_A.\boldsymbol{n}_B\right)(1+\boldsymbol{n}_A.\boldsymbol{n}_B)\right]\frac{\arccos\boldsymbol{n}_A.\boldsymbol{n}_B}{|\boldsymbol{n}_A\times\boldsymbol{n}_B|}\right\}$$
$$\left.+(1+\gamma)^3\frac{1-\boldsymbol{n}_A.\boldsymbol{n}_B}{1+\boldsymbol{n}_A.\boldsymbol{n}_B}\left[\left(1+\frac{r_B}{r_A}\right)^2\frac{2}{1+\boldsymbol{n}_A.\boldsymbol{n}_B}-\frac{r_B}{r_A}\right]\right\}\boldsymbol{P}_{AB}, \quad (9.15)$$

respectively.

Using (9.7), it is easily checked that equations (9.10) and (9.11) and (9.13) and (9.14) allow us to recover expressions (38*a*) and (38*b*) obtained in [5] for the expansion of $\hat{\underline{l}}_e$ and $\hat{\underline{l}}_r$ up to the second order in G. Of course, (9.13) and (9.14) are equivalent to (6.12) and (6.25), respectively. On the other hand, formulas (9.12) and (9.15) are new results.

10 Light ray emitted at infinity

A quasi-Minkowskian light ray coming from infinity in an initial direction defined by a given unit vector $\boldsymbol{N}_e$ and observed at a given point $\boldsymbol{x}_B$ is a relevant limiting case for modeling a lot of astrometric measurements. According to a notation introduced in [5], such a ray is denoted by $\Gamma_s(\boldsymbol{N}_e, \boldsymbol{x}_B)$. The corresponding null geodesic is assumed to be a perturbation in powers of G of the straight segment defined by the parametric equations

$$x^0_{(0)}(\lambda) = ct_B + \lambda r_c\,, \quad \boldsymbol{x}_{(0)}(\lambda) = \lambda r_c \boldsymbol{N}_e + \boldsymbol{x}_B\,, \quad -\infty < \lambda \le 0\,, \quad (10.1)$$

where

$$r_c = r_B|\boldsymbol{N}_e \times \boldsymbol{n}_B|\,. \quad (10.2)$$

Given the direction $\boldsymbol{N}_e$, $\boldsymbol{x}_B$ is supposed to satisfy the condition

$$|\lambda r_c \boldsymbol{N}_e + \boldsymbol{x}_B| > r_h \quad (10.3)$$

when $-\infty < \lambda \le 0$ in order to ensure that the straight segment coming from infinity in the direction $\boldsymbol{N}_e$ and ending at $\boldsymbol{x}_B$ is entirely lying in $\mathcal{D}_h$. Condition (10.3) is the extension of condition (5.13) when $\boldsymbol{x}_A$ is at infinity.

To apply the results of Section 9, let us consider a point $\boldsymbol{x}_A$ lying on $\Gamma_s(\boldsymbol{N}_e, \boldsymbol{x}_B)$. It is clear that the part of $\Gamma_s(\boldsymbol{N}_e, \boldsymbol{x}_B)$ joining $\boldsymbol{x}_A$ and $\boldsymbol{x}_B$ coincides with a quasi-Minkowskian null geodesic path $\Gamma_s(\boldsymbol{x}_A, \boldsymbol{x}_B)$. So, the impact parameters of $\Gamma_s(\boldsymbol{N}_e, \boldsymbol{x}_B)$ and $\Gamma_s(\boldsymbol{x}_A, \boldsymbol{x}_B)$ are equal. As a consequence, the coefficients q_1, q_2 and q_3 can be obtained as functions of $\boldsymbol{N}_e$ and $\boldsymbol{x}_B$ by taking the limit of equations (9.4)–(9.6) when $\boldsymbol{x}_A$ recedes toward the source of the light ray at infinity, i.e. when $r_A \to \infty$, $\boldsymbol{n}_A \to -\boldsymbol{N}_e$ and $\boldsymbol{N}_{AB} \to \boldsymbol{N}_e$. Using (10.2), and then taking into account that $\arccos \boldsymbol{n}_A.\boldsymbol{n}_B \to \pi - \arccos \boldsymbol{N}_e.\boldsymbol{n}_B$ when $\boldsymbol{n}_A \to -\boldsymbol{N}_e$, the following propositions can be stated.

Proposition 7. *Let $\boldsymbol{N}_e$ be a unit vector and $\boldsymbol{x}_B$ a point in $\mathcal{D}_h$ fulfilling condition* (10.3). *The impact parameter of a quasi-Minkowskian light ray emitted at infinity in the direction $\boldsymbol{N}_e$ and arriving at $\boldsymbol{x}_B$ is given by expansion* (7.8), *where r_c is expressed by* (10.2) *and the coefficients q_1, q_2, and q_3 are yielded by*[5]

$$q_1(\boldsymbol{N}_e, \boldsymbol{x}_B) = (1+\gamma)\frac{|\boldsymbol{N}_e \times \boldsymbol{n}_B|}{1 - \boldsymbol{N}_e.\boldsymbol{n}_B}, \tag{10.4}$$

$$q_2(\boldsymbol{N}_e, \boldsymbol{x}_B) = \kappa\left[1 + \boldsymbol{N}_e.\boldsymbol{n}_B \frac{\pi - \arccos \boldsymbol{N}_e.\boldsymbol{n}_B}{|\boldsymbol{N}_e \times \boldsymbol{n}_B|}\right] - (1+\gamma)^2 \frac{1 + \boldsymbol{N}_e.\boldsymbol{n}_B}{1 - \boldsymbol{N}_e.\boldsymbol{n}_B}, \tag{10.5}$$

$$\begin{aligned} q_3(\boldsymbol{N}_e, \boldsymbol{x}_B) = \frac{|\boldsymbol{N}_e \times \boldsymbol{n}_B|}{1 - \boldsymbol{N}_e.\boldsymbol{n}_B}\Bigg\{&\kappa_3\,(1 + \boldsymbol{N}_e.\boldsymbol{n}_B) + 2(1+\gamma)^3 \frac{1 + \boldsymbol{N}_e.\boldsymbol{n}_B}{1 - \boldsymbol{N}_e.\boldsymbol{n}_B} \\ &- (1+\gamma)\kappa\left[1 + (1 + 2\boldsymbol{N}_e.\boldsymbol{n}_B)\,\frac{\pi - \arccos \boldsymbol{N}_e.\boldsymbol{n}_B}{|\boldsymbol{N}_e \times \boldsymbol{n}_B|}\right]\Bigg\}. \end{aligned} \tag{10.6}$$

Proposition 8. *For the ray considered in Proposition 7, the light direction triple at point $\boldsymbol{x}_B$ is determined up to the third order by*

$$\hat{\underline{l}}_r^{(1)}(\boldsymbol{N}_e, \boldsymbol{x}_B) = -\frac{(1+\gamma)m}{r_B}\left[\boldsymbol{N}_e - \frac{|\boldsymbol{N}_e \times \boldsymbol{n}_B|}{1 - \boldsymbol{N}_e.\boldsymbol{n}_B}\boldsymbol{P}_e\right], \tag{10.7}$$

$$\begin{aligned} \hat{\underline{l}}_r^{(2)}(\boldsymbol{N}_e, \boldsymbol{x}_B) = &-\frac{m^2}{r_B^2}\left[\kappa - \frac{(1+\gamma)^2}{1 - \boldsymbol{N}_e.\boldsymbol{n}_B}\right]\boldsymbol{N}_e \\ &+ \frac{m^2}{r_B^2}\frac{1}{|\boldsymbol{N}_e \times \boldsymbol{n}_B|}\Bigg\{\kappa\left[\boldsymbol{N}_e.\boldsymbol{n}_B + \frac{\pi - \arccos \boldsymbol{N}_e.\boldsymbol{n}_B}{|\boldsymbol{N}_e \times \boldsymbol{n}_B|}\right] \\ &- (1+\gamma)^2 \frac{1 + \boldsymbol{N}_e.\boldsymbol{n}_B}{1 - \boldsymbol{N}_e.\boldsymbol{n}_B}\Bigg\}\boldsymbol{P}_e, \end{aligned} \tag{10.8}$$

$$\begin{aligned} \hat{\underline{l}}_r^{(3)}(\boldsymbol{N}_e, \boldsymbol{x}_B) = &-\frac{m^3}{r_B^3}\Bigg\{\kappa_3 - \frac{(1+\gamma)\kappa}{1 - \boldsymbol{N}_e.\boldsymbol{n}_B}\left[1 + \frac{\pi - \arccos \boldsymbol{N}_e.\boldsymbol{n}_B}{|\boldsymbol{N}_e \times \boldsymbol{n}_B|}\right] \\ &+ \frac{2(1+\gamma)^3}{(1 - \boldsymbol{N}_e.\boldsymbol{n}_B)^2}\Bigg\}\boldsymbol{N}_e \end{aligned}$$

5 Note that equation (114) yielding q_2 in [14] contains an extra factor $|\boldsymbol{N}_e \times \boldsymbol{n}_B|$ due to a typographic mistake. See the corrigendum quoted in [14].

$$+\frac{m^3}{r_B^3}\frac{1}{|N_e\times n_B|}\left\{\kappa_3\frac{1+N_e.n_B}{1-N_e.n_B}(2-N_e.n_B)\right.$$
$$-\frac{(1+\gamma)\kappa}{1-N_e.n_B}\left[N_e.n_B+(2+N_e.n_B)\frac{\pi-\arccos N_e.n_B}{|N_e\times n_B|}\right]$$
$$\left.+2(1+\gamma)^3\frac{1+N_e.n_B}{(1-N_e.n_B)^2}\right\}P_e, \tag{10.9}$$

where P_e is defined as

$$P_e=\left(\frac{N_e\times n_B}{|N_e\times n_B|}\right)\times N_e. \tag{10.10}$$

Exactly as in the case where the emission point is located at a finite distance from the origin, formula (10.9) is new, whereas (10.7) and (10.8) are equivalent to the expressions of the direction triples up to the second order derived in [5].

11 Enhanced terms in $\mathcal{T}^{(1)}$, $\mathcal{T}^{(2)}$, and $\mathcal{T}^{(3)}$

In the present chapter, the time transfer function $\mathcal{T}$ is obtained in the form of an asymptotic expansion in power series in G (or m) provided that condition (5.13) is met. However, it is clear that the physical reliability of this expansion requires that inequalities as follow:

$$\left|\mathcal{T}^{(n)}(x_A,x_B)\right| \ll \left|\mathcal{T}^{(n-1)}(x_A,x_B)\right| \tag{11.1}$$

are satisfied for any $n\geq 1$, with $\mathcal{T}^{(0)}(x_A,x_B)$ being conventionally defined as

$$\mathcal{T}^{(0)}(x_A,x_B)=\frac{1}{c}|x_B-x_A|\,.$$

The results obtained in the previous section enable us to find the conditions ensuring inequalities (11.1) for $n=1,2,3$. It is clear that the magnitude of the functions given by (6.6)–(6.8) may be extremely large when points x_A and x_B are located in almost opposite directions. This behavior corresponds to the "enhanced terms" determined up to G^2 for the deflection of light in [27] and up to G^3 for the time transfer function in [13]. Indeed, it is straightforwardly derived from (6.19) that

$$\frac{1}{1+n_A.n_B}\sim\frac{2r_A^2r_B^2}{(r_A+r_B)^2}\frac{1}{r_c^2} \tag{11.2}$$

when $1+n_A.n_B\to 0$. Using this relation to eliminate $1+n_A.n_B$, the following proposition is easily deduced from (6.6)–(6.8).

Proposition 9. *When x_A and x_B tend to be located in opposite directions (i.e. $1+n_A.n_B\to 0$), the first three perturbation terms in the time transfer function are enhanced ac-*

cording to the asymptotic expressions

$$\mathcal{T}^{(1)}_{enh}(\boldsymbol{x}_A, \boldsymbol{x}_B) \sim \frac{(1+\gamma)m}{c} \ln\left(\frac{4r_A r_B}{r_c^2}\right), \tag{11.3}$$

$$\mathcal{T}^{(2)}_{enh}(\boldsymbol{x}_A, \boldsymbol{x}_B) \sim -2\frac{(1+\gamma)^2 m^2}{c(r_A + r_B)} \frac{r_A r_B}{r_c^2}, \tag{11.4}$$

$$\mathcal{T}^{(3)}_{enh}(\boldsymbol{x}_A, \boldsymbol{x}_B) \sim 4\frac{(1+\gamma)^3 m^3}{c(r_A + r_B)^2} \left(\frac{r_A r_B}{r_c^2}\right)^2. \tag{11.5}$$

These expressions confirm the formulas obtained in [13] by a different method. It is worthy noticing that, at least up to G^3, γ is the only post-Newtonian parameter involved in the enhanced terms. When $\boldsymbol{x}_A$ and $\boldsymbol{x}_B$ tend to be located in opposite directions, the asymptotic behavior of each function $\mathcal{T}^{(n)}_{enh}$ is such that

$$\left|\mathcal{T}^{(n)}_{enh}(\boldsymbol{x}_A, \boldsymbol{x}_B)\right| \lesssim k_n \frac{2(1+\gamma)m}{r_A + r_B} \frac{r_A r_B}{r_c^2} \left|\mathcal{T}^{(n-1)}_{enh}(\boldsymbol{x}_A, \boldsymbol{x}_B)\right| \tag{11.6}$$

for $n = 1, 2, 3$, with $k_1 = 2$, $k_2 = k_3 = 1$ and $\mathcal{T}^{(0)}_{enh}(\boldsymbol{x}_A, \boldsymbol{x}_B) \sim r_A + r_B$. For $n = 3$, the formula (11.6) is straightforwardly derived from (11.4) and (11.5) (the symbol $\lesssim$ could be replaced by $\sim$). For $n = 1$, the formula results from the fact that $\ln x < x$ for any $x > 0$. Lastly, for $n = 2$, (11.6) obviously follows from the fact that $\ln(4r_A r_B / r_c^2) \to \infty$ when $1 + \boldsymbol{n}_A.\boldsymbol{n}_B \to 0$.

It results from (11.6) that inequalities (11.1) are satisfied for $n = 1, 2, 3$ as long as the zeroth-order distance of closest approach is such that a condition as follows

$$\frac{2m}{r_A + r_B} \frac{r_A r_B}{r_c^2} \ll 1 \tag{11.7}$$

is fulfilled. This inequality coincides with the condition ensuring the validity of the asymptotic expansions obtained in [13]. It may be expected that (11.7) is sufficient to ensure inequality (11.1) at any order.

It results from the definition of r_c that condition (11.7) is equivalent to

$$r_c \gg 2m \frac{|\boldsymbol{x}_B - \boldsymbol{x}_A|}{r_A + r_B} \frac{1}{|\boldsymbol{n}_A \times \boldsymbol{n}_B|}.$$

When $1 + \boldsymbol{n}_A.\boldsymbol{n}_B \to 0$, this inequality implies

$$r_c \gg \frac{m}{\sqrt{1 + \boldsymbol{n}_A.\boldsymbol{n}_B}}, \tag{11.8}$$

which in turn implies $r_c \gg m$. This last inequality means that condition (5.13) is met when inequality (11.7) is satisfied[6]. The full expressions of $\mathcal{T}^{(1)}$, $\mathcal{T}^{(2)}$, and $\mathcal{T}^{(3)}$ obtained

6 Note that the reciprocal is not true.

by the procedures developed in the present chapter can therefore be considered as reliable in a close superior conjunction as long as inequalities (11.1) hold.

To finish, it is worth noticing that condition (11.7) applied to a close superior conjunction is equivalent to an inequality as follows:

$$\pi - \arccos \boldsymbol{n}_A.\boldsymbol{n}_B \gg \sqrt{\frac{2m(r_A + r_B)}{r_A r_B}} \tag{11.9}$$

when (6.19) is taken into account. This last inequality clearly indicates that our procedures cannot be straightforwardly applied to the gravitational lensing configurations.

12 Application to some Solar System experiments

Condition (11.7) is fulfilled in experiments performed with photons exchanged between a spacecraft in the outer Solar System and a ground station. Indeed, noting that

$$\frac{m}{r_c}\frac{r_B}{r_c} < \frac{2m}{r_A + r_B}\frac{r_A r_B}{r_c^2} < 2\frac{m}{r_c}\frac{r_B}{r_c}$$

holds if $r_A > r_B$, replacing m by half the Schwarzschild radius of the Sun, $m_\odot$, and then putting $r_B = 1$ au, we find that inequalities

$$4.56 \times 10^{-4} \times \frac{R_\odot^2}{r_c^2} < \frac{2m_\odot}{r_A + r_B}\frac{r_A r_B}{r_c^2} < 9.12 \times 10^{-4} \times \frac{R_\odot^2}{r_c^2} \tag{12.1}$$

hold if $r_A > r_B$, with $R_\odot$ denoting the radius of the Sun. We put $R_\odot = 6.96 \times 10^8$ m. The other numerical parameters of the Sun used throughout this section are taken from [37].

Formulae (11.3)–(11.5) enable us to discuss the relevance of the terms $\mathcal{T}^{(2)}_{enh}$ and $\mathcal{T}^{(3)}_{enh}$ in a proposed mission like SAGAS, for instance. Indeed, this project plans to measure the parameter γ up to an accuracy reaching 10^{-8} with light rays traveling between

Table 1. Numerical values in ps of the main stationary contributions to the light travel time in the Solar System for various values of $r_c/R_\odot$. In each case, $r_A = 50$ au and $r_B = 1$ au. The parameters γ and κ are taken as $\gamma = 1$ and $\kappa = 15/4$, respectively. For the numerical estimates of $|\mathcal{T}^{(1)}_S|$ and $\mathcal{T}^{(1)}_{J_2}$, the light ray is assumed to propagate in the equatorial plane of the Sun. The dynamical effects due to the planetary perturbations are not taken into account.

$r_c/R_\odot$	$\lvert\mathcal{T}^{(1)}_S\rvert$	$\mathcal{T}^{(1)}_{J_2}$	$\mathcal{T}^{(2)}_{enh}$	$\mathcal{T}^{(2)}_{\kappa}$	$\mathcal{T}^{(3)}_{enh}$
1	10	2	−17 616	123	31.5
2	5	0.5	−4 404	61.5	2
5	2	0.08	−704.6	24.6	0.05

a spacecraft moving in the outer Solar System and the Earth. For $r_A = 50$ au and $r_B = 1$ au, the travel time of a ray passing in close proximity to the Sun (conjunction) is about 2.54×10^4 s. It follows from (11.3) that $\mathcal{T}^{(1)}$ is decreasing from 158 μs to 126 μs when r_c varies from $R_\odot$ to $5R_\odot$. As a consequence, reaching an accuracy of 10^{-8} on the measurement of γ requires to determine the light travel time with an accuracy of 0.7 ps. The numerical values of the respective contributions of $\mathcal{T}^{(2)}_{enh}$ and $\mathcal{T}^{(3)}_{enh}$ are indicated in Table 1. It is clear that the contribution of the enhanced term of order G^3 is larger than 2 ps when $r_c < 2R_\odot$. The same order of magnitude for $\mathcal{T}^{(3)}_{enh}$ may be expected in other proposed missions like ODYSSEY, LATOR, or ASTROD.

The above discussion also reveals that an experiment like SAGAS would enable to determine the post-Newtonian parameter κ with a relative precision amounting to 7×10^{-3}. In the Solar System, indeed, the term proportional to κ in (6.7) yields the asymptotic contribution

$$\mathcal{T}^{(2)}_{\kappa}(\boldsymbol{x}_A, \boldsymbol{x}_B) \sim \frac{\kappa \pi m_\odot^2}{c r_c} \tag{12.2}$$

when (11.2) holds. For a ray grazing the Sun ($r_c = R_\odot$), one has $\mathcal{T}^{(2)}_{\kappa} \approx 123$ ps if $\kappa = 15/4$. Hence the conclusion.

Before closing this study, it is worthy of note that the first-order contribution $\mathcal{T}^{(1)}_S$ to the time transfer function due to the gravitomagnetic effect of the solar rotation may be compared with the third-order-enhanced term. Indeed, it is easily inferred from equation (62) in [1] that for a ray traveling in the equatorial plane of the Sun

$$\left|\mathcal{T}^{(1)}_S(\boldsymbol{x}_A, \boldsymbol{x}_B)\right| \sim \frac{2(1+\gamma)GS_\odot}{c^4 r_c} \tag{12.3}$$

when (11.2) is checked, with $S_\odot$ being the angular momentum of the Sun. According to helioseismology, we can take $S_\odot \approx 2 \times 10^{41}$ kg m^2 s^{-1} (see, e.g. [38]). So, in the case where $r_c = R_\odot$, we have $|\mathcal{T}^{(1)}_S(\boldsymbol{x}_A, \boldsymbol{x}_B)| \approx 10$ ps. Furthermore, the contribution $\mathcal{T}^{(1)}_{J_2}$ due to the solar quadrupole moment $J_{2\odot}$ must also be considered for rays grazing the Sun. Using equation (24) in [4] for a ray traveling in the equatorial plane gives

$$\mathcal{T}^{(1)}_{J_2}(\boldsymbol{x}_A, \boldsymbol{x}_B) \sim \frac{(1+\gamma)m_\odot}{c} J_{2\odot} \frac{R_\odot^2}{r_c^2} \,. \tag{12.4}$$

Taking $J_{2\odot} \approx 2 \times 10^{-7}$ and putting $r_c = R_\odot$, (12.4) leads to $\mathcal{T}^{(1)}_{J_2} \approx 2$ ps.

13 Concluding remarks

Two methodologies enabling us to determine the time transfer function up to any given order of approximation in a static, spherically symmetric spacetime have been presented. The corresponding procedures are natively adapted to the case where both the emitter and the receiver of light rays are located at a finite distance from the origin of the spatial coordinates. These procedures lead to identical expressions for the time transfer function up to the order G^3 (see equations (6.6)–(6.8)). This coincidence is a new result. The reliability of the expression obtained for $\mathcal{T}^{(3)}$ in [14] is thus confirmed.

The procedure set out in Section 7 presents the advantage of exclusively involving elementary integrations which can be performed with any symbolic computer program, whatever the order of approximation. It is very likely that the procedure applied in Section 6 has the same property, even if it is not so easy to prove.

It must be emphasized that the explicit determination of the time transfer function up to the third order does not reduce to a purely mathematical improvement. This determination brings a rigorous proof of the existence of a third-order enhanced contribution to the time transfer function for light rays grazing the Sun. The enhanced term in $\mathcal{T}^{(3)}$ must be taken into account for determining the post-Newtonian parameter γ at a level of accuracy of 10^{-8} in Solar System experiments. It is worth noticing that for light rays almost grazing the Sun, this enhanced term is larger than the first-order Lense–Thirring effect due to the spinning of the Sun and than the first-order contribution due to the solar mass quadrupole.

The light direction triples $\widehat{\underline{l}}_e$ and $\widehat{\underline{l}}_r$ are now fully calculated up to the third order in G (see equations (9.10)–(9.15) for the generic case and (10.7)–(10.9) for a light ray emitted at infinity). As a consequence, the frequency shift between two observers could be determined up to the order G^3 by means of formula (3.9).

To finish, it may be noted that the calculations of the time transfer function performed here with the second procedure could be easily extended to quasi-Minkowskian light rays propagating in the equatorial plane of an axisymmetric, rotating body having a nonzero mass quadrupole. This methodology could be of interest for studying the light propagation in a Kerr metric, for example.

Acknowledgement: We are grateful to Bernard Linet for very helpful discussions and indebted to Olivier Minazzoli for his precious remarks.

References

[1] B. Linet and P. Teyssandier, *Phys. Rev. D* **66** 024045, 2002.

[2] C. Le Poncin-Lafitte, B. Linet and P. Teyssandier, *Class. Quantum Grav.* **21** 4463, 2004.

[3] P. Teyssandier, C. Le Poncin-Lafitte and B. Linet, A Universal Tool for Determining the Time Delay and the Frequency Shift of Light: Synge's World Function, *Lasers, Clocks and Drag-Free Control: Exploration of Relativistic Gravity in Space* (Springer series on *Astrophysics and Space Science Library* vol 349) ed. H. Dittus, C. Lammerzahl and S. G. Turyshev p 153 (*Preprint* arXiv:0711.0034), 2008.

[4] C. Le Poncin-Lafitte and P. Teyssandier, *Phys. Rev. D* **77** 044029, 2008.

[5] P. Teyssandier, *Class. Quantum Grav.* **29** 245010, 2012.

[6] S. Bertone, O. Minazzoli, M. Crosta, C. Le Poncin-Lafitte, A. Vecchiato and M.-C. Angonin, *Class. Quantum Grav.* **31** 015021, 2014.
[7] A. Hees, S. Bertone and C. Le Poncin-Lafitte, *Phys. Rev. D* **89** 064045, 2014.
[8] P. Wolf et al., *Exp. Astron.* **23** 651, 2009.
[9] B. Christophe et al., *Exp. Astron.* **23** 529, 2009.
[10] S. G. Turyshev et al., *Exp. Astron.* **27** 27, 2009.
[11] C. Braxmaier et al., *Exp. Astron.* **34** 181, 2012.
[12] O. Minazzoli and B. Chauvineau, *Class. Quantum Grav.* **28** 085010, 2011.
[13] N. Ashby and B. Bertotti, *Class. Quantum Grav.* **27** 145013, 2010.
[14] B. Linet and P. Teyssandier, *Class. Quantum Grav.* **30** 175008, 2013. Corrigendum: *Class. Quantum Grav.* **31** 079502, 2014.
[15] S. A. Klioner, *Sov. Astron.* **35** 523, 1991.
[16] S. M. Kopeikin and G. Schäfer, *Phys. Rev. D* **60** 124002, 1999.
[17] S. M. Kopeikin and B. Mashhoon, *Phys. Rev. D* **65** 064025, 2002.
[18] S. A. Klioner, *Astron. J.* **125** 1580, 2003.
[19] S. M. Kopeikin, P. Korobkov and A. Polnarev, *Class. Quantum Grav.* **23** 4299, 2006.
[20] M. Crosta, *Class. Quantum Grav.* **28** 235013, 2011.
[21] I. I. Shapiro, *Phys. Rev. Lett.* **13** 789, 1964.
[22] C. M. Will, *Theory and Experiment in Gravitational Physics*, 2nd edn. (Cambridge: Cambridge University Press), 1993.
[23] L. Blanchet, C. Salomon, P. Teyssandier and P. Wolf, *Astron. Astrophys.* **370** 320, 2001.
[24] G. W. Richter and R. A. Matzner, *Phys. Rev. D* **28** 3007, 1983.
[25] V. A. Brumberg, *Kinematika i Fisika Nebesnykh Tel* **3** 8 (in Russian); *Kinematics Phys. Celest. Bodies* **3** 6 (in English), 1987.
[26] V. A. Brumberg, *Essential Relativistic Celestial Mechanics* (Bristol: Adam Hilger), 1991.
[27] S. A. Klioner and S. Zschocke, *Class. Quantum Grav.* **27** 075015, 2010.
[28] R. W. John, *Exp. Tech. Phys.* **23** 127, 1975.
[29] P. Teyssandier and C. Le Poncin-Lafitte, *Class. Quantum Grav.* **25** 145020, 2008.
[30] A. F. Sarmiento, *Gen. Rel. Grav.* **14** 793, 1982.
[31] C. R. Keeton and A. O. Petters, *Phys. Rev. D* **72** 104006, 2005.
[32] J. L. Synge, *Relativity: The General Theory* (Amsterdam, North-Holland), 1964.
[33] C. Darwin, *Proc. Roy. Soc. London A* **249** 180, 1959.
[34] J.-P. Luminet, *Astron. Astrophys.* **75** 228, 1979.
[35] F. Giannoni, A. Masiello and P. Piccione, *Class. Quantum Grav.* **16** 731, 1999.
[36] S. Chandrasekhar, *The Mathematical Theory of Black Holes* (New York: Oxford University Press), 1983.
[37] IERS Conventions, *IERS Technical Note No 36*, G. Petit and B. Luzum (eds.) (Frankfurt am Main: Verlag des Bundesamts für Kartographie und Geodäsie, 2010), 2010.
[38] R. Komm, R. Howe, B. R. Durney and F. Hill, *Astrophys. J.* **586** 650, 2003.

Norbert Wex

Testing relativistic gravity with radio pulsars

1 Introduction

Next year, we will be celebrating the centenary of Einstein's general theory of relativity. On November 25, 1915, Einstein presented his field equations of gravitation (without cosmological term) to the Prussian Academy of Science [1]. With this publication, general relativity (GR) was finally completed as a logically consistent physical theory (*Damit ist endlich die allgemeine Relativitätstheorie als logisches Gebäude abgeschlossen.*). Already 1 week before, based on the vacuum form of his field equations, Einstein was able to show that his theory of gravitation naturally explains the anomalous perihelion advance of the planet Mercury [2]. While in hindsight, this can be seen as the first experimental test for GR, back in 1915 astronomers were still searching for a Newtonian explanation [3]. In his 1916 comprehensive summary of GR [4], Einstein proposed three experimental tests:

- Gravitational redshift (Einstein suggested to look for redshift in the spectral lines of stars).
- Light deflection (Einstein explicitly calculated the values for the Sun and Jupiter).
- Perihelion precession of planetary orbits (Einstein emphasized the agreement of GR, with the observed perihelion precession of Mercury with a reference to his calculations in [2]).

Gravitational redshift, a consequence of the equivalence principle, is common to all metric theories of gravity, and therefore in some respect its measurement has less discriminating power than the other two tests [5]. The first verification of gravitational light bending during the total eclipse on May 29, 1919, was far from being a high precision test, but clearly decided in favor of GR, against the Newtonian prediction, which is only half the GR value [6]. In the meantime this test has been greatly improved, in the optical with the astrometric satellite HIPPARCOS [7], and in the radio with very long baseline interferometry [8–10]. The deflection predicted by GR has been verified with a precision of 1.5×10^{-4}. An even better test for the curvature of spacetime in the vicinity of the Sun is based on the Shapiro delay, the so-called fourth test of GR [11]. A measurement of the frequency shift of radio signals exchanged with the Cassini spacecraft lead to a 10^{-5} confirmation of GR [12]. Apart from the four "classical" tests, GR has passed many other tests in the Solar System with flying colors: Lunar Laser Ranging tests for the strong equivalence principle and the de-Sitter precession of the Moon's

Norbert Wex: Max-Planck-Institut für Radioastronomie, Auf dem Hügel 69, D-53121, Bonn, Germany

orbit [13], the Gravity Probe B experiment for the relativistic spin precession of a gyroscope (geodetic and frame dragging) [14], and the Lense–Thirring effect in satellite orbits [15], just to name a few.

GR, being a theory where fields travel with finite speed, predicts the existence of gravitational waves that propagate with the speed of light [16] and extract energy from (nonaxisymmetric) material systems with accelerated masses [17]. This is also true for a self-gravitating system, where the acceleration of the masses is driven by gravity itself, a question which was settled in a fully satisfactory manner only several decades after Einstein's pioneering papers (see [18] for an excellent review). This fundamental property of GR could not be tested in the slow-motion environment of the Solar System, and the verification of the existence of gravitational waves had to wait until the discovery of the first binary pulsar in 1974 [19]. Also, all the experiments in the Solar System can only test the weak-field aspects of gravity. The spacetime of the Solar System is close to Minkowski space everywhere. To illustrate this: to first order (in standard coordinates) the spatial components of the spacetime metric can be written as $g_{ij} = (1 - 2\Phi/c^2)\delta_{ij}$, where Φ denotes the Newtonian gravitational potential. At the surface of the Sun one finds $\Phi/c^2 \sim -2 \times 10^{-6}$, while at the surface of a neutron star $\Phi/c^2 \sim -0.2$. Consequently, gravity experiments with binary pulsars, not only yielded the first tests of the radiative properties of gravity, they also took our gravity tests into a new regime of gravity.

To categorize gravity tests with pulsars and to put them into context with other gravity tests it is useful to introduce the following four gravity regimes:

G1 *Quasi-stationary weak-field regime*: The motion of the masses is slow compared to the speed of light ($v \ll c$) and spacetime is only very weakly curved, i.e. close to Minkowski spacetime everywhere. This is, for instance, the case in the Solar System.

G2 *Quasi-stationary strong-field regime*: The motion of the masses is slow compared to the speed of light ($v \ll c$), but one or more bodies of the system are strongly self-gravitating, i.e. spacetime in their vicinity deviates significantly from Minkowski space. Prime examples here are binary pulsars, consisting of two well-separated neutron stars.

G3 *Highly dynamical strong-field regime*: Masses move at a significant fraction of the speed of light ($v \sim c$) and spacetime is strongly curved and highly dynamical in the vicinity of the masses. This is the regime of merging neutron stars and black holes.

GW *Radiation regime*: Synonym for the collection of the radiative properties of gravity, most notably the generation of gravitational waves by material sources, the propagation speed of gravitational waves, and their polarization properties.

Figure 1 illustrates the different regimes. Gravity regime G1 is well tested in the Solar System. Binary pulsar experiments are presently our only precision experiments

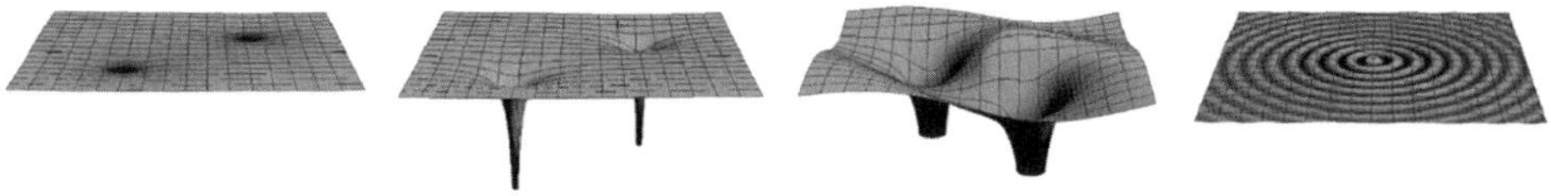

Fig. 1. Illustration of the different gravity regimes used in this review.

for gravity regime G2, and the best tests for the radiative properties of gravity (regime GW)[1]. In the near future, gravitational wave detectors will allow a direct detection of gravitational waves (regime GW) and probe the strong and highly dynamical spacetime of merging compact objects (regime G3). As we will discuss at the end of this review, pulsar timing arrays soon should give as direct access to the nano-Hz gravitational wave band and probe the properties of these ultra-low frequency gravitational waves (regime GW).

1.1 Radio pulsars and pulsar timing

Radio pulsars, i.e. rotating neutron stars with coherent radio emission along their magnetic poles, were discovered in 1967 by Jocelyn Bell and Antony Hewish [21]. Seven years later Russell Hulse and Joseph Taylor [19] discovered the first binary pulsar, a pulsar in orbit with a companion star. This discovery marked the beginning of gravity tests with radio pulsars. Presently, more than 2000 radio pulsars are known, out of which about 10% reside in binary systems [22]. The population of radio pulsars can be nicely presented in a diagram that gives the two main characteristics of a pulsar: the rotational period P and its temporal change $\dot{P}$ due to the loss of rotational energy (see Figure 2). Fast rotating pulsars with small $\dot{P}$ (millisecond pulsars) appear to be particularly stable in their rotation. On long time scales, some of them rival the best atomic clocks in their stability [23, 24]. This property makes them ideal tools for precision astrometry, and hence (most) gravity tests with pulsars are simply clock comparison experiments to probe the spacetime of the binary pulsar, where the "pulsar clock" is read off by counting the pulses in the pulsar signal (see Figure 3). As a result, a wide range of relativistic effects related to orbital binary dynamics, time dilation, and delays in the signal propagation can be tested. The technique used is the so-called *pulsar timing*, which basically consists of measuring the exact arrival time of pulses at the radio telescope on Earth, and fitting an appropriate *timing model* to these arrival times, to obtain a phase-connected solution. In the phase-connected approach lies the true strength of pulsar timing. The timing model has to account for every pulse over a time

1 Gravitational wave damping has also been observed in a double white-dwarf system, which has an orbital period of just 13 min [20]. This experiment combines gravity regimes G1 (note, $v/c \sim 3 \times 10^{-3}$) and GW of Figure 1.

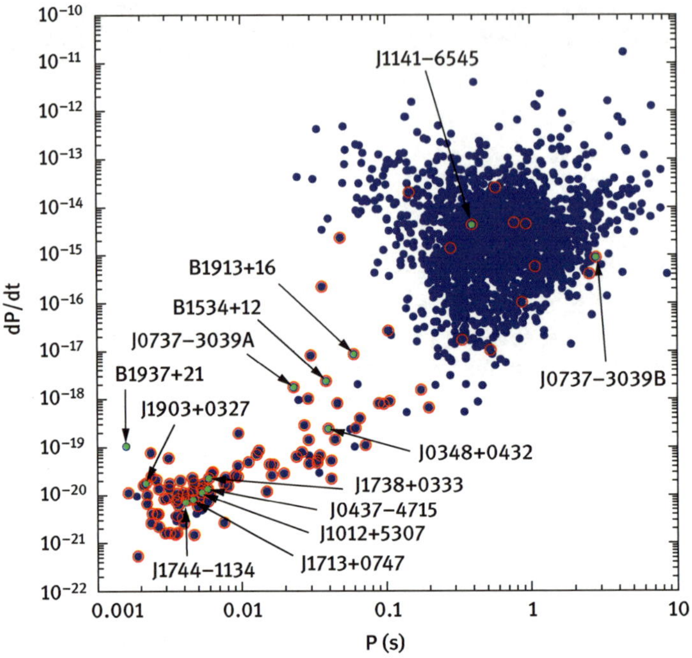

Fig. 2. The P–$\dot{P}$ diagram for radio pulsars. Binary pulsars are indicated by a red circle. Pulsars that play a particular role in this review are marked with a green dot and have their name as a label. The data are taken from the *ATNF Pulsar Catalogue* [22].

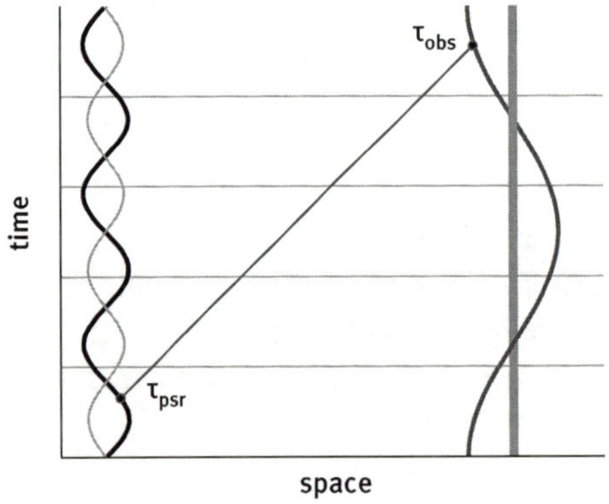

Fig. 3. Spacetime diagram illustration of pulsar timing. Pulsar timing connects the proper time of emission τ_{psr}, defined by the pulsar's intrinsic rotation, and the proper time of the observer on Earth τ_{obs}, measured by the atomic clock at the location of the radio telescope. The timing model, which expresses τ_{obs} as a function of τ_{psr}, accounts for various "relativistic effects" associated with the metric properties of the spacetime, i.e. the world line of the pulsar and the null-geodesic of the radio signal. In addition, it contains a number of terms related to the Earth motion and relativistic corrections in the Solar System, like time dilation and signal propagation delays (see [27] for details).

Table 1. Examples of precision measurements using pulsar timing. A number in bracket indicates the (one-sigma) uncertainty in the last digit of each value. The symbol $M_\odot$ stands for the Solar mass. (cf. Table 1 in [28]).

Rotational period:	$5.757451924362137(2)$ ms	[29]
Orbital period:	$0.102251562479(8)$ d	(Kramer et al., in preparation)
Small eccentricity:	$(3.5 \pm 1.1) \times 10^{-7}$	[30]
Distance:	$157(1)$ pc	[29]
Proper motion:	$140.915(1)$ mas yr^{-1}	[29]
Masses of neutron stars:	$m_p = 1.4398(2)\, M_\odot$	[31]
	$m_c = 1.3886(2)\, M_\odot$	[31]
Mass of millisecond pulsar:	$1.667(7)\, M_\odot$	[32]
Mass of white-dwarf companion:	$0.207(2)\, M_\odot$	[33]
Mass of Jupiter and moons:	$9.547921(2) \times 10^{-4}\, M_\odot$	[34]
Relativistic periastron advance:	$4.226598(5)$ deg yr^{-1}	[31]
Gravitational wave damping:	$0.504(3)$ pico – Hz yr^{-1}	(Kramer et al., in preparation)
GR validity (observed/GR):	$1.0000(5)$	(Kramer et al., in prepration)

scale of several years, in some cases even several decades. This makes pulsar timing extremely sensitive to even tiny deviations in the model parameters, and therefore vastly superior to a simple measurement of Doppler shifts in the pulse period. Table 1 illustrates the current precision capabilities of pulsar timing for various experiments, like mass determination, astrometry, and gravity tests. We will not go into the details of pulsar observations and pulsar timing here, since there are numerous excellent reviews on these topics, for instance [25, 26], just to mention two. In this review, we focus on the relativistic effects that play a role in pulsar timing observations, and how pulsar timing can be used to test gravitational phenomena in generic as well as theory-based frameworks.

1.2 Binary pulsar motion in gravity theories

While in Newtonian gravity there is an exact solution to the equations of motion of two point masses that interact gravitationally, no such exact analytic solution is known in GR. In GR, the two-body problem has to be solved numerically or on the basis of approximation methods. A particularly well established and successful approximation scheme to tackle the problem of motion of a system of well-separated bodies is the *post-Newtonian approximation*, which is based on the weak-field slow-motion assumption. However, to describe the motion and gravitational wave emission of binary pulsars, there are two main limitations of the post-Newtonian approximation that have to be overcome (cf. [35]):

(1) Near and inside the pulsar (and its companion, if it is as well a neutron star) the gravitational field is strong and the weak-field assumption no longer holds.

(2) When it comes to the generation of gravitational wave (wavelength λ_{GW}) and their back reaction on the orbital motion (size r and period P_b), the post-Newtonian approximation is only valid in the near zone ($r \ll \lambda_{\mathrm{GW}} = cP_b/2$), and breaks down in the radiation zone ($r > \lambda_{\mathrm{GW}}$) where gravitational waves propagate and boundary conditions are defined, like the "incoming radiation" condition.

In particular, the discovery of the Hulse–Taylor pulsar was a strong stimulus for the development of consistent approaches to compute the equations of motion for a binary system with strongly self-gravitating bodies (gravity regime G2). As a result, by now there are fully self-consistent derivations for the gravitational wave emission and the damping of the orbit due to the gravitational wave back reaction for such systems. In fact, in GR, there are several independent approaches that lead to the same result, giving equations of motion for a binary system with nonrotating components that include terms up to 3.5 post-Newtonian order (v^7/c^7) [36, 37]. For the relative acceleration in the center-of-mass frame one finds the general form

$$\ddot{\mathbf{r}} = -\frac{GM}{r^2} \times \left[(1 + A_2 + A_4 + A_5 + A_6 + A_7)\, \frac{\mathbf{r}}{r} + (B_2 + B_4 + B_5 + B_6 + B_7)\, \dot{\mathbf{r}} \right], \tag{1.1}$$

where the coefficients A_k and B_k are of the order c^{-k}, and are functions of $r \equiv |\mathbf{r}|$, $\dot{r}$, $v \equiv |\dot{\mathbf{r}}|$, and the masses (see [36] for explicit expressions). The quantity M denotes the total mass of the system. At this level of approximation, these equations-of-motion are also applicable to binaries containing strongly self-gravitating bodies, like neutron stars and black holes. This is a consequence of a remarkable property of Einstein's theory of gravity, the *effacement of the internal structure* [35, 38]. In GR, strong-field contributions are absorbed into the definition of the body's mass.

In GR's post-Newtonian approximation scheme, gravitational wave damping enters for the first time at the 2.5 post-Newtonian level (order v^5/c^5), as a term in the equations-of-motion that is not invariant against time reversal. The corresponding loss of orbital energy is given by the *quadrupole formula*, derived for the first time by Einstein within the linear approximation, for a material system where the gravitational interaction between the masses can be neglected [17]. As it turns out, the quadrupole formula is also applicable for gravity regime G2 of Figure 1, and therefore valid for binary pulsars as well (cf. [35]).

In alternative gravity theories, the gravitational wave back-reaction, generally, already enters at the 1.5 post-Newtonian level (order v^3/c^3). This is the result of the emission of dipolar gravitational waves, and adds terms A_3 and B_3 to equation (1.1) [5, 39]. Furthermore, one does no longer have an effacement of the internal structure of a compact body, meaning that the orbital dynamics, in addition to the mass, depends on the "sensitivity" of the body, a quantity that depends on its structure/compactness. Such modifications already enter at the "Newtonian" level, where the usual Newtonian gravitational constant G is replaced by a (body-dependent) effective gravitational constant $\mathcal{G}$. For alternative gravity theories, it therefore generally makes an important difference whether the pulsar companion is a compact neutron star or a much less compact white dwarf. In sum, alternative theories of gravity generally predict deviations from GR in both the quasi-stationary and the radiative properties of binary pulsars [40, 41].

At the first post-Newtonian level, for fully conservative gravity theories without preferred location effects, one can construct a generic *modified Einstein–Infeld–Hoffmann Lagrangian* for a system of two gravitationally interacting masses m_p (pulsar) and m_c (companion) at relative (coordinate) separation $r \equiv |\mathbf{x}_p - \mathbf{x}_c|$ and velocities $\mathbf{v}_p = \dot{\mathbf{x}}_p$ and $\mathbf{v}_c = \dot{\mathbf{x}}_c$:

$$
\begin{aligned}
L_{\mathrm{O}} = & -m_p c^2 \left(1 - \frac{\mathbf{v}_p^2}{2c^2} - \frac{\mathbf{v}_p^4}{8c^4}\right) - m_c c^2 \left(1 - \frac{\mathbf{v}_c^2}{2c^2} - \frac{\mathbf{v}_c^4}{8c^4}\right) \\
& + \frac{\mathcal{G} m_p m_c}{r} \left[1 - \frac{\mathbf{v}_p \cdot \mathbf{v}_c}{2c^2} - \frac{(\mathbf{r} \cdot \mathbf{v}_p)(\mathbf{r} \cdot \mathbf{v}_c)}{2c^2 r^2} + \varepsilon \frac{(\mathbf{v}_p - \mathbf{v}_c)^2}{2c^2}\right] \\
& - \xi \frac{\mathcal{G}^2 M m_p m_c}{2c^2 r^2},
\end{aligned}
\tag{1.2}
$$

where $M \equiv m_p + m_c$. The body-dependent quantities $\mathcal{G}$, ε, and ξ account for deviations from GR associated with the self-energy of the individual masses [5, 40]. In GR one simply finds $\mathcal{G} = G$, $\varepsilon = 3$, and $\xi = 1$. There are various analytical solutions to the dynamics of (1.2). The most widely used in pulsar astronomy is the quasi-Keplerian parametrization by Damour and Deruelle [42]. It forms the basis of pulsar-timing models for relativistic binary pulsars, as we will discuss in more details in Section 1.4.

Beyond the first post-Newtonian level there is no fully generic framework for the gravitational dynamics of a binary system. However, one can find equations-of-motion valid for a general class of gravity theories, like in [43] where a framework based on multi-scalar–tensor theories is introduced to discuss tests of relativistic gravity to the second post-Newtonian level, or in [44] where the explicit equations of motion for non-spinning compact objects to 2.5 post-Newtonian order for a general class of scalar–tensor theories of gravity are given.

1.3 Gravitational spin effects in binary pulsars

In relativistic gravity theories, in general, the proper rotation of the bodies of a binary system directly affects their orbital and spin dynamics. Equations of motion for spinning bodies in GR have been developed by numerous authors, and in the meantime go way beyond the leading order contributions (for reviews and references see, e.g., [35, 45–47]). For present-day pulsar timing experiments it is sufficient to have a look at the post-Newtonian leading order contributions. There one finds three contributions, the spin–orbit (SO) interaction between the pulsar spin $\mathbf{S}_p$ and the orbital angular momentum $\mathbf{L}$, the SO interaction between the companion spin $\mathbf{S}_c$ and the orbital angular momentum, and finally the spin–spin interaction between the spin of the pulsar and the spin of the companion [45].

Spin–spin interaction will remain negligible in binary pulsar experiments for the foreseeable future. They are many orders of magnitude below the second post-Newtonian and spin–orbit effects [48], and many orders of magnitude below the measurement precision of present timing experiments. For this reason, we will not further discuss spin–spin effects here.

For a boost invariant gravity theory, the (acceleration-dependent) Lagrangian for the spin–orbit interaction has the following general form (summation over spatial indices i, j):

$$L_{\mathrm{SO}}(\mathbf{x}_A, \mathbf{v}_A, \mathbf{a}_A) = \frac{1}{c^2} \sum_A S_A^{ij} \left[\frac{1}{2} v_A^i a_A^j + \sum_{B \neq A} \frac{\Gamma_A^B m_B}{r_{AB}^3} (v_A^i - v_B^i)(x_A^j - x_B^j) \right], \tag{1.3}$$

where $S_A^{ij} \equiv \varepsilon^{ijk} S_A^k$ is the antisymmetric spin tensor of body A [35, 40, 49]. The coupling function Γ_A^B can also account for strong-field effects in the spin–orbit coupling. In GR $\Gamma_A^B = 2G$. For bodies with negligible gravitational self-energy, one finds in the framework of the *parametrized post-Newtonian (PPN) formalism*[2] $\Gamma_A^B = (\gamma_{\mathrm{PPN}} + 1)G$, a quantity that is actually most tightly constrained by the light-bending and Shapiro-delay experiments in the Solar System, which test γ_{PPN} [8–10, 12].

In binary pulsars, spin–orbit coupling has two effects. On the one hand, it adds spin-dependent terms to the equations of motion (1.1), which cause a Lense–Thirring precession of the orbit (for GR see [45, 50]). So far this contribution could not be tested in binary pulsar experiments. Prospects of its measurement will be discussed in the future outlook in Section 8. On the other hand, it leads to secular changes in the orientation of the spins of the two bodies (geodetic precession), most importantly the observed pulsar in a pulsar binary [45, 51, 52]. As we discuss in more details in Section 3, a change in the rotational axis of the pulsar causes changes in the observed emission properties of the pulsar, as the line-of-sight gradually cuts through different regions of the magnetosphere.

As can be derived from (1.3), to first order in GR the geodetic precession of the pulsar, averaged over one orbit, is given by ($\hat{\mathbf{L}} \equiv \mathbf{L}/|\mathbf{L}|$)

$$\Omega_p^{\mathrm{SO}} = \frac{n_b}{1 - e^2} \left(2 + \frac{3m_c}{2m_p} \right) \frac{m_p m_c}{M^2} \frac{V_b^2}{c^2} \hat{\mathbf{L}}, \tag{1.4}$$

where $n_b \equiv 2\pi/P_b$ and $V_b \equiv (GMn_b)^{1/3}$.

2 The PPN formalism uses 10 parameters to parametrize in a generic way deviations from GR at the post-Newtonian level, within the class of metric gravity theories (see [5] for details).

It is expected that in alternative theories relativistic spin precession generally depends on self-gravitational effects, meaning, the actual precession may depend on the compactness of a self-gravitating body. For the class of theories that lead to the Lagrangian (1.3), equation (1.4) modifies to

$$\Omega_p^{\mathrm{SO}} = \frac{n_b}{1-e^2}\left[\frac{\Gamma_p^c}{\mathcal{G}} + \left(\frac{\Gamma_p^c}{\mathcal{G}} - \frac{1}{2}\right)\frac{m_c}{m_p}\right]\frac{m_p m_c}{M^2}\frac{\mathcal{V}_b^2}{c^2}\,\hat{\mathbf{L}}\,, \tag{1.5}$$

where $\mathcal{V}_b \equiv (\mathcal{G}Mn_b)^{1/3}$ is the strong-field generalization of V_b.

Effects from spin-induced quadrupole moments are negligible as well. For double neutron-star systems they are many orders of magnitude below the second post-Newtonian and spin–orbit effects, due to the small extension of the bodies [48]. If the companion is a more extended star, like a white dwarf or a main-sequence star, the rotationally induced quadrupole moment might become important. A prime example is PSR J0045-7319, where the quadrupole moment of the fast rotating companion causes a significant precession of the pulsar orbit [53]. For all the binary pulsars discussed here, the quadrupole moments of pulsar and companion are (currently) negligible.

Finally, certain gravitational phenomena, not present in GR, can even lead to a spin precession of isolated pulsars, for instance, a violation of the local Lorentz invariance and a violation of the local position invariance in the gravitational sector, as we will discuss in more details in Sections 5 and 6.

1.4 Phenomenological approach to relativistic effects in binary pulsar observations

For binary pulsar experiments that test the quasi-stationary strong-field regime (G2) and the gravitational wave damping (GW), a phenomenological parametrization, the so-called parametrized post-Keplerian (PPK) formalism, has been introduced by Damour [54] and extended by Damour and Taylor [40]. The PPK formalism parametrizes all the observable effects that can be extracted independently from binary pulsar timing and pulse-structure data. Consequently, the PPK formalism allows to obtain theory-independent information from binary pulsar observations by fitting for a set of *Keplerian* and *post-Keplerian parameters*.

The description of the orbital motion is based on the quasi-Keplerian parametrization of Damour and Deruelle, which is a solution to the first post-Newtonian equations of motion [42, 55]. The corresponding *Roemer delay* in the arrival time of the pulsar signals is

$$\Delta_{\mathrm{R}} = x\sin\omega\left[\cos U - e(1+\delta_r)\right] + x\cos\omega\left[1 - e^2(1+\delta_\theta)^2\right]^{1/2}\sin U\,, \tag{1.6}$$

where the eccentric anomaly U is linked to the proper time of the pulsar T via the Kepler equation

$$U - e\sin U = 2\pi\left[\left(\frac{T-T_0}{P_b}\right) - \frac{\dot{P}_b}{2}\left(\frac{T-T_0}{P_b}\right)^2\right] . \tag{1.7}$$

The five Keplerian parameters P_b, e, x, ω, and T_0 denote the orbital period, the orbital eccentricity, the projected semimajor axis of the pulsar orbit, the longitude of periastron, and the time of periastron passage, respectively. The post-Keplerian parameter δ_r is not separately measurable, i.e. it can be absorbed into other timing parameters, and the post-Keplerian parameter δ_θ has not been measured up to now in any of the binary pulsar systems. The relativistic precession of periastron changes the longitude of periastron ω according to

$$\omega = \omega_0 + \dot{\omega}\,\frac{P_b}{\pi}\,\arctan\left[\left(\frac{1+e}{1-e}\right)^{1/2}\tan\frac{U}{2}\right] , \tag{1.8}$$

meaning, that averaged over a full orbit, the location of periastron shifts by an angle $\dot{\omega}P_b$. The parameter $\dot{\omega}$ is the corresponding post-Keplerian parameter. A change in the orbital period, due to the emission of gravitational waves, is parametrized by the post-Keplerian parameter $\dot{P}_b$. Correspondingly, one has post-Keplerian parameters for the change in the orbital eccentricity and the projected semimajor axis:

$$e = e_0 + \dot{e}(T - T_0) , \tag{1.9}$$

$$x = x_0 + \dot{x}(T - T_0) . \tag{1.10}$$

Besides the Roemer delay Δ_{R}, there are two purely relativistic effects that play an important role in pulsar timing experiments. In an eccentric orbit, one has a changing time dilation of the "pulsar clock" due to a variation in the orbital velocity of the pulsar and a change of the gravitational redshift caused by the gravitational field of the companion. This so-called *Einstein delay* is a periodic effect, whose amplitude is given by the post-Keplerian parameter γ, and to the first order can be written as

$$\Delta_{\mathrm{E}} = \gamma \sin U . \tag{1.11}$$

For sufficiently edge-on and/or eccentric orbits the propagation delay suffered by the pulsar signals in the gravitational field of the companion becomes important. This so-called *Shapiro delay*, to first order, reads

$$\Delta_{\mathrm{S}} = -2r\ln\left[1 - e\cos U - s\sin\omega(\cos U - e) - s\cos\omega(1-e^2)^{1/2}\sin U\right] , \tag{1.12}$$

where the two post-Keplerian parameters r and s are called *range* and *shape* of the Shapiro delay, respectively. The latter is linked to the inclination of the orbit with respect to the line of sight, i, by $s = \sin i$. It is important to note, that for $i \to 90°$ equation (1.12) breaks down and higher order corrections are needed. But so far, equation (1.12) is fully sufficient for the timing observations of known pulsars [56].

Concerning the post-Keplerian parameters related to quasi-stationary effects, for the wide class of boost-invariant gravity theories one finds that they can be expressed as functions of the Keplerian parameters, the masses, and parameters generically accounting for gravitational self-field effects (cf. equation (1.2)) [5, 40]:

$$\dot{\omega} = \frac{n_b}{1-e^2}\left(\varepsilon - \frac{\xi}{2} + \frac{1}{2}\right)\frac{\mathcal{V}_b^2}{c^2}, \tag{1.13}$$

$$\gamma = \frac{e}{n_b}\left(\frac{G_{0c}}{\mathcal{G}} + \mathcal{K}_p^c + \frac{m_c}{M}\right)\frac{m_c}{M}\frac{\mathcal{V}_b^2}{c^2}, \tag{1.14}$$

$$r = \frac{1+\varepsilon_{0c}}{4}\frac{G_{0c}m_c}{c^3}, \tag{1.15}$$

$$s = x\,n_b\frac{M}{m_c}\frac{c}{\mathcal{V}_b}, \tag{1.16}$$

plus Ω^{SO} from equation (1.5). Here we have listed only those parameters that play a role in this review. For a complete list and a more detailed discussion, the reader is referred to [40]. The quantities G_{0c} and ε_{0c} are related to the interaction of the companion with a test particle or a photon. The parameter $\mathcal{K}_p^c$ accounts for a possible change in the moment of inertia of the pulsar due to a change in the local gravitational constant. In GR one finds $\mathcal{G} = G_{0c} = G$, $\varepsilon = \varepsilon_{0c} = 3$, $\xi = 1$ and $\mathcal{K}_p^c = 0$. Consequently

$$\dot{\omega}^{\mathrm{GR}} = \frac{3n_b}{1-e^2}\frac{V_b^2}{c^2}, \tag{1.17}$$

$$\gamma^{\mathrm{GR}} = \frac{e}{n_b}\left(1 + \frac{m_c}{M}\right)\frac{m_c}{M}\frac{V_b^2}{c^2}, \tag{1.18}$$

$$r^{\mathrm{GR}} = \frac{Gm_c}{c^3}, \tag{1.19}$$

$$s^{\mathrm{GR}} = x\,n_b\frac{M}{m_c}\frac{c}{V_b}. \tag{1.20}$$

These parameters are independent of the internal structure of the neutron star(s), due to the effacement of the internal structure, a property of GR [35, 38]. For most alternative gravity theories this will not be the case. For instance, in the mono-scalar–tensor theories $T_1(\alpha_0, \beta_0)$ of [57, 58], one finds[3]

$$\dot{\omega}^{T_1} = \frac{n_b}{1-e^2}\left(\frac{3-\alpha_p\alpha_c}{1+\alpha_p\alpha_c} - \frac{m_p\alpha_p^2\beta_c + m_c\alpha_c^2\beta_p}{2M(1+\alpha_p\alpha_c)^2}\right)\frac{\mathcal{V}_b^2}{c^2}, \tag{1.21}$$

$$\gamma^{T_1} = \frac{e}{n_b}\left(\frac{1+k_p\alpha_c}{1+\alpha_p\alpha_c} + \frac{m_c}{M}\right)\frac{m_c}{M}\frac{\mathcal{V}_b^2}{c^2}, \tag{1.22}$$

3 The mono-scalar–tensor theories $T_1(\alpha_0, \beta_0)$ of [57, 58] have a conformal coupling function $A(\varphi) = \alpha_0(\varphi - \varphi_0) + \beta_0(\varphi - \varphi_0)^2/2$. The Jordan–Fierz–Brans–Dicke gravity is the subclass with $\beta_0 = 0$, and $\alpha_0^2 = (2\omega_{\mathrm{BD}} + 3)^{-1}$.

$$r^{T_1} = \frac{G_* m_c}{c^3}\,, \tag{1.23}$$

$$s^{T_1} = x\, n_b \, \frac{M}{m_c} \frac{c}{\mathcal{V}_b}\,, \tag{1.24}$$

where $\mathcal{V}_b = [G_*(1 + \alpha_p \alpha_c) M n_b]^{1/3}$. The body-dependent quantities α_p and α_c denote the effective scalar coupling of pulsar and companion, respectively, and $\beta_A \equiv \partial\alpha_A/\partial\varphi_0$ where φ_0 denotes the asymptotic value of the scalar field at spatial infinity. The quantity k_p is related to the moment of inertia I_p of the pulsar via $k_p \equiv -\partial \ln I_p/\partial\varphi_0$. For a given equation of state, the parameters α_A, β_A, and k_A depend on the fundamental constants of the theory, e.g. α_0 and β_0 in $T_1(\alpha_0, \beta_0)$, and the mass of the body. As we will demonstrate later, these "gravitational form factors" can assume large values in the strong gravitational fields of neutron stars. Depending on the value of β_0, this is even the case for a vanishingly small α_0, where there are practically no measurable deviations from GR in the Solar System. In fact, even for $\alpha_0 = 0$, a neutron star, above a certain β_0-dependent critical mass, can have an effective scalar coupling α_A of order unity. This nonperturbative strong-field behavior, the so-called spontaneous scalarization of a neutron star, was discovered 20 years ago by Damour and Esposito-Farèse [57].

Finally, there is the post-Keplerian parameter $\dot{P}_b$, related to the damping of the orbit due to the emission of gravitational waves. We have seen above that in alternative gravity theories the back reaction from the gravitational wave emission might enter the equations of motion already at the 1.5 post-Newtonian level, giving rise to a $\dot{P}_b \propto \mathcal{V}_b^3/c^3$. To leading order one finds in mono-scalar–tensor gravity the dipolar contribution from the scalar field [58–60]:

$$\dot{P}_b = -2\pi \frac{m_p m_c}{M^2} \frac{1 + e^2/2}{(1 - e^2)^{5/2}} \frac{\mathcal{V}_b^3}{c^3} \frac{(\alpha_p - \alpha_c)^2}{1 + \alpha_p \alpha_c} + \mathcal{O}(\mathcal{V}_b^5/c^5)\,. \tag{1.25}$$

As one can see, the change in the orbital period due to dipolar radiation depends strongly on the difference in the effective scalar coupling α_A. Binary pulsar systems with a high degree of asymmetry in the compactness of their components are therefore ideal to test for dipolar radiation. An order unity difference in the effective scalar coupling would lead to a change in the binary orbit, which is several orders of magnitude ($\sim c^2/\mathcal{V}_b^2$) stronger than the quadrupolar damping predicted by GR.

At the 2.5 post-Newtonian level ($\propto \mathcal{V}_b^5/c^5$), in general, there are several contributions entering the $\dot{P}_b$ calculation:
- Monopolar waves for eccentric orbits.
- Higher order contributions to the dipolar wave damping.
- Quadrupolar waves from the tensor field, and the fields that are also responsible for the monopolar and/or dipolar waves.

For scalar–tensor gravity these expressions can be found in [61]. For GR, one finds the following expression, related to the *quadrupole formula* [62]:

$$\dot{P}_b^{\mathrm{GR}} = -\frac{192\pi}{5}\frac{m_p m_c}{M^2}\frac{1+73e^2/24+37e^4/96}{(1-e^2)^{7/2}}\frac{V_b^5}{c^5}\,. \tag{1.26}$$

Apart from a change in the orbital period, gravitational wave damping will also affect other post-Keplerian parameters. While gravitational waves carry away orbital energy and angular momentum, Keplerian parameters like the eccentricity and the semimajor axis of the pulsar orbit change as well. The corresponding post-Keplerian parameters are $\dot{e}$ and $\dot{x}$, respectively. However, these changes affect the arrival times of the pulsar signals much less than the $\dot{P}_b$, and therefore do (so far) not play any role in the radiative tests with binary pulsars.

As already mentioned in Section 1.2, there is no generic connection between the higher order gravitational wave damping effects and the parameters $\mathcal{G}$, ε, and ξ of the modified Einstein–Infeld–Hoffmann formalism. Such higher order, mixed radiative and strong-field effects depend in a complicated way on the structure of the gravity theory [40].

The post-Keplerian parameters are at the foundation of many of the gravity tests conducted with binary pulsars. As shown above, the exact functional dependence differs for given theories of gravity. A priori, the masses of pulsar and companion are undetermined, but they represent the only unknowns in this set of equations. Hence, once two post-Keplerian parameters are measured, the corresponding equations can be solved for the two masses, and the values for other post-Keplerian parameters can be predicted for an assumed theory of gravity. Any further post-Keplerian measurement must therefore be consistent with that prediction, otherwise the assumed theory has to be rejected. In other words, if $N \geq 3$ post-Keplerian parameters can be measured, a total of $N - 2$ independent tests can be performed. The method is very powerful, as any additionally measured post-Keplerian parameter is potentially able to fail the prediction and hence to falsify the tested theory of gravity. The standard graphical representation of such tests, as will become clear below, is the mass–mass diagram. Every measured post-Keplerian parameter defines a curve of certain width (given by the measurement uncertainty of the post-Keplerian parameter) in a m_p–m_c diagram. A theory has passed a binary pulsar test, if there is a region in the mass–mass diagram that agrees with all post-Keplerian parameter curves.

2 Gravitational wave damping

2.1 The Hulse–Taylor pulsar

The first binary pulsar to ever be discovered happened to be a rare double neutron star system. It was discovered by Russell Hulse and Joseph Taylor in summer 1974 [19].

The pulsar, PSR B1913+16, has a rotational period of 59 ms and is in a highly eccentric ($e = 0.62$) 7.75-h orbit around an unseen companion. Shortly after the discovery of PSR B1913+16, it has been realized that this system may allow the observation of gravitational wave damping within a time span of a few years [63, 64].

The first relativistic effect seen in the timing observations of the Hulse–Taylor pulsar was the secular advance of periastron $\dot{\omega}$. Thanks to its large value of 4.2 deg/yr, this effect was well measured already one year after the discovery [65]. Due to the, a priori, unknown masses of the system, this measurement could not be converted into a quantitative gravity test. However, assuming GR is correct, using equation (1.17) gives the total mass M of the system. From the modern value given in Table 2 one finds $M = m_p + m_c = 2.828378 \pm 0.000007\, M_\odot$ [31].[4]

It took a few more years to measure the Einstein delay (1.11) with good precision. In a single orbit this effect is exactly degenerate with the Roemer delay, and only due the relativistic precession of the orbit these two delays become separable [63, 67]. By the end of 1978, the timing of PSR B1913+16 yielded a measurement of the post-Keplerian parameter γ, which is the amplitude of the Einstein delay [68]. Together with the total mass from $\dot{\omega}^{\rm GR}$, equation (1.18) can now be used to calculate the individual masses. With the modern value for γ from Table 2, and the total mass given above, one finds the individual masses $m_p = 1.4398 \pm 0.0002\, M_\odot$ and $m_c = 1.3886 \pm 0.0002\, M_\odot$ for pulsar and companion respectively [31].

With the knowledge of the two masses, m_p and m_c, the binary system is fully determined in terms of the post-Keplerian parameters, and further GR effects can be calculated and compared with the observed values, providing an intrinsic consistency check of the theory. In fact, Taylor et al. [68] reported the measurement of a decrease

Table 2. Observed orbital timing parameters of PSR B1913+16, based on the Damour–Deruelle timing model (taken from [31]). Figures in parentheses represent estimated uncertainties in the last quoted digit.

T_0	Time of periastron passage (MJD)	52 144.90097841(4)
x	Projected semimajor axis of the pulsar orbit (s)	2.341782(3)
e	Orbital eccentricity	0.6171334(5)
P_b	Orbital period at T_0 (d)	0.322997448911(4)
ω_0	Longitude of periastron at T_0 (deg)	292.54472(6)
$\dot{\omega}$	Secular advance of periastron (deg/yr)	4.226598(5)
γ	Amplitude of Einstein delay (ms)	4.2992(8)
$\dot{P}_b$	Secular change of orbital period	$-2.423(1) \times 10^{-12}$

4 Strictly speaking, this is the total mass of the system scaled with an unknown Doppler factor D, i.e. $M^{\rm observed} = D^{-1} M^{\rm intrinsic}$ [40]. For typical velocities, $D - 1$ is expected to be of order 10^{-4}, see for instance [66]. In gravity tests based on post-Keplerian parameters, the factor D drops out and is therefore irrelevant [55].

in the orbital period $\dot{P}_b$, consistent with the quadrupole formula (1.26). This was the first proof for the existence of gravitational waves as predicted by GR. In the meantime the $\dot{P}_b$ is measured with a precision of 0.04% (see Table 2). However, this is not the precision with which the validity of the quadrupole formula is verified in the PSR B1913+16 system. The observed $\dot{P}_b$ needs to be corrected for extrinsic effects, most notably the differential Galactic acceleration and the Shklovskii effect, to obtain the intrinsic value caused by gravitational wave damping [69, 70]. The extrinsic contribution due to the Galactic gravitational field (acceleration **g**) and the proper motion (transverse angular velocity in the sky μ) are given by

$$\delta \dot{P}_b^{\rm ext} = \frac{P_b}{c}\left[\hat{\mathbf{K}}_0 \cdot (\mathbf{g}_{\rm PSR} - \mathbf{g}_\odot) + \mu^2 d\right] , \tag{2.1}$$

where $\hat{\mathbf{K}}_0$ is the unit vector pointing toward the pulsar, which is at a distance d from the Solar System. For PSR B1913+16, P_b and $\hat{\mathbf{K}}_0$ are measured with very high precision, and also μ is known with good precision (~ 8%). However, there is a large uncertainty in the distance d, which is also needed to calculate the Galactic acceleration of the PSR B1913+16 system, $\mathbf{g}_{\rm PSR}$, in equation (2.1). Due to its large distance, there is no direct

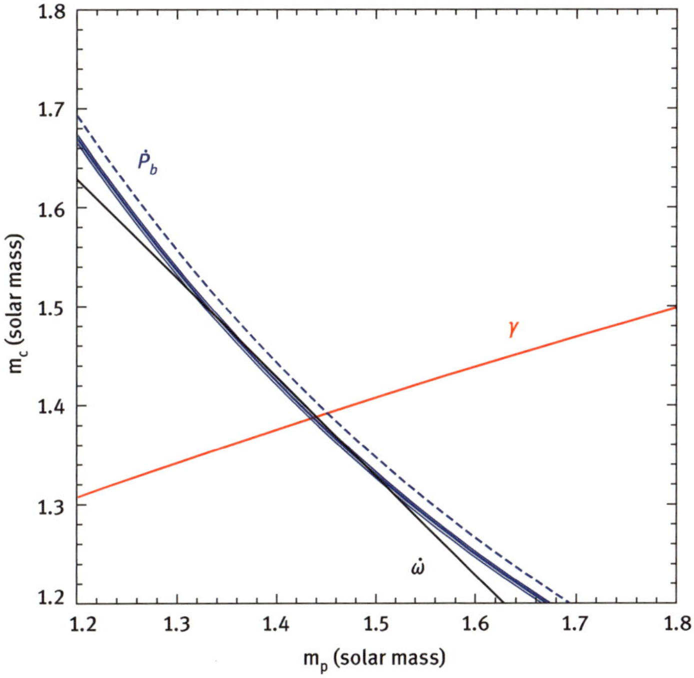

Fig. 4. Mass–mass diagram for PSR B1913+16 based on GR and the three observed post-Keplerian parameters $\dot{\omega}$ (black), γ (red), and $\dot{P}_b$ (blue). The dashed $\dot{P}_b$ curve is based on the observed $\dot{P}_b$, without corrections for Galactic and Shklovskii effects. The solid $\dot{P}_b$ curve is based on the corrected (intrinsic) $\dot{P}_b$, where the thin lines indicate the one-sigma boundaries. Values are taken from Table 2.

parallax measurement for d, and distance estimates are based on model-dependent methods, like the measured column density of free electrons between PSR B1913+16 and the Earth. Such methods are known to have large systematic uncertainties, and for this reason the distance to PSR B1913+16 is not well known: $d = 9.9 \pm 3.1$ kpc [31, 71]. In addition, there are further uncertainties, e.g. in the Galactic gravitational potential and the distance of the Earth to the Galactic center. Accounting for all these uncertainties leads to an agreement between $\dot{P}_b - \delta\dot{P}_b^{\rm ext}$ and $\dot{P}_b^{\rm GR}$ at the level of about 0.3% [31]. The corresponding mass–mass diagram is given in Figure 4. As the precision of the radiative test with PSR B1913+16 is limited by the model dependent uncertainties in equation (2.1), it is not expected that this test can be significantly improved in the near future.

Finally, besides the mass–mass diagram, there is a different way to illustrate the test of gravitational wave damping with PSR B1913+16. According to equation (1.7), the

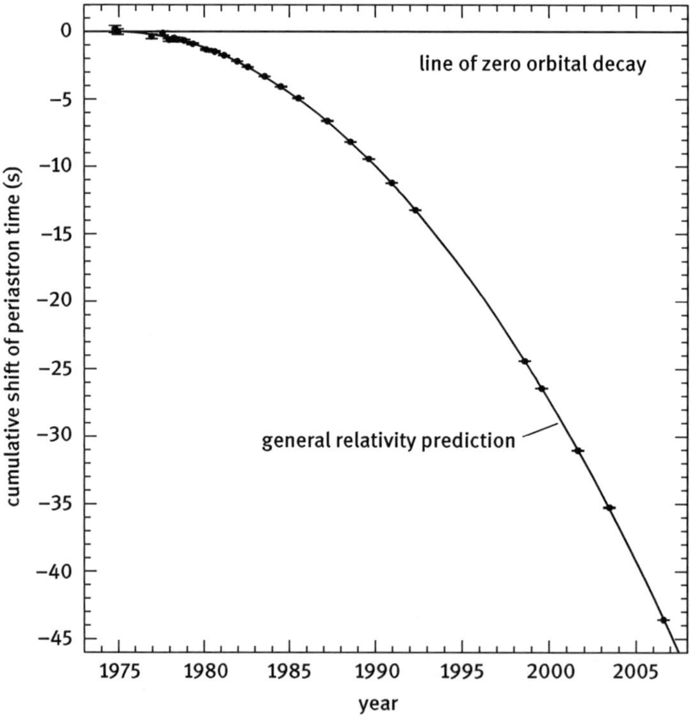

Fig. 5. Shift in the time of periastron passage of PSR B1913+16 due to gravitational wave damping. The parabola represents the GR prediction and the data points the timing measurements, with (vertical) error bars mostly too small to be resolved. The observed shift in periastron time is a direct measurement of the change in the world line of the pulsar due to the back reaction of the emitted gravitational waves (cf. Figure 3). The corresponding spatial shift amounts to about 20 000 km. Figure is taken from [31].

change in the orbital period, i.e. the post-Keplerian parameter $\dot{P}_b$, is measured from a shift in the time of periastron passage, where U is a multiple of 2π. One finds for the shift in periastron time, as compared to an orbit with zero decay

$$\Delta T = \frac{1}{2} P_b \dot{P}_b n^2 + \mathcal{O}(P_b \dot{P}_b^2 n^3)\,, \tag{2.2}$$

where $n = 0, 1, 2, \ldots$ denotes the number of the periastron passage, and is given by $n \simeq (T - T_0)/P_b$. Equation (2.2) represents a parabola in time, which can be calculated with high precision using the masses that come from $\dot{\omega}^{\mathrm{GR}}$ and γ^{GR} (see above). On the other hand, the observed cumulative shift in periastron can be extracted from the timing observations with high precision. A comparison of observed and predicted cumulative shift in the time of the periastron passage is given in Figure 5.

2.2 The Double Pulsar – The best test for Einstein's quadrupole formula, and more

In 2003, a binary system was discovered where at first one member was identified as a pulsar with a 23 ms period [72], before about half a year later the companion was also recognized as a radio pulsar with a period of 2.8 s [73]. Both pulsars, known as PSRs J0737-3039A and J0737-3039B, respectively, (or A and B hereafter), orbit each other in less than 2.5 h in a mildly eccentric ($e = 0.088$) orbit. As a result, the system is not only the first and only double neutron star system where both neutron stars are visible as active radio pulsars, but it is also the most relativistic binary pulsar laboratory for gravity known to date (see Figure 6). Just to give an example for the strength of relativistic effects, the advance of periastron, $\dot{\omega}$, is 17 degrees per year, meaning that the eccentric orbit does a full rotation in just 21 years. In this subsection, we briefly discuss the properties of this unique system, commonly referred to as the *Double Pulsar*, and highlight some of the gravity tests that are based on the radio observations of this system. For detailed reviews of the Double Pulsar see [74, 75].

In the case of the Double Pulsar a total of six post-Keplerian parameters have been measured by now. Five arise from four different relativistic effects visible in pulsar timing [76], while a sixth one can be determined from the effects of geodetic precession, which will be discussed in detail in Section 3.2. The relativistic precession of the orbit, $\dot{\omega}$, was measured within a few days after timing of the system commenced, and by 2006 was already known with a precision of 0.004% (see Table 3). At the same time the measurement of the amplitude of Einstein delay, γ, reached 0.7% (see Table 3). Due to the periastron precession of 17 deg per year, the Einstein delay was soon well separable from the Roemer delay. Two further post-Keplerian parameters came from the detection of the Shapiro delay, a propagation delay which depends on the companion mass and on the inclination angle of the orbit i, i.e. how close the pulsar signal passes the companion. These dependencies translate into the *shape* and *range* parameters s

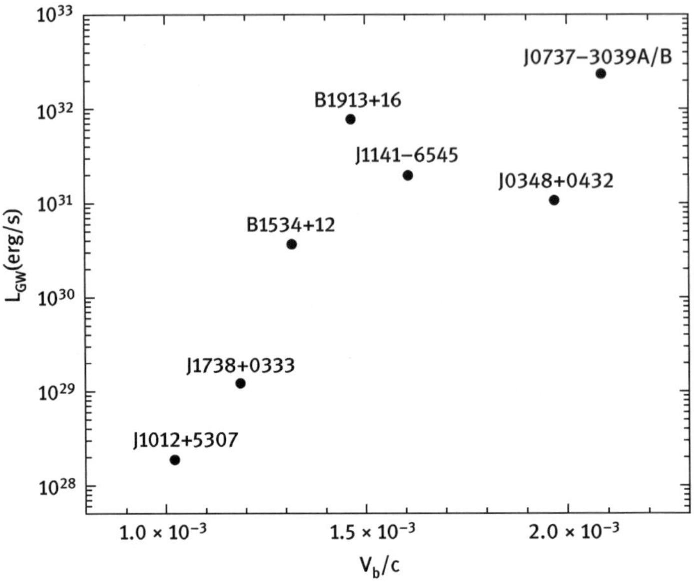

Fig. 6. Short-orbital-period ($P_b < 1$ day) binary pulsars used for gravity tests. The velocity V_b (divided by the speed of light c) is a direct measure for the strength of post-Newtonian effects in the orbital dynamics. The gravitational wave luminosity $L_{\rm GW}$ is an indicator for the strength of radiative effects that cause secular changes to the orbital elements due to gravitational wave damping.

and r. They were measured with a precision of 0.04% and 5%, respectively (see Table 3). From the measured value $s = \sin i = 0.99974^{+0.00016}_{-0.00039}$ ($i = 88.7^{\circ}{}^{+0.5^{\circ}}_{-0.8^{\circ}}$) one can already see how exceptionally edge-on this system is.[5] Finally, the decrease of the orbital period due to gravitational wave damping was measured with a precision of 1.4% just 3 years after the discovery of the system (see Table 3).

A unique feature of the Double Pulsar is its nature as a "dual-line source," i.e. we measure the orbits of both neutron stars at the same time. Obviously, the sizes of the two orbits are not independent from each other as they orbit a common center of mass. In GR, up to first post-Newtonian order the relative size of the orbits is identical to the inverse ratio of masses. Hence, by measuring the orbits of the two pulsars (relative to the center of mass), we obtain a precise measurement of the mass ratio. This ratio is directly observable, as the orbital inclination angle is obviously identical for both pulsars, i.e.

$$R \equiv \frac{m_{\rm A}}{m_{\rm B}} = \frac{a_{\rm B}}{a_{\rm A}} = \frac{a_{\rm B}\sin i/c}{a_{\rm A}\sin i/c} \equiv \frac{x_{\rm B}}{x_{\rm A}}\,. \tag{2.3}$$

5 The only binary pulsar known to be (most likely) even more edge-on is PSR J1614-2230 with $s = \sin i = 0.999894 \pm 0.000005$ ($i = 89.17^{\circ} \pm 0.02^{\circ}$) [77]. For this wide-orbit system ($P_b \approx 8.7$ d), however, no further post-Keplerian parameter is known that could be used in a gravity test.

Table 3. A selection of observed orbital timing parameters of the Double Pulsar, based on the Damour–Deruelle timing model (taken from [76]). All post-Keplerian parameters below are obtained from the timing of pulsar A. The timing precision for pulsar B is considerably lower, and allows only for a, in comparison, low precision measurement ($\sim$ 0.3%) of $\dot{\omega}$ [76]. Figures in parentheses represent estimated uncertainties in the last quoted digit.

$x_A \equiv a_A \sin i/c$	Projected semimajor axis of pulsar A (s)	1.415032(1)
$x_B \equiv a_B \sin i/c$	Projected semimajor axis of pulsar B (s)	1.5161(16)
e	Orbital eccentricity	0.0877775(9)
P_b	Orbital period (d)	0.10225156248(5)
$\dot{\omega}$	Secular advance of periastron (deg/yr)	16.89947(68)
γ	Amplitude of Einstein delay for A (ms)	0.3856(26)
$\dot{P}_b$	Secular change of orbital period	$-1.252(17) \times 10^{-12}$
s	*Shape* of Shapiro delay for A	$0.99974(-39, +16)$
r	*Range* of Shapiro delay for A (μs)	6.21(33)

This expression is not just limited to GR. In fact, it is valid up to post-Newtonian order and free of any explicit strong-field effects in any Lorentz-invariant theory of gravity (see [41] for a detailed discussion). Using the parameter values of Table 3, one finds that in the Double Pulsar the masses are nearly equal with $R = 1.0714 \pm 0.0011$.

As it turns out, all the post-Keplerian parameters measured from timing are consistent with GR. In addition, the region of allowed masses agrees well with the measured mass ratio R (see Figure 7). One has to keep in mind, that the test presented here is based on data published in 2006 [76]. In the meantime continued timing lead to a significant decrease in the uncertainties of the post-Keplerian parameters of the Double pulsar. This is especially the case for $\dot{P}_b$, for which the uncertainty typically decreases with $T_{\rm obs}^{-2.5}$ [78], $T_{\rm obs}$ being the total time span of timing observations. The new results will be published in an upcoming publication (Kramer et al., in preparation). As reported in [28], presently the Double Pulsar provides the best test for the GR quadrupole formalism for gravitational wave generation, with an uncertainty well below the 0.1% level. As discussed above, the Hulse–Taylor pulsar is presently limited by uncertainties in its distance. This raises the valid question, at which level such uncertainties will start to limit the radiative test with the Double Pulsar as well. Compared to the Hulse–Taylor pulsar, the Double Pulsar is much closer to Earth. Because of this, a direct distance estimate of $1.15^{+0.22}_{-0.16}$ kpc based on a parallax measurement with long-baseline interferometry was obtained [79]. Thus, with the current accuracy in the measurement of distance and transverse velocity, GR tests based on $\dot{P}_b$ can be taken to the 0.01% level. We will come back to this in Section 8, where we discuss some future tests with the Double Pulsar.

With the large number of post-Keplerian parameters and the known mass ratio, the Double Pulsar is the most over-constrained binary pulsar system. For this reason, one can do more than just testing specific gravity theories. The Double Pulsar allows for certain generic tests on the orbital dynamics, time dilation, and photon propaga-

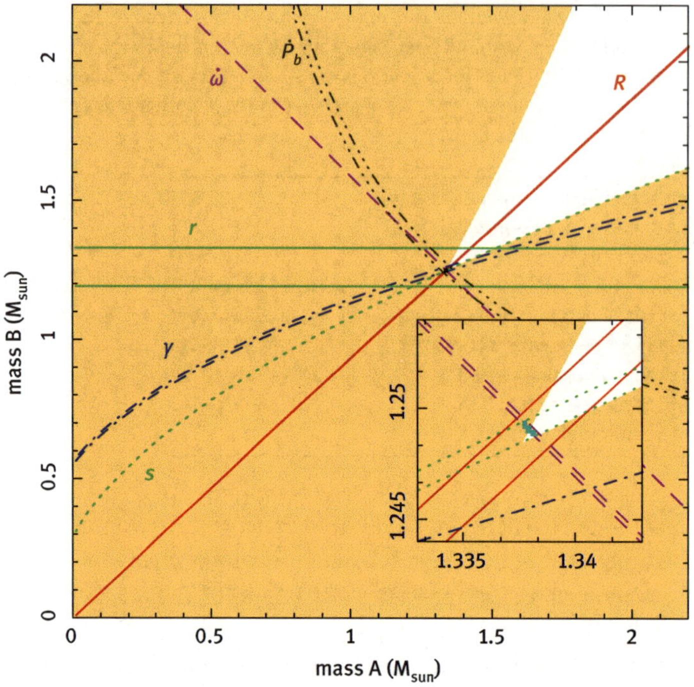

Fig. 7. GR mass–mass diagram based on timing observations of the Double Pulsar. The orange areas are excluded simply by the fact that $\sin i \leq 1$. The figure is taken from [75] (Ω_{SO} lines removed) and based on the timing solution published in [76].

tion of a spacetime with two strongly self-gravitating bodies [75]. First, the fact that the Double Pulsar gives access to the mass ratio, R, in any Lorentz-invariant theory of gravity, allows us to determine $m_A/M = R/(1+R) = 0.51724 \pm 0.00026$ and $m_B/M = 1/(1+R) = 0.48276 \pm 0.00026$. With this information at hand, the measurement of the shape of the Shapiro delay s can be used to determine $\mathcal{V}_b$ via equation (1.16): $\mathcal{V}_b/c = (2.0854 \pm 0.0014) \times 10^{-3}$. At this point, the measurement of the post-Keplerian parameters $\dot{\omega}$, γ, and r (equations (1.13), (1.14), (1.15)) can be used to impose restrictions on the "strong-field" parameters of Lagrangian (1.2) [75]:

$$\frac{2\varepsilon - \xi}{5} = 0.9995 \pm 0.0016\,, \tag{2.4}$$

$$\frac{G_{0B}}{\mathcal{G}} + \mathcal{K}_A^B = 1.005 \pm 0.010\,, \tag{2.5}$$

$$\frac{\varepsilon_{0B} + 1}{4}\,\frac{G_{0B}}{\mathcal{G}} = 1.009 \pm 0.054\,. \tag{2.6}$$

This is in full agreement with GR, which predicts one for all three of these expressions. Consequently, nature cannot deviate much from GR in the quasi-stationary strong-field regime of gravity (G2 in Figure 1).

2.3 PSR J1738+0333 – The best test for scalar–tensor gravity

The best “pulsar clocks” are found among the fully recycled millisecond pulsars, which have rotational periods less than about 10 ms (see e.g. [80]). A result of the stable mass transfer between companion and pulsar in the past – responsible for the recycling of the pulsar – is a very efficient circularization of the binary orbit, that leads to a pulsar-white dwarf system with very small residual eccentricity [81]. For such systems, the post-Keplerian parameters $\dot{\omega}$ and γ are generally not observable. There are a few cases, where the orbit is seen sufficiently edge-on, so that a measurement of the Shapiro delay gives access to the two post-Keplerian parameters r and s with good precision (see e.g. [82], which was the first detection of a Shapiro delay in a binary pulsar). With these two parameters the system is then fully determined, and in principle can be used for a gravity test in combination with a third measured (or constrained) post-Keplerian parameter (e.g. $\dot{P}_b$). Besides the Shapiro delay parameters, some of the circular binary pulsar systems offer a completely different access to their masses, which is not solely based on the timing observations in the radio frequencies. If the companion star is bright enough for optical spectroscopy, then we have a dual-line system, where the Doppler shifts in the spectral lines can be used, together with the timing observations of the pulsar, to determine the mass ratio R. Furthermore, if the companion is a white dwarf, the spectroscopic information in combination with models of the white dwarf and its atmosphere can be used to determine the mass of the white dwarf m_c, ultimately giving the mass of the pulsar via $m_p = R\,m_c$. As we will see in this and the following subsection, two of the best binary pulsar systems for gravity tests have their masses determined through such a combination of radio and optical astronomy.

PSR J1738+0333 was discovered in 2004 [83]. It has a spin period P of 5.85 ms and is a member of a low-eccentricity ($e < 4\times10^{-7}$) binary system with an orbital period P_b of just 8.5 h. The companion is an optically bright low-mass white dwarf (see Figure 8). Extensive timing observation over a period of 10 years allowed a determination of astrometric, spin, and orbital parameters with high precision [30], most notably

- A change in the orbital period of $(-17.0 \pm 3.1) \times 10^{-15}$.
- A timing parallax, which gives a model independent distance estimate of $d = 1.47 \pm 0.10$ kpc.

The latter is important to correct for the Shklovskii effect and the differential Galactic acceleration to obtain the intrinsic $\dot{P}_b$ (cf. equation (2.1)). Additional spectroscopic observations of the white dwarf gave the mass ratio $R = 8.1 \pm 0.2$ and the companion mass $m_c = 0.181^{+0.007}_{-0.005}\,M_\odot$, and consequently the pulsar mass $m_p = 1.47^{+0.07}_{-0.06}\,M_\odot$ [84]. It is important to note, that the mass determination for PSR B1738+0333 is free of any explicit strong-field contributions, since this is the case for the mass ratio [41], and certainly for the mass of the white dwarf, which is a weakly self-gravitating body, i.e. a gravity regime that has been well tested in the Solar System.

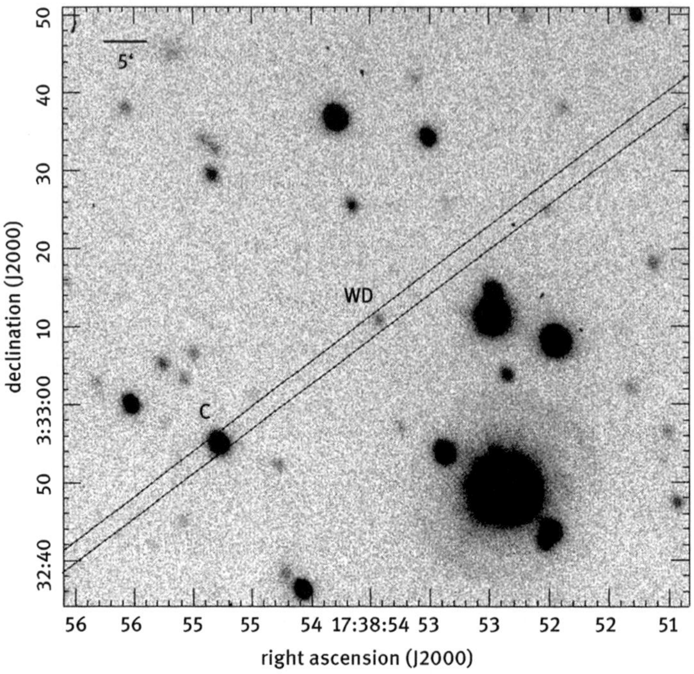

Fig. 8. Optical finding chart for the PSR J1738+0333 companion. Indicated are the white dwarf companion (WD), the slit orientation used during the observation and the comparison star (C) that was included in the slit. The white dwarf is sufficiently bright to allow for high signal-to-noise spectroscopy (see [84] for details, where this figure is taken from).

After using equation (2.1) to correct for the Shklovskii contribution, $\delta\dot{P}_b = P_b\mu^2 d/c = (8.3^{+0.6}_{-0.5}) \times 10^{-15}$, and the contribution from the Galactic differential acceleration, $\delta\dot{P}_b = (0.58^{+0.16}_{-0.14}) \times 10^{-15}$, one finds an intrinsic orbital period change due to gravitational wave damping of $\dot{P}_b^{\rm intr} = (-25.9 \pm 3.2) \times 10^{-15}$. This value agrees well with the prediction of GR, as can be seen in Figure 9.

The radiative test with PSR J1738+0333 represents a $\sim 15\%$ verification of GR's quadrupole formula. A comparison with the $< 0.1\%$ test from the Double Pulsar (see Section 2.2) raises the valid question of whether the PSR J1738+0333 experiment is teaching us something new about the nature of gravity and the validity of GR. To address this question, let us have a look at equation (1.25). Dipolar radiation can be a strong source of gravitational wave damping, if there is a sufficient difference between the effective coupling parameters α_p and α_c of pulsar and companion, respectively. For the Double Pulsar, where we have two neutron stars with $m_p \approx m_c$, one generally expects that $\alpha_p \approx \alpha_c$, and therefore the effect of dipolar radiation would be strongly suppressed. On the other hand, in the PSR J1738+0333 system there is a large difference in the compactness of the two bodies. For the weakly self-gravitating white-dwarf

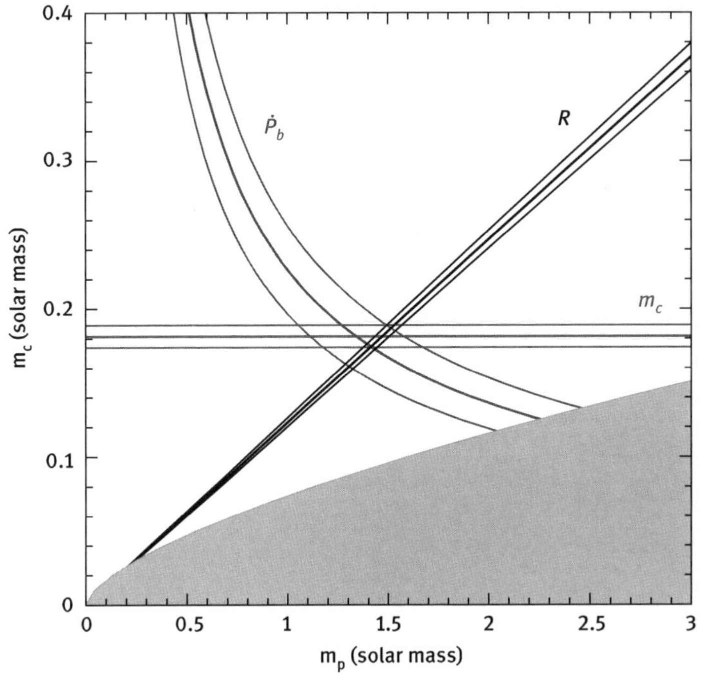

Fig. 9. GR mass–mass diagram based on the timing observations of PSR J1738+0333 and the optical observations of its white-dwarf companion, respectively. The thin lines indicate the one-sigma errors of the measured parameters. The gray area is excluded by the condition $\sin i \leq 1$.

companion $\alpha_c \simeq \alpha_0$, i.e. it assumes the weak-field value[6], while the strongly self-gravitating pulsar can have an α_p that significantly deviates from α_0. In fact, as discussed in Section 1.4, α_p can even be of oder unity in the presence of effects like strong-field scalarization. In the absence of nonperturbative strong-field effects one can do a first-order estimation $(\alpha_p - \alpha_c) \propto (\epsilon_p - \epsilon_c) + \mathcal{O}(\epsilon^2)$. For the Double Pulsar one finds $(\epsilon_p - \epsilon_c)^2 \approx 6 \times 10^{-5}$, which is significantly smaller than for the PSR J1738+0333 system, which has $(\epsilon_p - \epsilon_c)^2 \approx 0.012$.[7] As a consequence, the orbital decay of asymmetric systems like PSR J1738+0333 could still be dominated by dipolar radiation, even if the Double Pulsar agrees with GR. For this reason, PSR J1738+0333 is particularly useful to test gravity theories that violate the strong equivalence principle and therefore predict the emission of dipolar radiation. A well-known class of gravity theories, where this is the case, are scalar–tensor theories. As it turns out, PSR J1738+0333 is currently the best test system for these alternatives to GR (see Figure 10). In terms of equation (1.25),

6 From the Cassini experiment [12] one obtains $|\alpha_0| < 3 \times 10^{-3}$ (95% confidence).

7 These numbers are based on the equation of state MPA1 in [85]. Within GR, MPA1 has a maximum neutron-star mass of 2.46 $M_\odot$, which can also account for the high-mass candidates of [86–88].

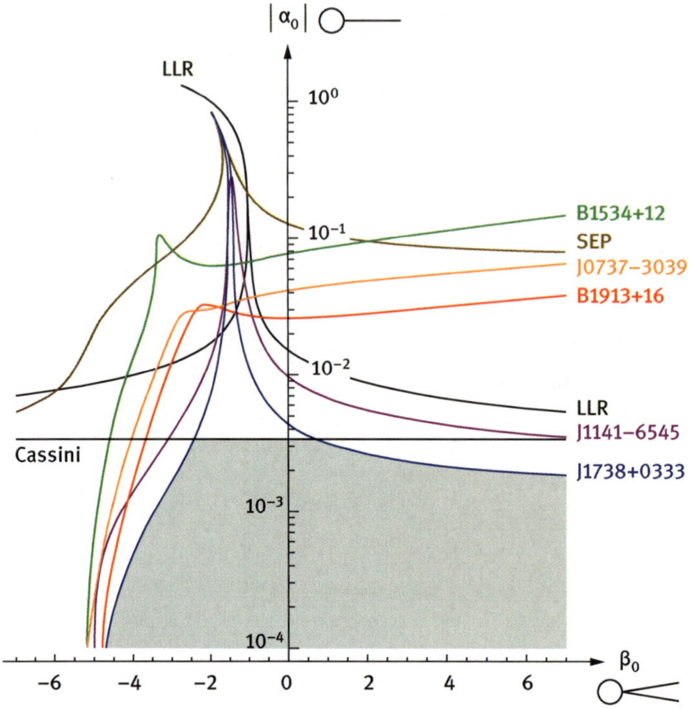

Fig. 10. Constraints on the class of $T_1(\alpha_0, \beta_0)$ scalar–tensor theories of [57, 58], from different binary pulsar and Solar System (Cassini and Lunar Laser Ranging) experiments. The gray area indicates the still allowed T_1 theories, and includes GR ($\alpha_0 = \beta_0 = 0$). It is obvious that PSR J1738+0333 is the most constraining experiment for most of the β_0 range, and is even competitive with Cassini in testing the Jordan–Fierz–Brans–Dicke theory ($\beta_0 = 0$). As can be clearly seen, the double neutron-star systems PSR B1534+12 [89], PSR B1913+16 (Hulse–Taylor pulsar) and PSR J0737-3039A/B (Double Pulsar) are considerably less constraining, as explained in the text. PSR J1141-6545 is also well suited for a dipolar radiation test [90], since it also has a white dwarf companion [91]. Figure is taken from [30].

one finds

$$|\alpha_p - \alpha_c| < 2 \times 10^{-3} \quad (95\% \text{ confidence}), \tag{2.7}$$

where for the weakly self-gravitating white dwarf companion $\alpha_c \simeq \alpha_0$. This limit can be interpreted as a generic limit on dipolar radiation, where $\alpha_p - \alpha_c$ is the difference of some hypothetical (scalar- or vector-like) "gravitational charges" [39].

2.4 PSR J0348+0432 – A massive pulsar in a relativistic orbit

PSR J0348+0432 was discovered in 2007 in a drift scan survey using the Green Bank radio telescope (GBT) [92, 93]. PSR J0348+0432 is a mildly recycled radio pulsar with a spin period of 39 ms. Soon it was found to be in a 2.46-h orbit with a low mass white dwarf companion. In fact, the orbital period is only 15 s longer than that of the Double Pulsar, which by itself makes this already an interesting system for gravity. Initial timing observations of the binary yielded an accurate astrometric position, which allowed for an optical identification of its companion [94]. As it turned out, the companion is a relatively bright white dwarf with a spectrum that shows deep Balmer lines. Like in the case of PSR J1738+0333, one could use high-resolution optical spectroscopy to determine the mass ratio $R = 11.70 \pm 0.13$ (see Figure 11) and the companion mass $m_c = 0.172 \pm 0.003\ M_\odot$. For the mass of the pulsar one then finds $m_p = R m_c = 2.01 \pm 0.04\ M_\odot$, which is presently the highest, well-determined neu-

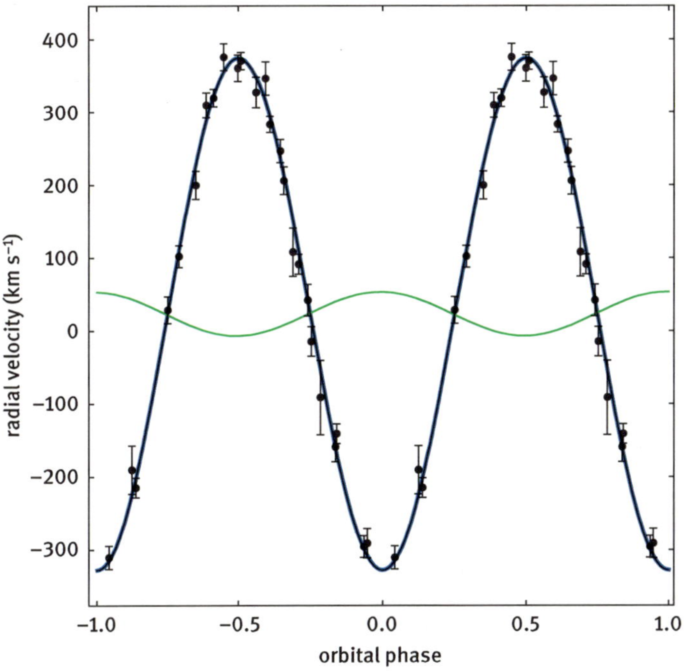

Fig. 11. Spectroscopically measured radial velocities for the white-dwarf companion of PSR J0348+0432. For illustration purposes the data are plotted twice. The fitted sinusoidal curve (blue) has an amplitude of 351 ± 4 km/s. As a comparison, the sinusoidal green line shows the radial velocity of the pulsar as derived from the timing solution. The amplitude of the green line is known with very high precision: 30.008235 ± 0.000016 km/s. The ratio of the amplitudes gives the mass ratio R. Figure is taken from [94].

tron star mass, and only the second neutron star with a well-determined mass close to $2M_\odot$.[8]

Since its discovery there have been regular timing observations of PSR J0348+0432 with three of the major radio telescopes in the world, the 100-m Green Bank Telescope, the 305-m radio telescope at the Arecibo Observatory, and the 100-m Effelsberg radio telescope. Based on the timing data, in 2013 Antoniadis et al. [94] reported the detection of a decrease in the orbital period of $\dot{P}_b = (-2.73 \pm 0.45) \pm 10^{-13}$ that is in full agreement with GR (see Figure 12). In numbers:

$$\dot{P}_b/\dot{P}_b^{\rm GR} = 1.05 \pm 0.18 \,. \tag{2.8}$$

As it turns out, using the distance inferred from the photometry of the white dwarf ($d \sim 2.1$ kpc) corrections due to the Shklovskii effect and differential acceleration in the Galactic potential (see equation (2.1)) are negligible compared to the measurement uncertainty in $\dot{P}_b$.

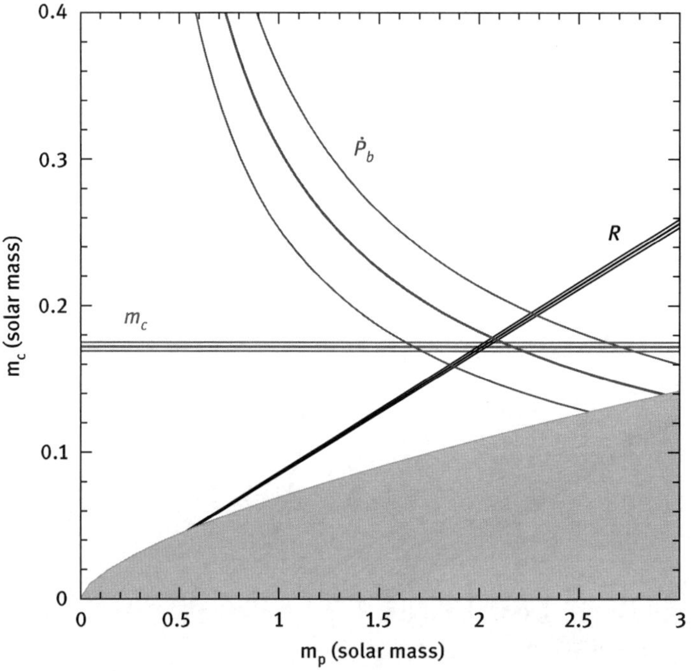

Fig. 12. GR mass–mass diagram based on timing and optical observations of the PSR J0348+0432 system. The thin lines indicate the one-sigma errors of the measured parameters. The gray area is excluded by the condition $\sin i \leq 1$.

---

8 The first well determined two Solar mass neutron star is PSR B1614-2230 [77], which is in a wide orbit and therefore does not provide any gravity test.

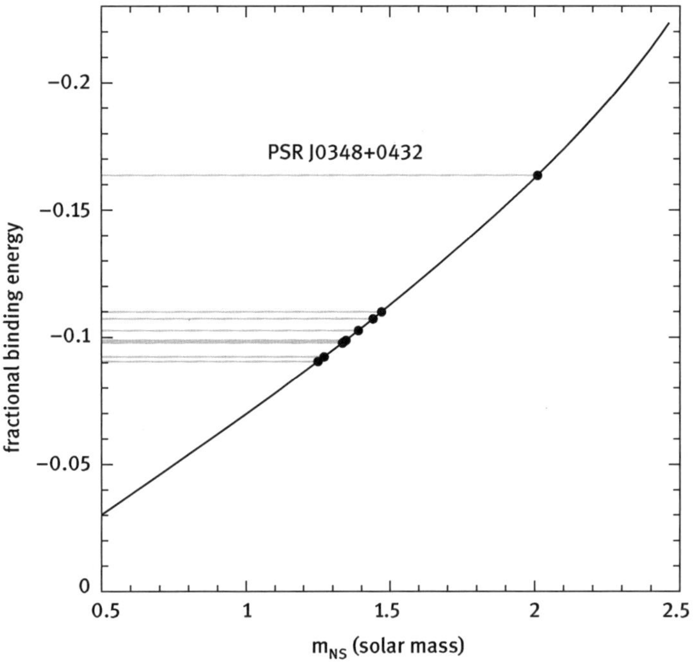

Fig. 13. Fractional gravitational binding energy of a neutron star as a function of its (inertial) mass, based on equation of state MPA1 [85]. The plot clearly shows the prominent position of PSR J0348+0432. The other dots indicate the neutron star masses of the individual test systems in Figure 10.

Like PSR 1738+0333, PSR J0348+0432 is a system with a large asymmetry in the compactness of the components, and therefore well suited for a dipolar radiation test. Using equation (1.25), the limit (2.8) can be converted into a limit on additional gravitational scalar or vector charges:

$$|\alpha_p - \alpha_0| < 5 \times 10^{-3} \quad (95\% \text{ confidence})\,. \tag{2.9}$$

This limit is certainly weaker than the limit (2.8), but it has a new quality as it tests a gravity regime in neutron stars that has not been tested before. Gravity tests before [94] were confined to "canonical" neutron star masses of $\sim 1.4\,M_\odot$. PSR J0348+0432 for the first time allows a test of the relativistic motion of a massive neutron star, which in terms of gravitational self-energy lies clearly outside the tested region (see Figure 13).

Although an increase in fractional binding energy of about 50% does not seem much, in the highly nonlinear gravity regime of neutron stars it could make a significant difference. To demonstrate this, [94] used the scalar–tensor gravity $T_1(\alpha_0, \beta_0)$ of [57, 58], which is known to behave strongly nonlinear in the gravitational fields of neutron stars, in particular for $\beta_0 < -4.0$. As shown in Figure 14, PSR J0348+0432 excludes a family of scalar–tensor theories that predict significant deviations from GR in

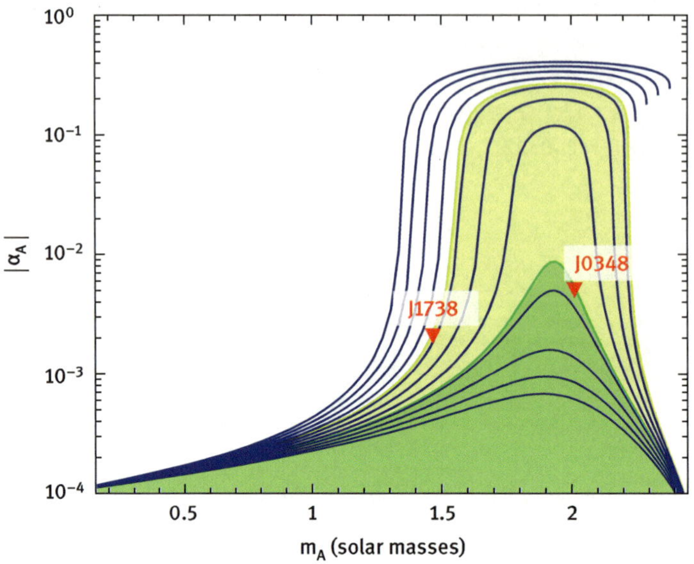

Fig. 14. Effective scalar coupling as a function of the neutron-star mass, in the $T_1(\alpha_0, \beta_0)$ mono-scalar–tensor gravity theory of [57, 58]. For the linear coupling of matter to the scalar field we have chosen $\alpha_0 = 10^{-4}$, a value well below the sensitivity of any near-future Solar System experiment, like GAIA [95]. The blue curves correspond to stable neutron-star configurations for different values of the quadratic coupling β_0: -5 to -4 (top to bottom) in steps of 0.1. The yellow area indicates the parameter space still allowed by the limit (2.7) [label "J1738"], whereas only the green area is in agreement with the limit (2.9) [label "J0348"]. The plot shows clearly how the massive pulsar PSR J0348+0432 probes deep into a new gravity regime. Neutron-star calculations are based on equation of state MPA1 [85] (see [94] for a different equation of state).

massive neutron stars and were not excluded by previous experiments, most notably the test done with PSR J1738+0333. To further illustrate this in a mass–mass diagram, Figure 15 shows a gravity theory with strong-field scalarization in massive neutron stars that passes the PSR J1738+0333 experiment, but is falsified by PSR J0348+0432.

With PSR J0348+0432, gravity tests now cover a range of neutron star masses from $1.25 M_\odot$ (PSR J0737-3039B) to $2 M_\odot$. No significant deviation from GR in the orbital motion of these neutron stars was found. These findings have interesting implications for the upcoming ground-based gravitational wave experiments, as we will briefly discuss in the next subsection.

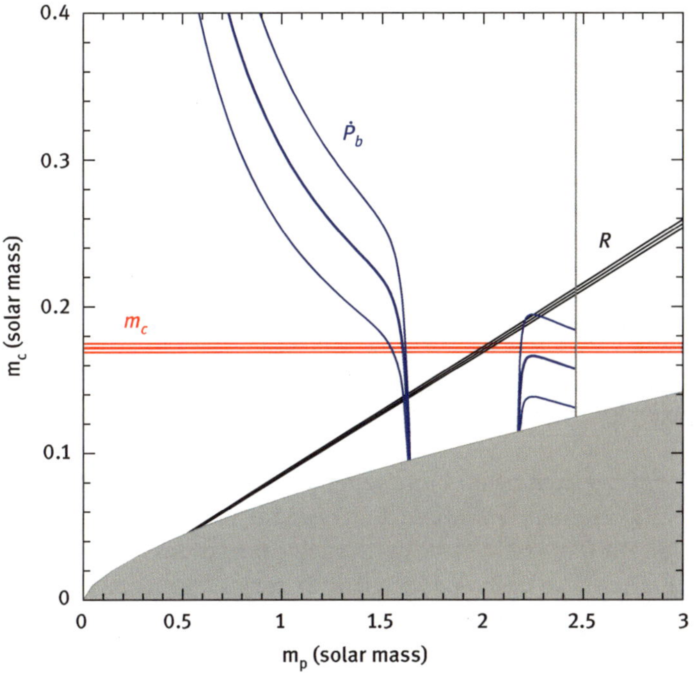

Fig. 15. Mass–mass diagram based on timing and optical observations of the PSR J0348+0432 system, for the mono-scalar–tensor gravity $T_1(10^{-4}, -4.5)$. The thin lines indicate the one-sigma errors of the measured parameters. The vertical gray line is at the maximum mass of a neutron star for the given theory and equation of state (MPA1). The gray area is excluded by the condition $\sin i \leq 1$. Obviously $T_1(10^{-4}, -4.5)$ is clearly falsified by this test, as there is no common region for the curves of the three parameters m_c, R and $\dot{P}_b$.

2.5 Implications for gravitational wave astronomy

The first detection of gravitational waves from astrophysical sources by ground-based laser interferometers, like LIGO[9] and VIRGO[10], will mark the beginning of a new era of gravitational wave astronomy [96]. One of the most promising sources for these detectors are merging compact binaries, consisting of neutron stars and black holes, whose orbits are decaying toward a final coalescence due to gravitational wave damping. While the signal sweeps in frequency f through the detectors' typical sensitive bandwidth $[f_{\text{in}}, f_{\text{out}}]$ from about 20 Hz to a few kHz, the gravitational wave signal will be deeply buried in the broadband noise of the detectors [96]. To detect the signal, one will have to apply a matched filtering technique, i.e. correlate the output of the detector with a template wave form. Consequently, it is crucial to know the binary's orbital

9 www.ligo.org
10 www.cascina.virgo.infn.it

Table 4. Contributions to the accumulated number of gravitational wave cycles in the frequency band of 20 Hz to 1350 Hz for a $2M_\odot/1.25M_\odot$ neutron-star pair (cf. equations (235) and (236) in [36]). The frequency of 1350 Hz corresponds to the innermost circular orbit of the merging binary system [36].

	Correction to LO	Number of cycles
LO (leading order)	–	4158.6
1pN	$(v/c)^2$	196.4
1.5pN	$(v/c)^3$	−123.6
2pN	$(v/c)^4$	7.2
2.5pN	$(v/c)^5$	−10.3
3pN	$(v/c)^6$	2.4
3.5pN	$(v/c)^7$	−0.9

phase with high accuracy for searching and analyzing the signals from in-spiraling compact binaries. Typically, one aims to lose less than one gravitational wave cycle in a signal with $\sim 10^4$ cycles. For this reason, within GR such calculations for the phase evolution of compact binaries have been conducted with great effort to cover many post-Newtonian orders including spin–orbit and spin–spin contributions (see [36, 97] for reviews). Table 4 illustrates the importance of the individual corrections to the number of cycles spent in the LIGO/VIRGO band[11] for two merging nonspinning neutron stars. For a later comparison, the two neutron-star masses are chosen to be $2M_\odot$ and $1.25M_\odot$, the highest and lowest neutron-star masses observed.

If the gravitational interaction between two compact masses is different from GR, the phase evolution over the last few thousand cycles, which fall into the bandwidth of the detectors, deviates from the (GR) template. This will degrade the ability to accurately determine the parameters of the merging binary, or in the worst case even prevent the detection of the signal. In scalar–tensor gravity, for instance, the evolution of the phase is modified because the system can now lose additional energy to dipolar waves [99, 100]. Depending on the difference between the effective scalar couplings of the two bodies, α_A and α_B, the 1.5 post-Newtonian dipolar contribution to the equations of motion could drive the gravitational wave signal many cycles away from the GR template. For this reason, it is desirable that potential deviations from GR in the interaction of two compact objects can be tested and constrained prior to the start of the advanced gravitational wave detectors. With its location at the high end of the

11 The advanced LIGO/VIRGO gravitational wave detectors are expected to have a lower end seismic noise cut-off at about 10 Hz [98]. For a low signal-to-noise ratio the low-frequency cut-off is considerably higher. In this review, we adapt a value of 20 Hz as the minimum frequency. The maximum frequency of a few kHz is not important here, since the frequency of the innermost circular orbit is well below the upper limit of the LIGO/VIRGO band.

measured neutron-star masses, PSR J0348+0432 with its limit (2.9) plays a particularly important role in such constraints.

The change in the number of cycles that fall into the frequency band of a gravitational wave detector due to a dipolar contribution is given, to leading order, by [99, 100]

$$\Delta N \approx -\frac{25}{21\,504\,\pi}\left(\frac{m_A m_B}{M^2}\right)^{2/5}(\alpha_A - \alpha_B)^2\left(u_{\rm in}^{-7/3} - u_{\rm out}^{-7/3}\right), \tag{2.10}$$

where $u \equiv \pi(G\mathcal{M}c^{-3})f$, and $\mathcal{M} \equiv (m_A m_B)^{3/5}M^{-1/5}$ is the chirp mass. Equation (2.10) is based on the assumption that α_0, α_A, and α_B are considerably smaller than unity, which is supported by binary pulsar experiments. For a $2/1.25M_\odot$ double neutron-star merger, one finds from equation (2.10) and the limit (2.9)

$$|\Delta N(f_{\rm in} = 20\,{\rm Hz}, f_{\rm out} = f_{\rm ICO})| < 0.4\,, \tag{2.11}$$

where $f_{\rm ICO} \approx 1350\,{\rm Hz}$ is the gravitational wave frequency of the innermost circular orbit (cf. [36]). The exact value of $f_{\rm ISCO}$ does not play an important role in equation (2.10), since $f_{\rm in} \ll f_{\rm ICO}$. This result is based on the extreme assumption, that the light neutron star has an effective scalar coupling which corresponds to the well-constrained weak-field limit, i.e. $\alpha_B = \alpha_0$. If the companion of the $2M_\odot$ neutron star is a $10M_\odot$ black hole, then the constraints on ΔN that can be derived from binary pulsar experiments are even tighter (see [94]). A comparison with Table 4 shows that the limit (2.11) is already below the contribution of the highest order correction calculated.

As explained in [94], binary pulsar experiment cannot exclude significant deviations associated with short-range fields (e.g. massive scalar fields), which could still impact the mergers for ground-based gravitational wave detectors. Nevertheless, the constraints on dipolar radiation obtained from binary pulsars provide added confidence in the use of elaborate GR templates to search for the signals of compact merging binaries in the LIGO/VIRGO data sets.

3 Geodetic precession

A few months after the discovery of the Hulse–Taylor pulsar, Damour and Ruffini [51] proposed a test for geodetic precession in that system. If the pulsar spin is sufficiently tilted with respect to the orbital angular momentum, the spin direction should gradually change over time (see Section 1.3). A change in the orientation of the spin axis of the pulsar with respect to the line-of-sight should lead to changes in the observed pulse profile. These pulse-profile changes manifest themselves in various forms [102], such as changes in the amplitude ratio or separation of pulse components [103, 104], the shape of the characteristic swing of the linear polarization [105], or the absolute value of the position angle of the polarization in the sky [75]. In principle, such changes could allow for a measurement of the precession rate and by this yield a test of GR. In practice, it turned out to be rather difficult to convert changes in the pulse profile into a quantitative test for the precession rate. And indeed, the Hulse–Taylor pulsar, in spite of prominent profile changes due to geodetic precession [103, 104], does not (yet) allow for a quantitative test of geodetic precession. This is mostly due to uncertainties in the orientation of the magnetic axis and the intrinsic beam shape [106].

Profile and polarization changes due to geodetic precession have been observed in other binary pulsars as well [107, 108], but again did not lead to a quantitative gravity test. A complete list of binary pulsars that up to date show signs of geodetic precession can be found in [28]. Out of the six pulsars listed in [28], so far only two allowed for quantitative constraints on their rate of geodetic precession. These two binary pulsars will be discussed in more details in the following.

3.1 PSR B1534+12

PSR B1534+12 is a 38 ms pulsar, which was discovered in 1991 [109]. It is a member of an eccentric ($e = 0.27$) double neutron-star system with an orbital period of about 10 h. Subsequent timing observations lead to the determination of five post-Keplerian parameters: $\dot{\omega}$, γ, $\dot{P}_b$, and r, s from the Shapiro delay [89]. The large uncertainty in the distance to this system still prevents its usage in a gravitational wave test, since the observed $\dot{P}_b$ has a large Shklovskii contribution, which one cannot properly correct for. The other four post-Keplerian parameters are nevertheless useful to test quasi-stationary strong-field effects. However, these tests are generally less constraining than tests from other pulsars (see, e.g. Figure 10).

Continued observations of PSR B1534+12 with the 305-m Arecibo radio telescope revealed systematic changes in the the observed pulsar profile by about 1% per year, as well as changes in the polarization properties of the pulsar [110]. As outlined above, such changes are expected from geodetic precession. Using equation (1.4) and the parameters from [89], one finds that GR predicts a precession rate of

$$\Omega^{\mathrm{SO}} = 0.51\ \mathrm{deg/yr} \tag{3.1}$$

for PSR B1534+12.

Besides the secular changes visible in the high signal-to-noise ratio pulse profile and polarization data of PSR B1534+12, Stairs et al. [105] reported the detection of special-relativistic aberration of the revolving pulsar beam due to orbital motion. Aberration periodically shifts the observed angle between the line of sight and spin axis of PSR B1534+12 by an amount that depends on the orientation of the pulsar spin, and therefore contains additional geometrical information. Combining these observations, Stairs et al. [105] were able to determine the system geometry, including the misalignment between the spin of PSR B1534+12 and the angular momentum of the binary motion, and constrain the rate of geodetic precession to

$$\begin{aligned} \Omega^{\mathrm{SO}} &= 0.44^{+0.48}_{-0.16}\ \mathrm{deg/yr} \quad (68\%\ \text{confidence})\,, \\ \Omega^{\mathrm{SO}} &= 0.44^{+4.6}_{-0.24}\ \mathrm{deg/yr} \quad (95\%\ \text{confidence})\,. \end{aligned} \tag{3.2}$$

Although the uncertainties are comparably large, these were the first beam-model-independent constraints on the geodetic precession rate of a binary pulsar. As can be seen, these model-independent constraints on the precession rate are consistent with the prediction by GR, as given in equation (3.1).

3.2 The Double Pulsar

In Section 2.2, we have seen the Double Pulsar as one of the most exciting "laboratories" for relativistic gravity, with a wealth of relativistic effects measured, allowing the determination of five post-Keplerian parameters from timing observations: $\dot{\omega}$, γ, $\dot{P}_b$, r, s. Calculating the inclination angle of the orbit i from $s = \sin i$, one finds that the line-of-sight is inclined with respect to the plane of the binary orbit by just about 1.3° [76]. As a consequence, during the superior conjunction the signals of pulsar A pass pulsar B at a distance of only 20 000 km. This is small compared to the extension of pulsar B's magnetosphere, which is roughly given by the radius of the

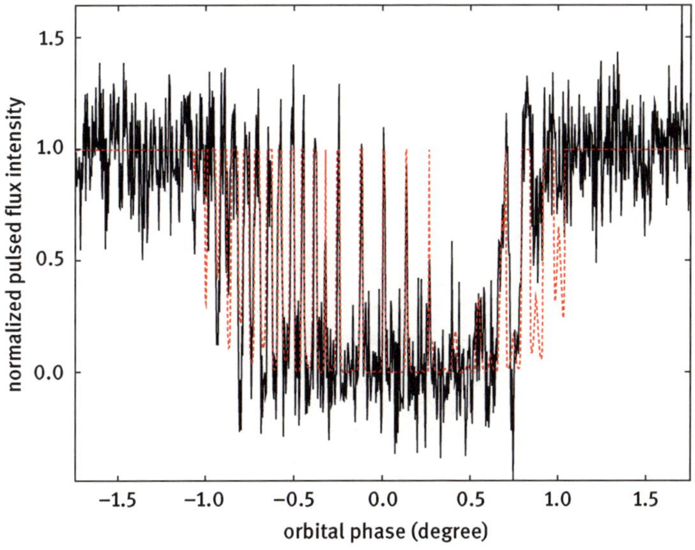

Fig. 16. Average eclipse profile of pulsar A observed at 820 MHz over a 5-day period around 11 April 2007 (black line). The model based on a co-rotating magnetosphere gives a good explanation of the eclipse profile (red dashed line). Figure is taken from [112].

light cylinder[12] $r_{\rm lc} \equiv cP/2\pi \sim 130\,000$ km. And indeed, at every superior conjunction pulsar A gets eclipsed for about 30 s due to absorption by the plasma in the magnetosphere of pulsar B [73]. A detailed analysis revealed that during every eclipse the light curve of pulsar A shows flux modulations that are spaced by half or integer numbers of pulsar B's rotational period [111] (see Figure 16). This pattern can be understood by absorbing plasma that co-rotates with pulsar B and is confined within the closed field lines of the magnetic dipole of pulsar B. As such, the orientation of pulsar B's spin is encoded in the observed light curve of pulsar A [112]. Over the course of several years, Breton et al. [112] observed characteristic shifts in the eclipse pattern, that can be directly related to a precession of the spin of pulsar B. From this analysis, Breton et al. were able to derive a precession rate of

$$\Omega^{\rm SO} = 4.77^{+0.66}_{-0.65}\ \text{deg/yr}\,. \tag{3.3}$$

The measured rate of precession is consistent with that predicted by GR ($\Omega^{\rm SO}_{\rm GR}$ = 5.07 deg/yr) within its one-sigma uncertainty. This is the sixth(!) post-Keplerian pa-

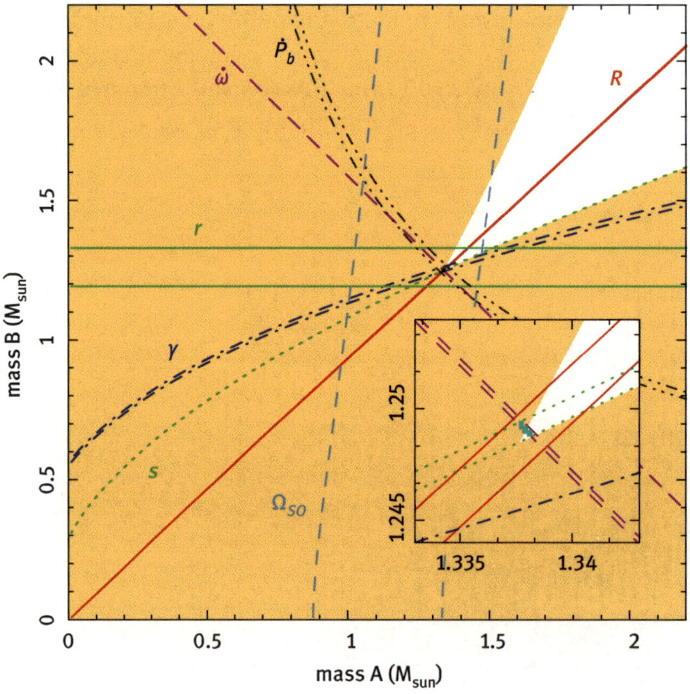

Fig. 17. GR mass–mass diagram for the Double Pulsar. Same as in Section 2.2 (Figure 7), plus the inclusion of the constraints from the geodetic precession of pulsar B ($\Omega_{\rm SO}$). Figure is taken from [75].

12 The light cylinder is defined as the surface where the co-rotating frame reaches the speed of light.

rameter measured in the Double-Pulsar system (see Figure 17). Furthermore, for the coupling function Γ_B^A, which parametrizes strong-field deviation in alternative gravity theories (see equation (1.5)), one finds

$$\Gamma_B^A/\mathcal{G} = 1.90 \pm 0.22\,, \tag{3.4}$$

which agrees with the GR value $\Gamma_B^A/G = 2$. Although the geodetic precession of a gyroscope was confirmed to better than 0.3% by the Gravity Probe B experiment [14], the clearly less precise test with Double Pulsar B (13%) for the first time gives a good measurement of this effect for a strongly self-gravitating "gyro," and by this represents a qualitatively different test.

The geodetic precession of pulsar B not only changes the pattern of the flux modulations observed during the eclipse of pulsar A, it also changes the orientation of pulsar B's emission beam with respect to our line of sight. As a result of this, geodetic precession has by now turned pulsar B in such a way, that since 2009 it is no longer seen by radio telescopes on Earth [113]. From their model, Perera et al. [113] predicted that the reappearance of pulsar A is expected to happen around 2035 with the same part of the beam, but could be as early as 2014 if one assumes a symmetric beam shape.

Finally, for pulsar A GR predicts a precession rate of 4.78 deg/yr, which is comparable to that of pulsar B. However, since the light-cylinder radius of pulsar A ($\sim$ 1000 km) is considerably smaller than that of pulsar B, there are no eclipses that could give insight into the orientation of its spin. Moreover, long-term pulse profile observations indicate that the misalignment between the spin of pulsar A and the orbital angular momentum is less than 3.2° (95% confidence) [114]. For such a close alignment, geodetic precession does not expected to cause any significant changes in the spin direction (cf. equations (1.4) and (1.5)). This, on the other hand, is good news for tests based on timing observations. One does not expect a complication in the analysis of the pulse arrival times due to additional modeling of a changing pulse profile, like this is, for instance, the case in PSR J1141-6545 [90].

4 The strong equivalence principle

The strong equivalence principle (SEP) extends the weak equivalence principle (WEP) to the universality of free fall (UFF) of self-gravitating bodies. In GR, WEP, and SEP are fulfilled, i.e. in GR the world line of a body is independent of its chemical composition and gravitational binding energy. Therefore, a detection of a SEP violation would directly falsify GR. On the other hand, alternative theories of gravity generally violate SEP. This is also the case for most metric theories of gravity [5]. For a weakly self-gravitating body in a weak external gravitational field one can simply express a violation of SEP as a difference between inertial and gravitational mass that is proportional to the gravitational binding energy E_{grav} of the mass:

$$\frac{m_G}{m_I} \simeq 1 + \eta \frac{E_{\text{grav}}}{m_I c^2} \equiv 1 + \eta\,\epsilon\,. \tag{4.1}$$

The Nordtvedt parameter η is a theory-dependent constant. In the parameterized post-Newtonian (PPN) framework, η is given as a combination of different PPN parameters (see [5] for details). As a consequence of (4.1), the Earth ($\epsilon \approx -5\times10^{-10}$) and the Moon ($\epsilon \approx -2\times10^{-11}$) would fall differently in the gravitational field of the Sun (Nordtvedt effect [115]). The parameter η is therefore tightly constrained by the lunar-laser-ranging (LLR) experiments to $\eta = (3.0 \pm 3.6) \times 10^{-4}$, which is in perfect agreement with GR where $\eta = 0$ [116].

In view of the smallness of the self-gravity of Solar System bodies, the LLR experiment says nothing about strong-field aspects of SEP. SEP could still be violated in extremely compact objects, like neutron stars, meaning that a neutron star would feel a different acceleration in an external gravitational field than weakly self-gravitating bodies. For such a strong-field SEP violation, the best current limits come from millisecond pulsar-white dwarf systems with wide orbits. If there is a violation of UFF by neutron stars, then the gravitational field of the Milky Way would polarize the binary orbit [117]. In comparison with the LLR experiment, such tests have two disadvantages: (i) the much weaker polarizing external field ($|\mathbf{g}| \sim 2\times10^{-8}$ cm/s^2, as compared to the ~ 0.6 cm/s^2 of the Solar gravitational field at the location of the Earth–Moon system), and (ii) the significantly lower precision in the ranging, which is of the order of a few 10^3 cm for the best pulsar experiments ($\sim$ 1 cm for LLR). This is almost completely counterbalanced by the gravitational binding energy of the neutron star, which is a large fraction of its total inertial mass energy ($\epsilon \sim -0.1$) and more than eight orders of magnitude larger than that of the Earth. This results in experiments with comparable limits on a SEP violation, which nonetheless are complementary since they probe different regimes of binding energy. The recent discovery of a millisecond pulsar in a hierarchical triple (see [118] and Ransom et al., in preparation) might allow for a significant improvement in testing SEP, as it combines a strong external field $\mathbf{g}$ with a large fractional binding energy ϵ.

Since beyond the first post-Newtonian approximation there is no general PPN formalism available, discussions of gravity tests in this regime are done in various theory-specific frameworks. A particularly suitable example for a framework that allows a detailed investigation of higher order/strong-field deviations from GR, is the above mentioned two-parameter class of mono-scalar–tensor theories $T_1(\alpha_0, \beta_0)$ of [57, 58], which for certain values of β_0 exhibit significant strong-field deviations from GR, and a correspondingly strong violation of SEP for neutron stars. To illustrate this violation of SEP, it is sufficient to look at the leading "Newtonian" terms in the equations-of-motion of a three-body system with masses m_a ($a = 1, 2, 3$) [61]:

$$\ddot{\mathbf{x}}_a = -\sum_{b\neq a} \mathcal{G}_{ab} m_b \frac{\mathbf{x}_a - \mathbf{x}_b}{|\mathbf{x}_a - \mathbf{x}_b|^3}\,, \tag{4.2}$$

where the body-dependent effective gravitational constant $\mathcal{G}_{ab}$ is related to the bare gravitational constant G_* by

$$\mathcal{G}_{ab} = G_*(1 + \alpha_a \alpha_b) \,. \tag{4.3}$$

As mentioned above, for a neutron star α_a can significantly deviate from the weak-field value $\alpha_0 \ll 1$. The structure dependence of the effective gravitational constant $\mathcal{G}_{ab}$ has the consequence that the pulsar does not fall in the same way as its companion, in the gravitational field of our Galaxy. For a binary pulsar with a noncompact companion, e.g. a white dwarf, that effect should be most prominent. Since both the white dwarf and the Galaxy are weakly self-gravitating bodies, their effective scalar coupling can be approximated by α_0, and one finds from equation (4.2)

$$\ddot{\mathbf{x}}_{\rm PSR} - \ddot{\mathbf{x}}_{\rm WD} \simeq -G(1 + \delta_{\rm P00})M \frac{\mathbf{x}_{\rm PSR} - \mathbf{x}_{\rm WD}}{|\mathbf{x}_{\rm PSR} - \mathbf{x}_{\rm WD}|^3} + \delta_{\rm P00}\, \mathbf{g} + \mathbf{a}_{\rm PN} \,, \tag{4.4}$$

where $\delta_{\rm P00} \equiv (\alpha_{\rm PSR} - \alpha_0)\alpha_0$, and where $\mathbf{g}$ is the gravitational acceleration caused by the Galaxy at the location of the binary pulsar.[13] Also, the contribution from post-Newtonian dynamics, term $\mathbf{a}_{\rm PN}$, has been added, whose most important consequence is the secular precession of periastron, $\dot{\omega}_{\rm PN}$. The $\mathbf{g}$-related term reflects the violation of SEP, which modifies the orbital dynamics of binary pulsars. This can be confronted with pulsar observations to test for a violation of SEP. In the following we briefly discuss different tests of SEP with binary pulsars. For a more complete review of the topic of this section see [119]. The discussion below is not specific to scalar–tensor gravity, and the quantity $\delta_{\rm P00}$ can be generically seen as the difference between inertial and gravitational mass.

4.1 The Damour–Schäfer test

In 1991, when Damour and Schäfer first investigated the orbital dynamics of a binary pulsar under the influence of a SEP violation [117], only four binary pulsars were known in the Galactic disk. Two of these (PSR B1913+16 and PSR B1957+20) were clearly inadequate for that test, not only because of the compactness of their orbits, but also because PSR B1913+16 is member of a double neutron star system that lacks the required amount of asymmetry in the binding energy, necessary for a stringent test of a SEP violation, and PSR B1957+20 is a so-called black-widow pulsar, where the companion suffers significant irregular mass losses, due to the irradiation by the pulsar. The remaining systems were PSR B1855+09 [82] and PSR B1953+29 [120]. Both of these systems have wide orbits with small eccentricities, $e = 2.2 \times 10^{-5}$ and $e = 3.3 \times 10^{-4}$, respectively.

13 Here we used $G \equiv G^*(1 + \alpha_0^2)$, and we dropped terms of order α_0^3 and smaller.

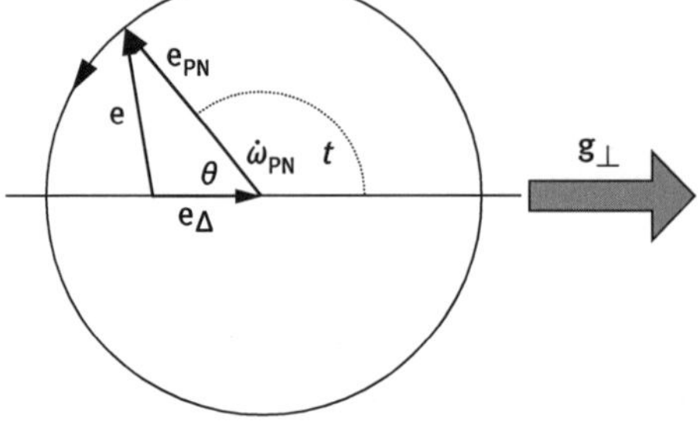

Fig. 18. Time evolution of the orbital eccentricity vector $\mathbf{e} = \mathbf{e}_{\mathrm{PN}} + \mathbf{e}_{\Delta}$ for a small-eccentricity binary, in the presence of a SEP violation. The vector $\mathbf{g}_{\perp}$ represents the projection of the external acceleration in the orbital plane.

Damour and Schäfer found for small-eccentricity binary systems that a violation of SEP leads to a characteristic polarization of the orbit, which is best represented by a vector addition where the end-point of the observed eccentricity vector $\mathbf{e}(t)$ evolves along a circle in an eccentric way (see Figure 18). The polarizing eccentricity $\mathbf{e}_{\Delta}$ is proportional to δ_{P00} and therefore, a limit on $|\mathbf{e}_{\Delta}|$ would directly pose a limit on δ_{P00}. Unfortunately neither e_{PN} nor θ in Figure 18 are measurable quantities. Also, a direct test for a change in $\mathbf{e}(t)$ is in many cases not feasible, as the expected changes are much too small compared to the available measurement precision from timing (we will discuss exceptions below). In their test, Damour and Schäfer realized if one excludes small θ values with a given probability, one can set an upper limit on $|\mathbf{e}_{\Delta}|$ without knowing e_{PN}. This is the basic idea behind the Damour–Schäfer test. The angle θ can be assumed to have a uniform probability distribution in the range $[0°, 360°)$ if the following two conditions are met:

(DS1) The system should have a sufficient age, so that one can assume that the relativistic precession of the orbit will have caused the eccentricity vector to have made many turns since the system's birth, thereby effectively randomizing the relative orientation $\theta \equiv \pi - \dot{\omega}_{\mathrm{PN}}t$ (cf. equation (5.3) below).

(DS2) The rate of periastron advance $\dot{\omega}_{\mathrm{PN}}$ should be appreciably larger than the angular velocity of the pulsar's rotation of the Galaxy with which $\mathbf{g}$ rotates in the reference frame of the binary system. As a result, the projection of the Galactic acceleration vector onto the orbit can be considered constant.

Only if these conditions are met, the Damour–Schäfer test can be applied. Both of the systems considered in [117] fulfill these two criteria. Damour and Schäfer derived 90% confidence limits of $|\delta_{\mathrm{P00}}| < 5.6 \times 10^{-2}$ and $|\delta_{\mathrm{P00}}| < 1.1 \times 10^{-2}$, from PSR B1855+09 and PSR B1953+29 respectively.

Once the eccentricity of a wide binary pulsar system is measured, there is generally little one can do to improve the Damour–Schäfer test with that system. Significant improvement of the Damour–Schäfer test has to come from the discovery of a new system. The larger the orbital period and the smaller the orbital eccentricity, the tighter a limit can be derived from a Damour–Schäfer test. In fact, the figure-of-merit for this test is P_b^2/e. In the meantime, quite a few suitable systems have been discovered (see e.g. Table 4 in [121]). However, one cannot just pick the one with the best figure-of-merit from that ensemble, as this introduces a selection bias, since it is possible that the small eccentricity of the selected sytem is actually the result of a SEP violation where by chance θ is small. In fact, if one has a large number of systems, there is a high probability that there are systems with small θ [119, 122]. In this case, one has to properly combine all the systems in a statistical test. The latest results based on a proper statistical treatment can be found in [121, 123] which give 95% confidence limits of order 5×10^{-3}. The slightly better limit in [121] has a caveat by including PSR J1711-4322, which has a large figure-of-merit but neither fulfills condition DS1 nor DS2. Concerning DS2, as shown in [124], PSR J1711-4322 is at a location in the Galactic plane where $\dot{\omega}_{\rm PN}$ is close, or even equal to the Galactic rotation. This can lead to a highly nonuniform evolution of $\theta(t)$.

4.2 Direct tests

There is an underlying assumption in the Damour–Schäfer test for multiple systems, which is related to the mass dependence of a SEP violation. Constraining a $\delta_{\rm P00}$ from a set of pulsar-white dwarf systems in a generic way, requires the assumption that $\delta_{\rm P00}$ is practically independent of the mass of the neutron star, as these systems have different pulsar masses. Even in the absence of a nonperturbative behavior, where to first order $\delta_{\rm P00}$ is proportional to ϵ (cf. equation (4.1)), we can have deviations from that assumption of oder 30% along the range of observed neutron star masses. And in the presence of nonperturbative strong-field effects, like the spontaneous scalarization mentioned above, this assumption is strongly violated. For this reason, it is desirable to have direct tests, based on long-term timing observations of individual systems, used to directly constrain $\dot{e}$ (see [119] for details). As it requires a number of conditions to be met, like high timing precision and knowledge on the orbital orientation, only few systems turn out to be suitable at present. In [119] two binary pulsar systems have been identified as particularly suitable for a direct test of a SEP violation: PSR J1713+0747 and PSR J1903+0327. While the work on PSR J1713+0747 is still in progress, preliminary results for PSR J1903+0327 have been published in [119]. PSR J1903+0327 is a millisecond pulsar with good timing precision in a wide ($x = 105.593$ lt-s), highly eccentric ($e = 0.44$) orbit [32, 125]. The pulsar is comparably massive ($m_p = 1.67\ M_\odot$) and the companion is a Sun-like main sequence star. At present, the limit from PSR J1903+0327 (~ few %) cannot compete with the results of the Damour–Schäfer test mentioned above. But these limits are expected to improve with time, just by continuous timing observations.

In summary, there are several advantages of a direct test [119] compared to the Damour–Schäfer test:

- The tests are conducted for specific neutron star masses, and therefore are meaningful even in the presence of a nonperturbative strong-field behavior.
- The test no longer requires probabilistic consideration for unknown angles, and therefore cannot only set an upper limit, but also has the potential to detect a SEP violation.
- There are hardly any relevant additional effects, that lead to a nonzero $\dot{e}$. For wide binary orbits, $\dot{e}$ from gravitational wave damping is absolutely negligible. Therefore the precision of this test is expected to just keep on improving with time, at least for the foreseeable future.
- We do not need to restrict our sample to systems with small eccentricities. In fact, in an eccentric system the violation of SEP would not only cause a change in the orbital eccentricity ($\dot{e}$) but also, depending on the orientation, change the inclination of the orbital plane, which leads to a change in the projected semimajor axis ($\dot{x}$). This allows for a unique cross-check, since the ratio $\dot{x}/(x\dot{e})$ for a SEP violation only depends on the orientation and the eccentricity of the pulsar orbit [119].

5 Local Lorentz invariance of gravity

Some alternative gravity theories allow the Universal matter distribution to single out the existence of a preferred frame, which breaks the symmetry of local Lorentz invariance (LLI) for the gravitational interaction. In the post-Newtonian parametrization of semi-conservative gravity theories, LLI violation is characterized by two parameters, α_1 and α_2 [5]. Nonvanishing α_1 and α_2 modify the dynamics of self-gravitating systems that move with respect to the preferred frame (preferred-frame effects). In GR one finds $\alpha_1 = \alpha_2 = 0$.

As the most natural preferred frame, generally one chooses the frame associated with the isotropic cosmic microwave background (CMB), meaning that the preferred frame is assumed to be fixed by the global matter distribution of the Universe. From the five-year Wilkinson Microwave Anisotropy Probe (WMAP) satellite experiment, a CMB dipole measurement with high precision was obtained [126]. The CMB dipole corresponds to a motion of the Solar System with respect to the CMB with a velocity of 369.0 ± 0.9 km/s in direction of Galactic longitude and latitude $(l, b) = (263.99° \pm 0.14°, 48.26° \pm 0.03°)$. The numbers quoted in the next two subsections, will be with respect to the CMB frame. A generalization to other frames is straightforward, and was done in some of the references cited below.

The most important (weak-field) constraints on preferred-frame effects do come form Lunar Laser Ranging (LLR) [127],

$$\alpha_1 = (-0.7 \pm 1.8) \times 10^{-4} \quad (95\%\ \mathrm{CL})\,, \tag{5.1}$$

and the alignment of the Sun's spin with the total angular momentum of the planets in the Solar System [128],

$$|\alpha_2| < 2.4 \times 10^{-7}\,. \tag{5.2}$$

5.1 Constraints on $\hat{\alpha}_1$ from binary pulsars

In binary pulsars, the isotropic violation of Lorentz invariance in the gravitational sector should lead to characteristic preferred frame effects in the binary dynamics, if the barycenter of the binary is moving relative to the preferred frame with a velocity $\mathbf{w}$. For small-eccentricity binaries, the effects induced by $\hat{\alpha}_1$ and $\hat{\alpha}_2$ (the hat indicates possible modifications by strong-field effects) decouple, and can therefore be tested independently [129, 130].

In case of a nonvanishing $\hat{\alpha}_1$, the observed eccentricity vector $\mathbf{e}$ of a small-eccentricity binary pulsar is a vectorial superposition of a "rotating eccentricity" $\mathbf{e}_R(t)$ and a fixed "forced eccentricity" $\mathbf{e}_F$: $\mathbf{e}(t) = \mathbf{e}_F + \mathbf{e}_R(t)$ [129]. The rotating eccentricity has a constant length e_R, and rotates with the relativistic precession of periastron, $\dot{\omega}$, in the orbital plane. This is identical to the dynamics caused by a violation of the strong equivalence principle (cf. Section 4), with the forced eccentricity this time pointing into the direction of $\hat{\mathbf{L}} \times \mathbf{w}$. As a consequence, the binary orbit changes from a less to a more eccentric configuration and back on a time scale of

$$T_{\dot{\omega}} \equiv \frac{2\pi}{\dot{\omega}} \simeq (1140\,\mathrm{yr}) \left(\frac{P_b}{1\ \mathrm{day}}\right)^{5/3} \left(\frac{M}{2M_\odot}\right)^{-2/3}\,, \tag{5.3}$$

where we have assumed that the true $\dot{\omega}$ does not deviate significantly from the one predicted by GR (equation (1.17)), an assumption that is well justified by other binary pulsar experiments, like the generic tests in the Double Pulsar (cf. Section 2.2).

The forced eccentricity $\mathbf{e}_F$ is determined by the strength of the preferred frame effect. Its magnitude is approximately given by

$$e_F \simeq 0.093\,\hat{\alpha}_1 \frac{m_p - m_c}{M} \left(\frac{M}{2M_\odot}\right)^{-1/3} \left(\frac{P_b}{1\ \mathrm{day}}\right)^{1/3} \left(\frac{w \sin\psi}{300\ \mathrm{km/s}}\right)\,, \tag{5.4}$$

where ψ is the angle between $\mathbf{w}$ and $\hat{\mathbf{L}}$ (see [129] for a detailed expression). The observation of small eccentricities in binary pulsars, like $e \sim 10^{-7}$ for PSR J1738+0333 does not directly constrain $\hat{\alpha}_1$. The orientation of the a priory unknown intrinsic $\mathbf{e}_R$ could be such, that it compensates for a large $\mathbf{e}_F$. If the system is sufficiently old, one can assume a uniform probability distribution in $[0°, 360°)$ for $\theta(t)$. Like in the

Damour–Schäfer test for SEP, one can now set a probabilistic upper limit on e_F, and by this on $\hat{\alpha}_1$, by excluding θ values close to alignment of $\mathbf{e}_R$ and $\mathbf{e}_F$. Based on this method, [129] found a limit of $|\hat{\alpha}_1| < 5 \times 10^{-4}$ with 90% confidence.

But even if θ happens to be close to 0°, due to the relativistic precession it will not remain there, and a large e_F cannot remain hidden for ever. In fact, if $\dot{\omega}$ is sufficiently large (greater than $\sim$ 1° per year) a significant change in the orbital eccentricity should become observable over time scales of a few years, even if at the start of the observation there was a complete cancelation between $\mathbf{e}_R$ and $\mathbf{e}_F$. This can be used to constrain $\hat{\alpha}_1$ [130]. Hence, in contrast to the SEP test of Section 4.1, one now looks for suitable binary pulsars with short orbital periods. The best such test comes from PSR J1738+0333 (see Section 2.3). This binary pulsar is ideal for this test for several reasons:

- The orbit has an extremely small, well-constrained eccentricity of $\sim 10^{-7}$ [30].
- The (calculated) relativistic precession of periastron is about 1.6°/yr, and the binary has been observed by now for about 10 years [30]. Hence, $\theta(t)$ has covered an angle of 16° in that time.
- The 3D velocity with respect to the Solar System is known with good precision from timing and optical observations, meaning that one can compute $\mathbf{w}$ [30, 84].
- The orientation of the system is such, that the unknown angle of the ascending node Ω has little influence on the $\hat{\alpha}_1$ limit, hence, there is no need for probabilistic considerations to exclude certain values of Ω [130].

Consequently, PSR J1738+0333 leads to the best constraints of α_1-like violations of the local Lorentz invariance of gravity, giving [130]

$$\hat{\alpha}_1 = -0.4^{+3.7}_{-3.1} \times 10^{-5} \quad \text{(95\% confidence)}\,. \tag{5.5}$$

This limit is not only five times better than the current most stringent limit on α_1 obtained in the Solar System (cf. equation (5.1)), it is also sensitive to potential deviations related to the strong self-gravity of the pulsar. For nonperturbative deviations one can, for illustration purposes, do an expansion with respect to the fractional binding energy ϵ of the neutron star,

$$\hat{\alpha}_1 = \alpha_1 + \mathcal{C}_1\epsilon + \mathcal{O}(\epsilon^2)\,. \tag{5.6}$$

Since $\epsilon \sim -0.1$ for PSR J1738+0333, we get tight constraints for $\mathcal{C}_1$, a parameter that is virtually unconstraint by the LLR experiment, since $\epsilon \sim -5 \times 10^{-10}$ for the Earth.

5.2 Constraints on $\hat{\alpha}_2$ from binary and solitary pulsars

In the presence of a nonvanishing $\hat{\alpha}_2$, a small-eccentricity binary system experiences a precession of the orbital angular momentum around the fixed direction $\mathbf{w}$ with an angular frequency

$$\Omega_{\hat{\alpha}_2}^{\text{prec}} = -\hat{\alpha}_2 \frac{\pi}{P_b} \left(\frac{w}{c}\right)^2 \cos\psi$$
$$\simeq -(0.066^\circ/\text{yr})\,\hat{\alpha}_2 \left(\frac{P_b}{1\,\text{day}}\right)^{-1} \left(\frac{w}{300\,\text{km/s}}\right)^2 \cos\psi\,, \tag{5.7}$$

where ψ is the angle between the orbital angular momentum and $\mathbf{w}$ [130]. In binary pulsars, such a precession should become visible as a secular change in the projected semimajor axis of the pulsar orbit, $\dot{x}$, which is an observable timing parameter. The two binary pulsars PSRs J1012+5307 and J1738+0333 turn out to be particularly useful for such a test, since both of them have optically bright white dwarf companions, which allowed the determination of the masses in the system, and the 3D systemic velocity with respect to the preferred frame [84, 131, 132].

Unfortunately, in general, the orientation of a binary pulsar orbit with respect to $\mathbf{w}$ and the line-of-sight cannot be fully determined from timing observation. As a consequence, one cannot directly test $\hat{\alpha}_2$ from observed constraints for $\dot{x}$. In fact, since the longitude of the ascending node Ω is not measured, neither for PSR J1012+5307 nor for PSR J1738+0333, the orientation of these systems could in principle be such, that an $\hat{\alpha}_2$-induced precession would not lead to a significant $\dot{x}$. Assuming a random distribution of Ω in the interval $[0^\circ, 360^\circ)$, one can use probabilistic considerations to exclude such unfavorable orientations. A detailed discussion of this test can be found in [130], where the following 95% confidence limits are derived:

$$\begin{aligned} |\hat{\alpha}_2| &< 3.6 \times 10^{-4} \quad \text{from PSR J1012+5307}\,, \\ |\hat{\alpha}_2| &< 2.9 \times 10^{-4} \quad \text{from PSR J1738+0333}\,, \\ |\hat{\alpha}_2| &< 1.8 \times 10^{-4} \quad \text{from PSRs J1012+5307 and J1738+0333 combined}\,. \end{aligned} \tag{5.8}$$

It is important to note, that for the last limit, based on the statistical combination of the two systems, one has to assume that $\hat{\alpha}_2$ has only a weak functional dependence on the neutron-star mass in the range of 1.3–2.0 $M_\odot$.

The limit for $\hat{\alpha}_2$ obtained from binary pulsars are still several orders of magnitude weaker than the α_2 limit which Nordtvedt derived in 1987 from the alignment of the Sun's spin with the orbital planes of the planets [128]. In the same paper, Nordtvedt pointed out that solitary fast rotating pulsars could be used in a similar way to obtain tight constraints for α_2. This can be directly seen from equation (5.7), which holds for a rotating self-gravitating star if P_b is replaced by the rotational period P of the star. While the five-billion-year base-line for the Solar experiment is typically a factor of $\sim 10^9$ longer than the observational time-span T_{obs} of pulsars, for millisecond pulsars P is $\sim 10^9$ shorter than the rotational period of the Sun. In fact, the first millisecond pulsar PSR B1937+21, discovered in 1982 [133], by now has a figure of merit T_{obs}/P that is ~ 10 times larger than that of the Sun.

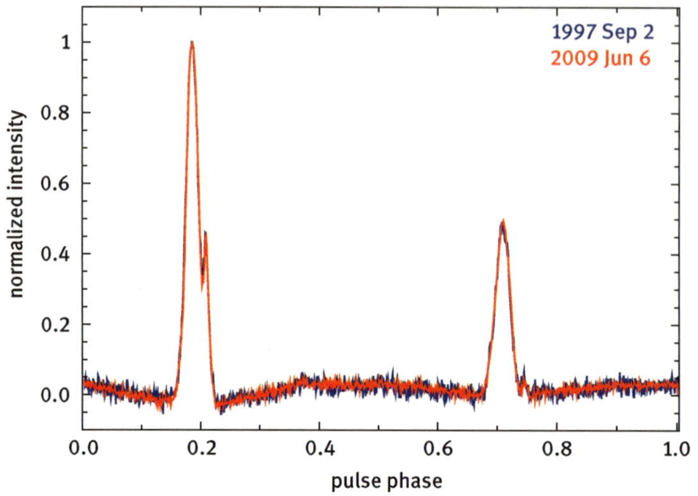

Fig. 19. Comparison of two pulse profiles of PSR B1937+21 obtained at two different epochs. The blue one was obtained on September 2, 1997, while the red one was obtained on June 6, 2009. The main peak is aligned and scaled to have the same intensity. There exists no visible difference within the noise level. Profiles were taken from [134].

The precession of a solitary pulsar due to a nonvanishing $\hat{\alpha}_2$ would lead to characteristic changes in the observed pulse profile over time scales of years, just like in the case of binary pulsars that experience geodetic precession (cf. Section 3). Consequently, a nondetection of such changes can be converted into constraints for $\hat{\alpha}_2$. Recently, Shao et al. [134] used the two solitary millisecond pulsars PSRs B1937+21 and J1744-1134 for such an experiment. For both pulsars they utilized a consistent set of data, taken over a time span of approximately 15 years with the same observing system at the 100-m Effelsberg radio telescope. The continuity in the observing system was key for an optimal comparison of the high signal-to-noise ratio profiles over time. As it turns out, both pulsars, PSRs B1937+21 and J1744-1134, do not show any detectable profile evolution in the last 15 years. As an example of such a nondetection see Figure 19, which shows two pulse profiles of PSR B1937+21 obtained at different epochs.

Similarly to the $\hat{\alpha}_2$ test with the binary pulsars, there are unknown angles in the orientation of the pulsar spin, for which certain values have to be excluded based on probabilistic considerations. From extensive Monte Carlo simulations Shao et al. found with 95% confidence

$$
\begin{aligned}
|\hat{\alpha}_2| &< 2.5 \times 10^{-8} \quad \text{from PSR B1937+21}\,, \\
|\hat{\alpha}_2| &< 1.5 \times 10^{-8} \quad \text{from PSR J1744-1134}\,, \\
|\hat{\alpha}_2| &< 1.6 \times 10^{-9} \quad \text{from PSRs B1937+21 and J1744-1134 combined}\,.
\end{aligned}
\tag{5.9}
$$

These limits are significantly tighter than the α_2 limit from the Sun's spin orientation. Like in the case of the $\hat{\alpha}_1$ test (previous subsection), this test also covers potential deviations related to the strong self-gravity of the pulsar, and in the combination of the two pulsars, makes the assumption that $\hat{\alpha}_2$ depends only weakly on the neutron-star mass.

An important difference to the aforementioned tests with binary pulsars is, that for solitary pulsars one cannot determine the radial velocity. It enters the determination of **w** as a free parameter. However, as shown in [134] the unknown radial velocities for PSRs B1937+21 and J1744-1134 only have a marginal effect on the limits. For the limits above it was assumed that both pulsars are gravitationally bound in the Galactic potential. But even if one relaxes this assumption and allows for unphysically large radial velocities, exceeding 1000 km/s, the limits get weaker by at most $\sim 40\%$.

As a final remark, a combined test of $\hat{\alpha}_1$ and $\hat{\alpha}_2$ for the gravitational interaction of two strongly self-gravitating objects using the Double Pulsar, has been proposed in [135]. At the time of that publication, however, the observational time span was not long enough to disentangle potential preferred-frame effects from other orbital contributions. The large correlations with orbital parameters in the timing solution, lead to rather weak limits on $\hat{\alpha}_1$ and $\hat{\alpha}_2$ in the Double Pulsar.

5.3 Constraints on $\hat{\alpha}_3$ from binary pulsars

In nonconservative gravity theories, there is a third PPN parameter related to preferred-frame effects, denoted by α_3, which identically vanishes in GR. Besides its association with preferred-frame effects, α_3 is also associated with a violation of the conservation of total momentum, the key feature of nonconservative gravity theories. Because of this, a nonzero α_3 leads to a self-acceleration for a self-gravitating rotating body. The acceleration is perpendicular to the angular rotation of the body, Ω and its motion with respect to the preferred frame, **w**. The α_3-induced acceleration is given by [5]

$$\mathbf{a}_{\alpha_3} = -\frac{\alpha_3}{3}\,\epsilon\,\mathbf{w}\times\Omega\,. \tag{5.10}$$

The quantity ϵ is the fractional binding energy of the body, as defined in equation (4.1). Due to their high binding energy ($\epsilon \sim -0.1$) and their fast rotation ($\Omega \sim 10^3$ rad/s), millisecond pulsars are ideal objects to probe for α_3 related effects. In fact, as shown in [136], millisecond pulsars in binary systems with a large orbital period P_b and a small eccentricity e are the best test systems for $\hat{\alpha}_3$, where the hat indicates a strong-field generalization. As shown in [136], $\mathbf{a}_{\alpha_3}$ has a polarizing effect on the binary orbit, analogous to the one induced by a SEP violation (cf. Section 4). Consequently a Damour–Schäfer test can be applied to constrain α_3. The same requirements as in the SEP test (cf. DS1 and DS2 in Section 4) have to be met. DS2 is important, since the binary system should not move appreciably in the Galaxy during the build up of the polarization induced by $\mathbf{a}_{\alpha_3}$, otherwise the equations given in [136] do not apply. The latest limit, based on a proper statistical analysis of a large sample of known binary pulsars, comes from [121]: $|\hat{\alpha}_3| < 5.5 \times 10^{-20}$ (95% confidence). PSR J1711-4322, which violates DS1 and DS2, has also been included in this analysis. However, due to its slow rotation ($P = 103$ ms) it plays only a minor part in that test. The limit of [121] is more than a factor of $\sim 10^{13}$ better than the best Solar System limit [5], and by far the tightest limit on any of the PPN parameters.

6 Local position invariance of gravity

The local position invariance (LPI) of gravity states that the outcome of any local gravitational experiment is independent of where and when it is performed. If the LPI is violated in the gravitational sector, the gravitational interaction of a localized self-gravitating system depends on the direction of its acceleration in the gravitational field of an external mass [5]. In such a scenario, the dynamics of the Solar System or a binary system depends on the overall matter distribution in our Galaxy, and one would experience a directional dependence in the locally measured gravitational constant. Within the PPN formalism, such an anisotropy is described by the *Whitehead parameter* ξ (not to be confused with the parameter ξ in equation (1.2)) [5]. It is interesting to note, that even for fully conservative theories of gravity one may have $\xi \neq 0$. In GR the gravitational interaction fulfills LPI and therefore GR has $\xi = 0$.

For small-eccentricity binaries, ξ primarily induces a precession of the orbital angular momentum around the direction of the external gravitational field, $\mathbf{n}_\mathrm{G}$, with the angular velocity

$$\Omega^{\mathrm{prec}} = -\xi \, \frac{2\pi}{P_b} \, \frac{\Phi_\mathrm{G}}{c^2} \, \cos\psi \,, \tag{6.1}$$

where Φ_G is the Newtonian galactic potential at the position of the system, and ψ is the angle between the orbital angular momentum and $\mathbf{n}_\mathrm{G}$. Due to the analogy with equation (5.7), one immediately sees that the same kind of analysis, as outlined in Section 5.2 for testing $\hat{\alpha}_2$, can be performed to constrain $\hat{\xi}$. Like in the previous section, the hat indicates the strong-field generalization of the PPN parameter. From a combined analysis of PSRs J1012+5307 and J1738+0333 one obtains [137]

$$|\hat{\xi}| < 3.1 \times 10^{-4} \quad \text{(95\% confidence)} \,. \tag{6.2}$$

This limit surpasses the weak-field limit on ξ obtained from the nondetection of anomalous Earth tides by about one order of magnitude.

A nonvanishing ξ would also affect an isolated rotating body [128]. Like for a binary system, the angular momentum, i.e., the spin, of a self-gravitating object with internal equilibrium should precess around $\mathbf{n}_\mathrm{G}$. The precessional frequency is given by equation (6.1), if P_b is replaced by the rotational period P of the isolated body. Again, one has the analogy to the $\hat{\alpha}_2$ tests of Section 5.2). Consequently, the same data and method used in the $\hat{\alpha}_2$ test with solitary pulsars can be used to constrain $|\hat{\xi}|$. One obtains with 95% confidence [138]:

$$\begin{aligned} |\hat{\xi}| &< 2.2 \times 10^{-8} \quad \text{from PSR B1937+21}\,, \\ |\hat{\xi}| &< 1.2 \times 10^{-7} \quad \text{from PSR J1744-1134}\,, \\ |\hat{\xi}| &< 3.9 \times 10^{-9} \quad \text{from PSRs B1937+21 and J1744-1134 combined}\,. \end{aligned} \tag{6.3}$$

These limits are significantly (up to three orders of magnitude) better than the limit obtained from the Solar spin [128, 138].

As mentioned above, a violation of the LPI for gravity is directly related to a directional dependence of the local gravitational constant G. Consequently, the limits (6.3) can straightforwardly be converted into limits on an anisotropy of G. Corresponding to the combined limit from PSRs B1937+21 and J1744-1134 in (6.3), one finds

$$\left|\frac{\Delta G}{G}\right|^{\text{anisotropy}} < 4 \times 10^{-16} \quad (95\%\ \text{confidence})\,, \tag{6.4}$$

which is the most constraining limit on the anisotropy of G [138].

7 A varying gravitational constant

The locally measured Newtonian gravitational constant G may vary with time as the Universe evolves. In fact, this is expected for most alternatives to GR that violate the strong equivalence principle [5]. By now there are various tests to constrain a change in the gravitational constant on different time scales. Some tests probe a change over the cosmological history, i.e. $G(t)$, others a present change, i.e. today's $\dot{G}$ (see [139] for a review).

In a binary system, a time variation of G changes the orbital period P_b. If the gravitational binding energy of the masses is small, like for Solar System bodies, this change is to first order given by [140]

$$\frac{\dot{P}_b}{P_b} = -\frac{\dot{n}_b}{n_b} = -2\,\frac{\dot{G}}{G}\,, \tag{7.1}$$

and the semimajor axis of the relative motion changes according to

$$\frac{\dot{a}}{a} = -\frac{\dot{G}}{G}\,. \tag{7.2}$$

In the Solar System, the Lunar Laser Ranging (LLR) experiment gives the best limit. Based on 39 years of LLR data, [141] derived a limit of

$$\frac{\dot{G}}{G} = (-0.7 \pm 3.8) \times 10^{-13}\,\text{yr}^{-1} = (-0.001 \pm 0.005)\,H_0\,. \tag{7.3}$$

The value for the Hubble constant $H_0 = 67.8\,\text{km/s/Mpc}$ is taken from [142].

Equation (7.1) is not applicable to binary pulsars. Contrary to weakly self-gravitating bodies, in binary pulsars the dependence on the gravitational self-energy cannot be neglected [143]. A change in G changes the gravitational binding energy of a self-gravitating body, and by this its mass. While such a change is negligible in the Earth–Moon system, since the fractional binding energy is very small for these bodies ($\epsilon_{\mathrm{Earth}} \approx -5 \times 10^{-10}$), it is significant for neutron stars, where the gravitational self-energy accounts for a significant fraction of the mass ($\epsilon_{\mathrm{NS}} \sim -0.1$). A detailed calculation can be found in [143], which shows that for a binary pulsar system equation (7.1) has to be modified to

$$\frac{\dot{P}_b}{P_b} = -2\frac{\dot{G}}{G}\left[1 - \left(1 + \frac{m_c}{2M}\right)s_p - \left(1 + \frac{m_p}{2M}\right)s_c\right], \tag{7.4}$$

where the "sensitivity"

$$s_A \equiv -\left.\frac{\partial(\ln m_A)}{\partial(\ln G)}\right|_N \tag{7.5}$$

measures how the mass of body A changes with a change of the local gravitational constant G, for a fixed baryon number N (see [5] for details). For a given mass, the sensitivity of a neutron-star depends on the equation of state and on the specifics of the gravity theory. Figure 20 shows the sensitivity for Jordan–Fierz–Brans–Dicke gravity, for different α_0 (i.e. ω_{BD}) and two different equations of state. If the companion of

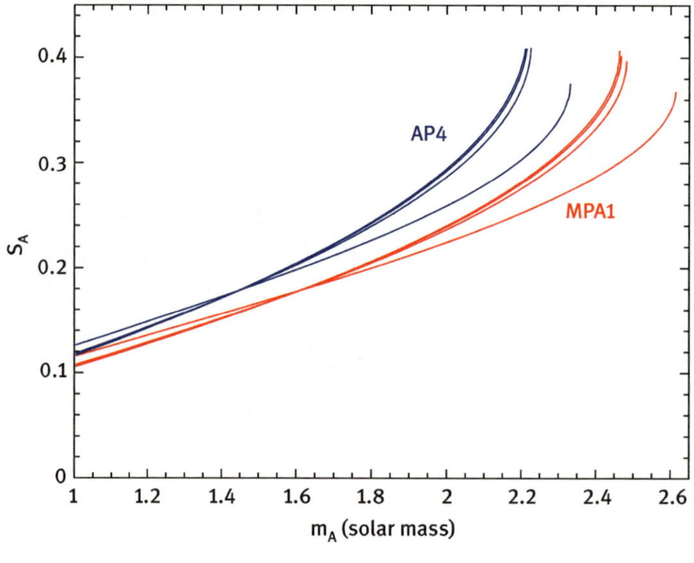

Fig. 20. Sensitivity s_A for Jordan–Fierz–Brans–Dicke theory, i.e. $T_1(\alpha_0, 0)$, and for two different equations of state (red: MPA1 [85], blue: AP4 [144]). For each equation of state four lines have been calculated, corresponding to $|\alpha_0| = 0.5, 0.2, 0.1, 0.01$ as the maximum mass decreases. For $|\alpha_0| < 0.01$, s_A is practically independent of α_0.

the pulsar is a weakly self-gravitating star, like a white dwarf, s_c becomes negligible and equation (7.4) simplifies to

$$\frac{\dot{P}_b}{P_b} \simeq -2\frac{\dot{G}}{G}\left[1-\left(1+\frac{m_c}{2M}\right)s_p\right] . \tag{7.6}$$

The currently best pulsar limit for a change in the gravitational constant comes from the pulsar-white dwarf system PSR J0437-4715 ($P_b = 5.74$ d). A direct confrontation of equation (7.6) with the timing observations of that pulsar yields [145]

$$\frac{\dot{G}}{G} = \frac{(-5 \pm 26)\times 10^{-13}}{1-1.1s_p}\,\mathrm{yr}^{-1} \quad (95\%\ \text{confidence}) . \tag{7.7}$$

The factor $(1 - 1.1s_p)$ weakens the limit by typically 30%, and has been neglected in [145]. As pointed out in [132], the limit (7.7) has the following caveat. It is generally expected that a gravity theory with a varying gravitational constant also predicts the existence of dipolar gravitational waves, that modify $\dot{P}_b$, and could in principle even balance a significant part of a decrease in G. In fact, just to give an example, in Jordan–Fierz–Brans–Dicke theory $\dot{P}_b^{\dot{G}} \sim -\dot{P}_b^{\mathrm{dipole}}$ for binary pulsar-white dwarf systems that have orbital periods of $\sim$ 10 d, like PSR J0437-4715. In the absence of nonperturbative strong-field effects one finds for the change in the orbital period of a pulsar-white dwarf system in a small-eccentricity orbit the combined expression (cf. equation (1.25))

$$\frac{\dot{P}_b - \dot{P}_b^{\mathrm{GR}}}{P_b} \simeq -2\frac{\dot{G}}{G}\left[1-\left(1+\frac{m_c}{2M}\right)s_p\right] - \frac{4\pi^2}{P_b^2}\frac{Gm_pm_c}{c^3M}\kappa_D s_p^2 + \mathcal{O}(s_p^3) . \tag{7.8}$$

The constant κ_D is a theory-dependent constant, which is a priori unknown in generic test, where no specific gravity theory is applied. As proposed in [132], it is now possible to combine two pulsars with a sufficiently large difference in their orbital periods P_b to constrain $\dot{G}$ and κ_D simultaneously. In [30], the best pulsar for testing dipolar radiation, PSR J1738+0333 (see Section 2.3), and the best pulsar for a $\dot{G}$ test, PSR J0437-4715, have been combined to give joint constraints for a variation in G and dipolar radiation (see Figure 21). In this generic test, one has to make certain reasonable assumptions about s_A and how it changes with mass m_A, since PSR J1738+0333 and PSR J0437-4715 have different masses (see [30] for details). As one can see from Figure 21, the pulsar limit on $\dot{G}$ is still somewhat weaker than the one from LLR (7.9), but obtained with a completely independent method.

Apart from providing an independent test for a varying gravitational constant, binary pulsar experiments can test for strong-field enhancements of $\dot{G}$. To illustrate this, we use scalar–tensor gravity, where a change in the locally measured gravitational constant G is the result of a change in the scalar field(s). More specifically, in the $T_1(\alpha_0, \beta_0)$ theory of [57, 58], LLR tests for a variation of the gravitational constant that is given by

$$\frac{\dot{G}}{G} = 2\left[1+\frac{\beta_0}{1+\alpha_0^2}\right]\alpha_0\dot{\varphi}_0 \tag{7.9}$$

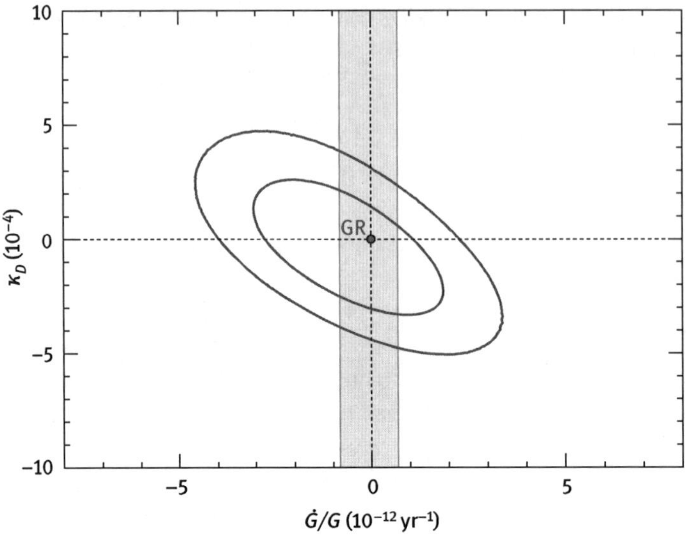

Fig. 21. A joint $\dot{G}$–κ_D test based on PSRs J1738+0333 and PSR J0437-4715. The inner contour includes 68.3% and the outer contour 95.4% of all probability. GR ($\dot{G} = \kappa_D = 0$) is well within the inner contour and close to the peak of probability density. The vertical gray band includes regions consistent with the one-sigma constraints for $\dot{G}/G$ from LLR (equation (7.3)). Generally only the upper half of the diagram has physical meaning, as the radiation of dipolar gravitational waves is expected to make the system lose orbital energy. Figure is taken from [30].

(see equation (167) of [139]). For the effective gravitational constant between two strongly self-gravitating bodies (as measured in the physical Jordan-frame), equation (7.9) changes to

$$\frac{\dot{\mathcal{G}}}{\mathcal{G}} = 2\left[1 + \frac{\alpha_A\beta_B + \alpha_B\beta_A}{2\alpha_0(1 + \alpha_A\alpha_B)}\right]\alpha_0\dot{\varphi}_0 \,. \tag{7.10}$$

In the presence of significant scalarization effects in the strong gravitational fields of neutron stars, the expression in square brackets of equation (7.10) can be considerably larger than the corresponding one in equation (7.9), even for β_0 values which are not yet excluded by binary pulsar experiments (see Figure 22). As a conclusion, $\dot{G}$ tests with binary pulsars can be more sensitive than LLR tests in situations where a change in the gravitational constant gets enhanced by strong-field effects in neutron stars. The details depend on the specifics of the gravity theory and the mass of the neutron star. Also, a complete analysis needs to account for corresponding changes in the masses, i.e. the analog to equation (7.4). We will not go into these details here.

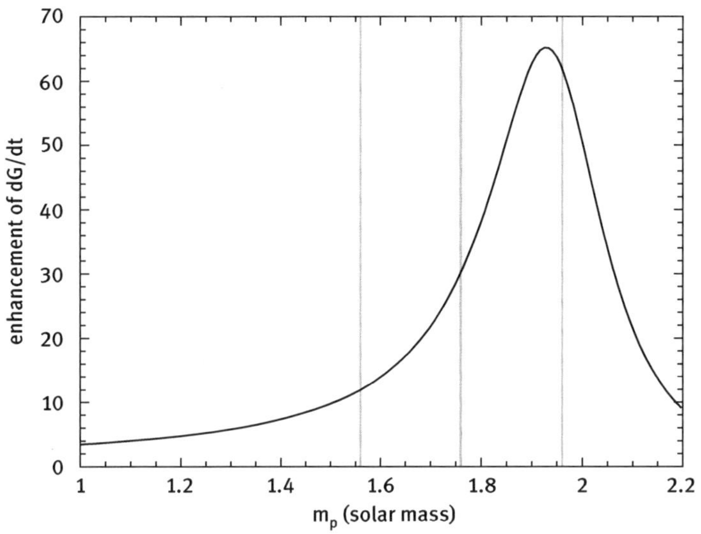

Fig. 22. Enhancement of $\dot{G}$ in a pulsar-white dwarf system as a function of the pulsar mass m_p. Figure shows the ratio $(\dot{\mathcal{G}}/\mathcal{G})$ / $(\dot{G}/G)$ as given in equations (7.9) and (7.10) for $T_1(10^{-4}, -4.3)$ gravity, a theory which still passes the PSR J0348+0432 test (see Figure 14). The gray vertical lines indicate the mass range for PSR J0437-4715 (mean and one-sigma uncertainties) [29]. PSR J1614-2230, with its mass of $1.97 \pm 0.04\, M_\odot$ and orbital period of 8.7 days [77], seems to be a promising future test for $\dot{G}$, once tight constraints on the intrinsic $\dot{P}_b$ can be derived from the timing observations.

8 Summary and outlook

With their discovery of the first binary pulsar four decades ago, Joseph Taylor and Russell Hulse opened a new field of experimental gravity, which has been an active field of research ever since. Besides the Hulse–Taylor pulsar, which led to the first confirmation of the existence of gravitational waves, astronomy has seen the discovery of many new binary pulsars suitable for precision gravity tests. Arguably the most exiting discovery was the Double Pulsar in 2003, which by now provides the best test for GR's quadrupole formalism of gravitational wave generation ($< 0.1\%$ uncertainty), and the best test for the relativistic spin precession of a strongly self-gravitating body. In addition to this, it is the binary pulsar with the most post-Keplerian parameters measured, allowing for a number of generic constraints on strong-field deviations from GR. For certain aspects of gravity, binary pulsars with white dwarf companions have proven to be even better "test laboratories" than the Double Pulsar. These are gravitational phenomena, predicted by alternatives to GR, that depend on the difference in the compactness/binding energy of the two components, like gravitational dipolar radiation and a violation of the strong equivalence principle. By now, pulsar-white dwarf systems, like PSR J1738+0333, set quite stringent limits (coupling strength less than about 10^{-3}) on the existence of any additional "gravitational charges" associated with light or massless fields. And the recent discovery of a massive pulsar in a relativistic binary system (PSR J0348+0432), for the first time allowed to test the orbital motion of a neutron star that is significantly more compact than pulsars of previous gravity tests. For certain aspects of gravity, solitary pulsars turned out to be ideal probes. The current best limits on the PPN parameters α_2, related to the existence of a preferred frame for gravity, and ξ, related to a violation of local position invariance of the gravitational interaction, do come from pulse-profile observations of two solitary millisecond pulsars. In all these tests, pulsars go beyond Solar System tests, since they are also sensitive to deviations that occur only in the strong-field environment of neutron stars.

So far, GR has passed all these tests with flying colors. Will this continue for ever? Is GR our final answer to the macroscopic description of gravity? Pulsar astronomy will certainly continue to investigate this question. Many of the tests mentioned here will simply improve by continued timing observations of the known pulsars. In fact, the measurement precision for some of the post-Keplerian parameters increases fast with time. For instance, in regular observations (with the same hardware) the uncertainty in the change of the orbital period $\dot{P}_b$ decreases with $T_{\rm obs}^{-2.5}$, $T_{\rm obs}$ denoting the observing time span. Improvements in the hardware, like new broad-band receivers (e.g. [146]), will further boost the timing precision. For pulsars like PSR J1738+0333 and PSR J0348+0432 soon the modeling of the white dwarf will be the limiting factor, while for the Double Pulsar the corrections of the external contributions to $\dot{P}_b$ will be the challenging bit, in particular if one wants to reach the $\sim 10^{-5}$ level at which higher oder contributions to $\dot{P}_b$ [147, 148] and the Lense–Thirring contribution to the orbital dynamics [45, 50] become relevant (see [75] for a detailed discussion). The upcoming next generation of radio telescopes, like the The five-hundred-meter Aperture Spherical radio Telescope (FAST) [149] and the The Square Kilometre Array (SKA) [150], will certainly promise a big step toward this goal. With SKA, for many pulsars one can hope for a factor of 100 improvement in timing precision [151]. The SKA also promises to provide excellent direct distance measurements to pulsars, either directly by utilizing the long baselines of the SKA to form high angular resolution images, or by fitting for the timing parallax in the arrival times of the pulsar signals [152]. In combination with new models for the gravitational potential of our Galaxy, in particular after new missions like GAIA [153], one will be able to accurately determine the extrinsic "contaminations" of $\dot{P}_b$ via equation (2.1), and by this know the intrinsic $\dot{P}_b$. This is key for any high precision gravitational wave test with binary pulsars, but also crucial to measure the Lense–Thirring drag in the Double Pulsar [75].

Reducing the parameter uncertainties for known pulsars is one way to push gravity tests forward, finding new, more relativistic systems is the other. Presently there are a number of pulsar surveys underway that promise the discovery of many new pulsars. New techniques, like acceleration searches [154] and high-performance computing, e.g. Einstein@Home [155], promise the detection of pulsars in tight orbits, which generally cannot be found with traditional methods. There is considerable hope among pulsar astronomers, that this will finally also lead to the discovery of a pulsar-black hole system, occasionally called the "holy grail" of pulsar astronomy. Such a system is expected to provide a superb new probe of relativistic gravity and black hole properties, like the dragging of spacetime by the rotation of the black hole [156–158]. According to GR, for an astrophysical black hole (Kerr solution) there is an upper limit for its spin, given by $S_{max} = GM^2/c$. It would pose an interesting challenge to GR, if the timing of a pulsar-black hole system indicates a spin $S > S_{max}$. But even for gravity theories that predict the same properties for black holes as in GR, a pulsar–black hole system would constitute an excellent test system, due to the high grade of asymmetry in the strong-field properties of these two components (see [158] for simulations based on $T_1(\alpha_0, \beta_0)$ scalar–tensor theories). A pulsar in a close orbit ($P_b < 1$ yr) around the super-massive black hole ($m_{BH} \approx 4 \times 10^6\,M_\odot$) in the center of our Galaxy would be the ultimate test system, in that context. According to the mock data analysis in [159], for such a system a precise measurement of the quadrupole moment of the black hole, and therefore a test of the no-hair theorem, should be possible, provided that the environment of the pulsar orbit is sufficiently clean. Finding and timing a pulsar in the center of our Galaxy is certainly challenging. A promising result in that direction is the very recent detection of radio signals from a magnetar near the Galactic center black hole [160], even if this pulsar is still too far away from the super-massive black hole (~ 0.1 pc) to probe its spacetime.

Until now, all gravitational wave tests are based on probing the near-zone of a binary pulsar's spacetime by measuring how the back reaction of the gravitational radiation changes the world lines of the source masses. As outlined above, with the Double Pulsar this test has reached a precision of better than 0.1%. Presently there are considerable efforts to achieve a direct detection of gravitational waves, i.e. measure the properties of the far field of such radiative spacetimes by using appropriate test masses. Ground based laser interferometric gravitational wave observatories, like LIGO and VIRGO, have mirrors with separations of a few kilometers. Their sensitivity is in the range from 10 Hz to few 10^3 Hz. Planned space-based detectors, like eLISA[14], will have three drag-free satellites as test masses with a typical separation of $\sim 10^6$ km, and should be sensitive to gravitational waves from about 10^{-4} to 0.1 Hz. For the ultra-low frequency band (few nano-Hz) pulsar timing arrays are currently the most promising detectors [161]. In these experiments the Earth/Solar System and a collection of very stable pulsars act as the test masses. A gravitational wave becomes apparent in a pulsar timing array by the changes it causes in the arrival times of the pulsar signals. Due to the fitting of the rotational frequency ν and its time derivative $\dot{\nu}$ for every pulsar, such a detector is only sensitive to wavelengths up to $\sim c\,T_{obs}$.[15] This

14 www.elisascience.org/

15 It has been suggested to use the orbital period of binary pulsars to test for gravitational waves of considerably longer wavelength [162, 163].

leads to the special situation that the length of the "detector arms" is much larger than the wavelength. As a consequence, the observed timing signal contains two contributions, the so-called *pulsar term*, related to the impact of the gravitational wave on the pulsar when the radio signal is emitted, and the *Earth term* corresponding to the impact of the gravitational wave on the Earth during the arrival of the radio signal at the telescope [164, 165]. The most promising source in the nano-Hz frequency band is a stochastic gravitational wave background, as a result of many mergers of super-massive black hole binaries in the past history of the Universe [166, 167]. With the large number of "detector arms," pulsar timing arrays have enough information to explore the properties of the nano-Hz gravitational wave background in details, once its signal is clearly detected in the data. Are there more than the two Einsteinian polarization modes (alternative metric theories can have up to six)? Is the propagation speed of nano-Hz gravitational waves frequency depended? Does the graviton carry mass? These are some of the main questions that can be addressed with pulsar timing arrays [168, 169]. The isolation of a single source in the pulsar timing array data would give us a unique opportunity to study the merger evolution of a super-massive black hole binary, since the signal in the Earth term and the signal in the pulsar term show two different states of the system, which are typically several thousand years apart [170]. For these kind of gravity experiments, however, we might have to wait till the full SKA has collected a few years of data, which probably brings us close to the year 2030.

Acknowledgement: I would like to thank John Antoniadis and Lijing Shao for carefully reading the manuscript and their valuable comments. I am grateful to Jim Lattimer for providing the tables for equations-of-state AP4 and MPA1.

References

[1] A. Einstein. Die Feldgleichungen der Gravitation. *Sitzungsberichte der Königlich Preußischen Akademie der Wissenschaften (Berlin)*, pages 844–847, 1915.

[2] A. Einstein. Erklärung der Perihelbewegung des Merkur aus der allgemeinen Relativitätstheorie. *Sitzungsberichte der Königlich Preußischen Akademie der Wissenschaften (Berlin)*, pages 831–839, 1915.

[3] H. Seeliger. Über die Anomalien in der Bewegung der innern Planeten. *Astronomische Nachrichten*, 201:273, 1915.

[4] A. Einstein. Die Grundlage der allgemeinen Relativitätstheorie. *Annalen der Physik*, 354:769–822, 1916.

[5] C. M. Will. *Theory and Experiment in Gravitational Physics*. Cambridge University Press, Cambridge, 1993.

[6] F. W. Dyson, A. S. Eddington and C. Davidson. A determination of the deflection of light by the Sun's gravitational field, from observations made at the total eclipse of May 29, 1919. *Royal Society of London Philosophical Transactions A Series*, 220:291–333, 1920.

[7] M. Froeschle, F. Mignard and F. Arenou. Determination of the PPN Parameter gamma with the HIPPARCOS Data. In R. M. Bonnet, E. Høg, P. L. Bernacca, L. Emiliani, A. Blaauw, C. Turon, J. Kovalevsky, L. Lindegren, H. Hassan, M. Bouffard, B. Strim, D. Heger, M. A. C. Perryman and L. Woltjer, editors, *Hipparcos – Venice '97*, Vol. 402 of *ESA Special Publication*, pages 49–52, 1997.

[8] D. E. Lebach, B. E. Corey, I. I. Shapiro, M. I. Ratner, J. C. Webber, A. E. E. Rogers, J. L. Davis and T. A. Herring. Measurement of the solar gravitational deflection of radio waves using very-long-baseline interferometry. *Phys. Rev. Lett.*, 75:1439–1442, 1995.

[9] S. S. Shapiro, J. L. Davis, D. E. Lebach and J. S. Gregory. Measurement of the Solar Gravitational Deflection of Radio Waves using Geodetic Very-Long-Baseline Interferometry Data, 1979–1999. *Phys. Rev. Lett.*, 92:121101, 2004.

[10] E. Fomalont, S. Kopeikin, G. Lanyi and J. Benson. Progress in measurements of the gravitational bending of radio waves using the VLBA. *Astrophys. J.*, 699:1395–1402, 2009.

[11] I. I. Shapiro. Fourth test of general relativity. *Phys. Rev. Lett.*, 13:789–791, 1964.

[12] B. Bertotti, L. Iess and P. Tortora. A test of general relativity using radio links with the Cassini spacecraft. *Nature*, 425:374–376, 2003.

[13] K. Nordtvedt. 30 years of lunar laser ranging and the gravitational interaction. *Class. Quantum Grav.*, 16:A101–A112, 1999.

[14] C. W. F. Everitt, D. B. Debra, B. W. Parkinson, J. P. Turneaure, J. W. Conklin, M. I. Heifetz, G. M. Keiser, A. S. Silbergleit, T. Holmes, J. Kolodziejczak, M. Al-Meshari, J. C. Mester, B. Muhlfelder, V. G. Solomonik, K. Stahl, P. W. Worden, Jr., W. Bencze, S. Buchman, B. Clarke, A. Al-Jadaan, H. Al-Jibreen, J. Li, J. A. Lipa, J. M. Lockhart, B. Al-Suwaidan, M. Taber and S. Wang. Gravity Probe B: Final results of a space experiment to test general relativity. *Phys. Rev. Lett.*, 106:221101, 2011.

[15] I. Ciufolini and E. C. Pavlis. A confirmation of the general relativistic prediction of the Lense–Thirring effect. *Nature*, 431:958–960, 2004.

[16] A. Einstein. Näherungsweise Integration der Feldgleichungen der Gravitation. *Sitzungsberichte der Königlich Preußischen Akademie der Wissenschaften (Berlin)*, pages 688–696, 1916.

[17] A. Einstein. Über Gravitationswellen. *Sitzungsberichte der Königlich Preußischen Akademie der Wissenschaften (Berlin)*, pages 154–167, 1918.

[18] D. Kennefick. *Traveling at the Speed of Thought: Einstein and the Quest for Gravitational Waves*. Princeton University Press, Princeton and Oxford, 2007.

[19] R. A. Hulse and J. H. Taylor. Discovery of a pulsar in a binary system. *Astrophys. J.*, 195:L51–L53, 1975.

[20] J. J. Hermes, M. Kilic, W. R. Brown, D. E. Winget, C. Allende Prieto, A. Gianninas, A. S. Mukadam, A. Cabrera-Lavers and S. J. Kenyon. Rapid Orbital Decay in the 12.75-minute Binary White Dwarf J0651+2844. *Astrophys. J.*, 757:L21, 2012.

[21] A. Hewish, S. J. Bell, J. D. H. Pilkington, P. F. Scott and R. A. Collins. Observation of a rapidly pulsating radio source. *Nature*, 217:709–713, 1968.

[22] R. N. Manchester, G. B. Hobbs, A. Teoh and M. Hobbs. The Australia telescope national facility pulsar catalogue. *Astrophys. J.*, 129:1993–2006, 2005. www.atnf.csiro.au/research/pulsar/psrcat/.

[23] Duncan R. Lorimer. Binary and millisecond pulsars. *Living Rev. Relativity*, 8, 2005. www.livingreviews.org/lrr-2005-7.

[24] G. Hobbs, W. Coles, R. N. Manchester, M. J. Keith, R. M. Shannon, D. Chen, M. Bailes, N. D. R. Bhat, S. Burke-Spolaor, D. Champion, A. Chaudhary, A. Hotan, J. Khoo, J. Kocz, Y. Levin, S. Oslowski, B. Preisig, V. Ravi, J. E. Reynolds, J. Sarkissian, W. van Straten, J. P. W. Verbiest, D. Yardley and X. P. You. Development of a pulsar-based time-scale. *Mon. Not. R. Astron. Soc.*, 427:2780–2787, 2012.

[25] I. H. Stairs. Testing General Relativity with Pulsar Timing. *Living Rev. Relativity*, 6, 2003. www.livingreviews.org/lrr-2003-5.

[26] D. R. Lorimer and M. Kramer. *Handbook of Pulsar Astronomy*. Cambridge University Press, Cambridge, 2004.

[27] R. T. Edwards, G. B. Hobbs and R. N. Manchester. TEMPO2, a new pulsar timing package – II. The timing model and precision estimates. *Mon. Not. R. Astron. Soc.*, 372:1549–1574, 2006.

[28] M. Kramer. Probing gravitation with pulsars. *Proceedings of the International Astronomical Union*, 8:19–26, 2012.

[29] J. P. W. Verbiest, M. Bailes, W. van Straten, G. B. Hobbs, R. T. Edwards, R. N. Manchester, N. D. R. Bhat, J. M. Sarkissian, B. A. Jacoby and S. R. Kulkarni. Precision Timing of PSR J0437-4715: An Accurate Pulsar Distance, a High Pulsar Mass, and a Limit on the Variation of Newton's Gravitational Constant. *Astrophys. J.*, 679:675–680, 2008.

[30] P. C. C. Freire, N. Wex, G. Esposito-Farèse, J. P. W. Verbiest, M. Bailes, B. A. Jacoby, M. Kramer, I. H. Stairs, J. Antoniadis and G. H. Janssen. The relativistic pulsar-white dwarf binary PSR J1738+0333 – II. The most stringent test of scalar-tensor gravity. *Mon. Not. R. Astron. Soc.*, 423:3328–3343, 2012.

[31] J. M. Weisberg, D. J. Nice and J. H. Taylor. Timing Measurements of the Relativistic Binary Pulsar PSR B1913+16. *Astrophys. J.*, 722:1030–1034, 2010.

[32] P. C. C. Freire, C. G. Bassa, N. Wex, I. H. Stairs, D. J. Champion, S. M. Ransom, P. Lazarus, V. M. Kaspi, J. W. T. Hessels, M. Kramer, J. M. Cordes, J. P. W. Verbiest, P. Podsiadlowski, D. J. Nice, J. S. Deneva, D. R. Lorimer, B. W. Stappers, M. A. McLaughlin and F. Camilo. On the nature and evolution of the unique binary pulsar J1903+0327. *Mon. Not. R. Astron. Soc.*, 412:2763–2780, 2011.

[33] A. W. Hotan, M. Bailes and S. M. Ord. High-precision baseband timing of 15 millisecond pulsars. *Mon. Not. R. Astron. Soc.*, 369:1502–1520, 2006.

[34] D. J. Champion, G. B. Hobbs, R. N. Manchester, R. T. Edwards, D. C. Backer, M. Bailes, N. D. R. Bhat, S. Burke-Spolaor, W. Coles, P. B. Demorest, R. D. Ferdman, W. M. Folkner, A. W. Hotan, M. Kramer, A. N. Lommen, D. J. Nice, M. B. Purver, J. M. Sarkissian, I. H. Stairs, W. van Straten, J. P. W. Verbiest and D. R. B. Yardley. Measuring the mass of Solar System planets using pulsar timing. *Astrophys. J.*, 720:L201–L205, 2010.

[35] T. Damour. The problem of motion in Newtonian and Einsteinian gravity. In S. W. Hawking and W. Israel, editors, *Three Hundred Years of Gravitation*, pages 128–198. Cambridge University Press, Cambridge; New York, 1987.

[36] L. Blanchet. Gravitational Radiation from Post-Newtonian Sources and Inspiralling Compact Binaries. *Living Rev. Relativity*, 9, 2006.

[37] T. Futamase and Y. Itoh. The Post-Newtonian Approximation for Relativistic Compact Binaries. *Living Rev. Relativity*, 10, 2007. www.livingreviews.org/lrr-2007-2.

[38] T. Damour. Gravitational radiation and the motion of compact bodies. In N. Deruelle and T. Piran, editors, *Gravitational Radiation*, pages 59–144, Amsterdam, Netherlands; New York, U.S.A., 1983. North-Holland; Elsevier.

[39] J.-M. Gérard and Y. Wiaux. Gravitational dipole radiations from binary systems. *Phys. Rev. D*, 66:024040, 2002.

[40] T. Damour and J. H. Taylor. Strong-field tests of relativistic gravity and binary pulsars. *Phys. Rev. D*, 45:1840–1868, 1992.

[41] T. Damour. Binary Systems as Test-Beds of Gravity Theories. In M. Colpi, P. Casella, V. Gorini, U. Moschella and A. Possenti, editors, *Astrophysics and Space Science Library*, Vol. 359 of *Astrophysics and Space Science Library*, pages 1–41, 2009.

[42] T. Damour and N. Deruelle. General Relativistic Celestial Mechanics of Binary Systems. I. The post-Newtonian motion. *Ann. Inst. Henri Poincaré Phys. Théor.*, 43:107–132, 1985.

[43] T. Damour and G. Esposito-Farèse. Testing gravity to second post-Newtonian order: A field-theory approach. *Phys. Rev. D*, 53:5541–5578, 1996.

[44] S. Mirshekari and C. M. Will. Compact binary systems in scalar-tensor gravity: Equations of motion to 2.5 post-Newtonian order. *Phys. Rev. D*, 87:084070, 2013.

[45] B. M. Barker and R. F. O'Connell. Gravitational two-body problem with arbitrary masses, spins, and quadrupole moments. *Phys. Rev. D*, 12:329–335, 1975.

[46] V. A. Brumberg. *Essential relativistic celestial mechanics*. Adam Hilger, Bristol, England and New York, 1991.

[47] J. Hartung, J. Steinhoff and G. Schäfer. Next-to-next-to-leading order post-Newtonian linear-in-spin binary Hamiltonians. *Annalen der Physik*, 525:359–394, 2013.

[48] N. Wex. The second post-Newtonian motion of compact binary-star systems with spin. *Class. Quantum Grav.*, 12:983–1005, 1995.

[49] T. Damour. Problème des deux corps et freinage de rayonnement en relativité générale. *C. R. Acad. Sci. Paris, Série II*, 294:1355–1357, 1982.

[50] T. Damour and G. Schäfer. Higher order relativistic periastron advances in binary pulsars. *Nuovo Cimento B*, 101:127, 1988.

[51] T. Damour and R. Ruffini. Certain new verifications of general relativity made possible by the discovery of a pulsar belonging to a binary system. *Academie des Sciences Paris Comptes Rendus Serie Sciences Mathematiques*, 279:971–973, 1974.

[52] G. Börner, J. Ehlers and E. Rudolph. Relativistic spin precession in two-body systems. *Astron. Astrophys.*, 44:417–420, 1975.

[53] V. M. Kaspi, M. Bailes, R. N. Manchester, B. W. Stappers and J. F. Bell. Evidence from a precessing pulsar orbit for a neutron-star birth kick. *Nature*, 381:584–586, 1996.

[54] T. Damour. Strong-field tests of general relativity and the binary pulsar. In A. Coley, C. Dyer and T. Tupper, editors, *Proceedings of the 2nd Canadian Conference on General Relativity and Relativistic Astrophysics*, pages 315–334, 1988.

[55] T. Damour and N. Deruelle. General relativistic celestial mechanics of binary systems. II. The post-Newtonian timing formula. *Ann. Inst. Henri Poincaré Phys. Théor.*, 44:263–292, 1986.

[56] M. Ali. Relativistic Propagation Effects in Binary Pulsar Signals for Next Generation Radio Telescopes. Master Thesis, University of Bonn, Germany, 2011.

[57] T. Damour and G. Esposito-Farèse. Nonperturbative strong-field effects in tensor-scalar theories of gravitation. *Phys. Rev. Lett.*, 70:2220–2223, 1993.

[58] T. Damour and G. Esposito-Farèse. Tensor-scalar gravity and binary-pulsar experiments. *Phys. Rev. D*, 54:1474–1491, 1996.

[59] D. M. Eardley. Observable effects of a scalar gravitational field in a binary pulsar. *Astrophys. J.*, 196:L59–L62, 1975.

[60] C. M. Will. Gravitational radiation from binary systems in alternative metric theories of gravity – Dipole radiation and the binary pulsar. *Astrophys. J.*, 214:826–839, 1977.

[61] T. Damour and G. Esposito-Farèse. Tensor-multi-scalar theories of gravitation. *Class. Quantum Grav.*, 9:2093–2176, September 1992.

[62] P. C. Peters. Gravitational radiation and the motion of two point masses. *Phys. Rev.*, 136:B1224–B1232, 1964.

[63] V. A. Brumberg, I. B. Zeldovich, I. D. Novikov and N. I. Shakura. Component masses and inclination of binary systems containing a pulsar, determined from relativistic effects. *Sov. Astronomy Lett.*, 1:2–4, 1975.

[64] R. V. Wagoner. Test for the existence of gravitational radiation. *Astrophys. J.*, 196:L63–L65, 1975.

[65] J. H. Taylor, R. A. Hulse, L. A. Fowler, G. E. Gullahorn and J. M. Rankin. Further observations of the binary pulsar PSR 1913+16. *Astrophys. J.*, 206:L53–L58, 1976.
[66] N. Wex, V. Kalogera and M. Kramer. Constraints on supernova kicks from the double neutron star system PSR B1913+16. *Astrophys. J.*, 528:401–409, 2000.
[67] R. Blandford and S. A. Teukolsky. On the measurement of the mass of PSR 1913+16. *Astrophys. J.*, 198:L27–L29, 1975.
[68] J. H. Taylor, L. A. Fowler and P. M. McCulloch. Measurements of general relativistic effects in the binary pulsar PSR 1913+16. *Nature*, 277:437–440, 1979.
[69] I. S. Shklovskii. Possible causes of the secular increase in pulsar periods. *Sov. Astronomy*, 13:562–565, 1970.
[70] T. Damour and J. H. Taylor. On the orbital period change of the binary pulsar PSR 1913+16. *Astrophys. J.*, 366:501–511, 1991.
[71] J. M. Weisberg, S. Stanimirović, K. Xilouris, A. Hedden, A. de la Fuente, S. B. Anderson and F. A. Jenet. Arecibo H I absorption measurements of pulsars and the electron density at intermediate longitudes in the first galactic quadrant. *Astrophys. J.*, 674:286–294, 2008.
[72] M. Burgay, N. D'Amico, A. Possenti, R. N. Manchester, A. G. Lyne, B. C. Joshi, M. A. McLaughlin, M. Kramer, J. M. Sarkissian, F. Camilo, V. Kalogera, C. Kim and D. R. Lorimer. An increased estimate of the merger rate of double neutron stars from observations of a highly relativistic system. *Nature*, 426:531–533, 2003.
[73] A. G. Lyne, M. Burgay, M. Kramer, A. Possenti, R. N. Manchester, F. Camilo, M. A. McLaughlin, D. R. Lorimer, N. D'Amico, B. C. Joshi, J. Reynolds and P. C. C. Freire. A Double-Pulsar system: A rare laboratory for relativistic gravity and plasma physics. *Science*, 303:1153–1157, 2004.
[74] M. Kramer and I. H. Stairs. The Double Pulsar. *Annual Review of Astronomy and Astrophysics*, 46:541–572, 2008.
[75] M. Kramer and N. Wex. TOPICAL REVIEW: The double pulsar system: a unique laboratory for gravity. *Class. Quantum Grav.*, 26:073001, 2009.
[76] M. Kramer, I. H. Stairs, R. N. Manchester, M. A. McLaughlin, A. G. Lyne, R. D. Ferdman, M. Burgay, D. R. Lorimer, A. Possenti, N. D'Amico, J. M. Sarkissian, G. B. Hobbs, J. E. Reynolds, P. C. C. Freire and F. Camilo. Tests of general relativity from timing the Double Pulsar. *Science*, 314:97–102, 2006.
[77] P. B. Demorest, T. Pennucci, S. M. Ransom, M. S. E. Roberts and J. W. T. Hessels. A two-solar-mass neutron star measured using Shapiro delay. *Nature*, 467:1081–1083, 2010.
[78] R. Blandford and S. A. Teukolsky. Arrival-time analysis for a pulsar in a binary system. *Astrophys. J.*, 205:580–591, 1976.
[79] A. T. Deller, M. Bailes and S. J. Tingay. Implications of a VLBI Distance to the Double Pulsar J0737-3039A/B. *Science*, 323:1327–1329, 2009.
[80] J. P. W. Verbiest, M. Bailes, W. A. Coles, G. B. Hobbs, W. van Straten, D. J. Champion, F. A. Jenet, R. N. Manchester, N. D. R. Bhat, J. M. Sarkissian, D. Yardley, S. Burke-Spolaor, A. W. Hotan and X. P. You. Timing stability of millisecond pulsars and prospects for gravitational-wave detection. *Mon. Not. R. Astron. Soc.*, 400:951–968, 2009.
[81] E. S. Phinney. Pulsars as probes of Newtonian dynamical systems. *Royal Society of London Philosophical Transactions Series A*, 341:39–75, 1992.
[82] M. F. Ryba and J. H. Taylor. High-precision timing of millisecond pulsars. I – Astrometry and masses of the PSR 1855+09 system. *Astrophys. J.*, 371:739–748, 1991.
[83] B. A. Jacoby. *Recycled pulsars*. PhD thesis, California Institute of Technology, California, USA, 2005.
[84] J. Antoniadis, M. H. van Kerkwijk, D. Koester, P. C. C. Freire, N. Wex, T. M. Tauris, M. Kramer and C. G. Bassa. The relativistic pulsar-white dwarf binary PSR J1738+0333 – I. Mass determination and evolutionary history. *Mon. Not. R. Astron. Soc.*, 423:3316–3327, 2012.

[85] H. Müther, M. Prakash and T. L. Ainsworth. The nuclear symmetry energy in relativistic Brueckner–Hartree–Fock calculations. *Physics Letters B*, 199:469–474, 1987.

[86] P. C. C. Freire, S. M. Ransom, S. Bégin, I. H. Stairs, J. W. T. Hessels, L. H. Frey and F. Camilo. Eight New Millisecond Pulsars in NGC 6440 and NGC 6441. *Astrophys. J.*, 675:670–682, 2008.

[87] M. H. van Kerkwijk, R. P. Breton and S. R. Kulkarni. Evidence for a massive neutron star from a radial-velocity study of the companion to the black-widow pulsar PSR B1957+20. *Astrophys. J.*, 728:95, 2011.

[88] R. W. Romani, A. V. Filippenko, J. M. Silverman, S. B. Cenko, J. Greiner, A. Rau, J. Elliott and H. J. Pletsch. PSR J1311-3430: A heavyweight neutron star with a flyweight helium companion. *Astrophys. J.*, 760:L36, 2012.

[89] I. H. Stairs, S. E. Thorsett, J. H. Taylor and A. Wolszczan. Studies of the relativistic binary pulsar PSR B1534+12. I. Timing analysis. *Astrophys. J.*, 581:501–508, 2002.

[90] N. D. R. Bhat, M. Bailes and J. P. W. Verbiest. Gravitational-radiation losses from the pulsar white-dwarf binary PSR J1141-6545. *Phys. Rev. D*, 77(12):124017, 2008.

[91] J. Antoniadis, C. G. Bassa, N. Wex, M. Kramer and R. Napiwotzki. A white dwarf companion to the relativistic pulsar PSR J1141-6545. *Mon. Not. R. Astron. Soc.*, 412:580–584, 2011.

[92] J. Boyles, R. S. Lynch, S. M. Ransom, I. H. Stairs, D. R. Lorimer, M. A. McLaughlin, J. W. T. Hessels, V. M. Kaspi, V. I. Kondratiev, A. Archibald, A. Berndsen, R. F. Cardoso, A. Cherry, C. R. Epstein, C. Karako-Argaman, C. A. McPhee, T. Pennucci, M. S. E. Roberts, K. Stovall and J. van Leeuwen. The Green Bank Telescope 350 MHz Drift-scan survey. I. Survey observations and the discovery of 13 pulsars. *Astrophys. J.*, 763:80, 2013.

[93] R. S. Lynch, J. Boyles, S. M. Ransom, I. H. Stairs, D. R. Lorimer, M. A. McLaughlin, J. W. T. Hessels, V. M. Kaspi, V. I. Kondratiev, A. M. Archibald, A. Berndsen, R. F. Cardoso, A. Cherry, C. R. Epstein, C. Karako-Argaman, C. A. McPhee, T. Pennucci, M. S. E. Roberts, K. Stovall and J. van Leeuwen. The Green Bank Telescope 350 MHz Drift-scan Survey II: Data Analysis and the Timing of 10 New Pulsars, Including a Relativistic Binary. *Astrophys. J.*, 763:81, 2013.

[94] J. Antoniadis, P. C. C. Freire, N. Wex, T. M. Tauris, R. S. Lynch, M. H. van Kerkwijk, M. Kramer, C. Bassa, V. S. Dhillon, T. Driebe, J. W. T. Hessels, V. M. Kaspi, V. I. Kondratiev, N. Langer, T. R. Marsh, M. A. McLaughlin, T. T. Pennucci, S. M. Ransom, I. H. Stairs, J. van Leeuwen, J. P. W. Verbiest and D. G. Whelan. A massive pulsar in a compact relativistic binary. *Science*, 340:448, April 2013.

[95] D. Hobbs, B. Holl, L. Lindegren, F. Raison, S. Klioner and A. Butkevich. Determining PPN γ with Gaia's astrometric core solution. *Proceedings of the International Astronomical Union*, 5:315–319, 2009.

[96] B. S. Sathyaprakash and B. F. Schutz. Physics, astrophysics and cosmology with gravitational waves. *Living Rev. Relativity*, 12, 2009. www.livingreviews.org/lrr-2009-2.

[97] L Blanchet. Post-newtonian theory and the two-body problem. In L. Blanchet, A. Spallicci and B. Whiting, editors, *Mass and Motion in General Relativity*, Vol. 162 of *Fundamental Theories of Physics*, pages 125–166. Springer Netherlands, 2011.

[98] LIGO Scientific Collaboration, Virgo Collaboration, J. Aasi, J. Abadie, B. P. Abbott, R. Abbott, T. D. Abbott, M. Abernathy, T. Accadia and F. Acernese, et al. Prospects for Localization of Gravitational Wave Transients by the Advanced LIGO and Advanced Virgo Observatories. *ArXiv e-prints*, 1304.0670, 2013.

[99] C. M. Will. Testing scalar-tensor gravity with gravitational-wave observations of inspiralling compact binaries. *Phys. Rev. D*, 50:6058–6067, 1994.

[100] T. Damour and G. Esposito-Farèse. Gravitational-wave versus binary-pulsar tests of strong-field gravity. *Phys. Rev. D*, 58(4):042001, 1998.

[101] E. Barausse, C. Palenzuela, M. Ponce and L. Lehner. Neutron-star mergers in scalar-tensor theories of gravity. *Phys. Rev. D*, 87:081506, 2013.

[102] N. D. H. Dass and V. Radhakrishnan. The new binary pulsar and the observation of gravitational spin precession. *Astrophys. J.*, 16:L135–L139, 1975.

[103] J. M. Weisberg, R. W. Romani and J. H. Taylor. Evidence for geodetic spin precession in the binary pulsar 1913 + 16. *Astrophys. J.*, 347:1030–1033, 1989.

[104] M. Kramer. Determination of the geometry of the PSR B1913+16 system by geodetic precession. *Astrophys. J.*, 509:856–860, 1998.

[105] I. H. Stairs, S. E. Thorsett and Z. Arzoumanian. Measurement of gravitational spin–orbit coupling in a binary-pulsar system. *Phys. Rev. Lett.*, 93:141101, 2004.

[106] J. M. Weisberg and J. H. Taylor. General relativistic geodetic spin precession in binary pulsar B1913+16: Mapping the emission Beam in two dimensions. *Astrophys. J.*, 576:942–949, 2002

[107] R. N. Manchester, M. Kramer, I. H. Stairs, M. Burgay, F. Camilo, G. B. Hobbs, D. R. Lorimer, A. G. Lyne, M. A. McLaughlin, C. A. McPhee, A. Possenti, J. E. Reynolds and W. van Straten. Observations and modeling of relativistic spin precession in PSR J1141-6545. *Astrophys. J.*, 710:1694–1709, 2010.

[108] G. Desvignes, M. Kramer, I. Cognard, L. Kasian, J. van Leeuwen, I. Stairs and G. Theureau. PSR J1906+0746: From relativistic spin-precession to beam modeling. *Proceedings of the International Astronomical Union*, 8:199–202, 2012.

[109] A. Wolszczan. A nearby 37.9-ms radio pulsar in a relativistic binary system. *Nature*, 350:688–690, 1991.

[110] I. H. Stairs, S. E. Thorsett, J. H. Taylor and Z. Arzoumanian. Geodetic Precession in PSR B1534+12. In M. Kramer, N. Wex and R. Wielebinski, editors, *IAU Colloq. 177: Pulsar Astronomy – 2000 and Beyond*, Vol. 202 of *Astronomical Society of the Pacific Conference Series*, pages 121–124, 2000.

[111] M. A. McLaughlin, A. G. Lyne, D. R. Lorimer, A. Possenti, R. N. Manchester, F. Camilo, I. H. Stairs, M. Kramer, M. Burgay, N. D'Amico, P. C. C. Freire, B. C. Joshi and N. D. R. Bhat. The Double Pulsar System J0737-3039: Modulation of A by B at Eclipse. *Astrophys. J.*, 616:L131–L134, 2004.

[112] R. P. Breton, V. M. Kaspi, M. Kramer, M. A. McLaughlin, M. Lyutikov, S. M. Ransom, I. H. Stairs, R. D. Ferdman, F. Camilo and A. Possenti. Relativistic spin precession in the double pulsar. *Science*, 321:104–107, 2008.

[113] B. B. P. Perera, M. A. McLaughlin, M. Kramer, I. H. Stairs, R. D. Ferdman, P. C. C. Freire, A. Possenti, R. P. Breton, R. N. Manchester, M. Burgay, A. G. Lyne and F. Camilo. The Evolution of PSR J0737-3039B and a Model for Relativistic Spin Precession. *Astrophys. J.*, 721:1193–1205, 2010.

[114] R. D. Ferdman, I. H. Stairs, M. Kramer, R. P. Breton, M. A. McLaughlin, P. C. C. Freire, A. Possenti, B. W. Stappers, V. M. Kaspi, R. N. Manchester and A. G. Lyne. The Double Pulsar: Evidence for neutron star formation without an iron core-collapse supernova. *Astrophys. J.*, 767:85, 2013.

[115] K. Nordtvedt. Testing relativity with laser ranging to the moon. *Phys. Rev.*, 170:1186–1187, 1968.

[116] J. Müller, F. Hofmann and L. Biskupek. Testing various facets of the equivalence principle using lunar laser ranging. *Class. Quantum Grav.*, 29:184006, 2012.

[117] T. Damour and G. Schäfer. New tests of the strong equivalence principle using binary-pulsar data. *Phys. Rev. Lett.*, 66:2549–2552, 1991.

[118] R. Lynch. The hunt for new pulsars with the Green Bank Telescope. *Proceedings of the International Astronomical Union*, 8:41–46, 2012.

[119] P. C. C. Freire, M. Kramer and N. Wex. Tests of the universality of free fall for strongly self-gravitating bodies with radio pulsars. *Class. Quantum Grav.*, 29:184007, 2012.

[120] V. Boriakoff, R. Buccheri and F. Fauci. Discovery of a 6.1-ms binary pulsar PSR1953+29. *Nature*, 304:417–419, 1983.

[121] M. E. Gonzalez, I. H. Stairs, R. D. Ferdman, P. C. C. Freire, D. J. Nice, P. B. Demorest, S. M. Ransom, M. Kramer, F. Camilo, G. Hobbs, R. N. Manchester and A. G. Lyne. High-precision timing of five millisecond pulsars: Space velocities, binary evolution, and equivalence principles. *Astrophys. J.*, 743:102, 2011.

[122] N. Wex. Small-eccentricity binary pulsars and relativistic gravity. In M. Kramer, N. Wex and R. Wielebinski, editors, *IAU Colloq. 177: Pulsar Astronomy – 2000 and Beyond*, Vol. 202 of *Astronomical Society of the Pacific Conference Series*, pages 113–116, 2000.

[123] I. H. Stairs, A. J. Faulkner, A. G. Lyne, M. Kramer, D. R. Lorimer, M. A. McLaughlin, R. N. Manchester, G. B. Hobbs, F. Camilo, A. Possenti, M. Burgay, N. D'Amico, P. C. Freire and P. C. Gregory. Discovery of three wide-orbit binary pulsars: Implications for binary evolution and equivalence principles. *Astrophys. J.*, 632:1060–1068, 2005.

[124] M. Kehl and A. Krieger. Auswirkungen der Verletzung des Starken Äquivalenzprinzips auf die Dynamik von Binärpulsaren im Gravitationsfeld der Milchstraße und in Kugelsternhaufen. Bachelor thesis, University of Bonn, Germany, 2012.

[125] D. J. Champion, S. M. Ransom, P. Lazarus, F. Camilo, C. Bassa, V. M. Kaspi, D. J. Nice, P. C. C. Freire, I. H. Stairs, J. van Leeuwen, B. W. Stappers, J. M. Cordes, J. W. T. Hessels, D. R. Lorimer, Z. Arzoumanian, D. C. Backer, N. D. R. Bhat, S. Chatterjee, I. Cognard, J. S. Deneva, C.-A. Faucher-Giguère, B. M. Gaensler, J. Han, F. A. Jenet, L. Kasian, V. I. Kondratiev, M. Kramer, J. Lazio, M. A. McLaughlin, A. Venkataraman and W. Vlemmings. An eccentric binary millisecond pulsar in the galactic plane. *Science*, 320:1309–1312, 2008.

[126] G. Hinshaw, J. L. Weiland, R. S. Hill, N. Odegard, D. Larson, C. L. Bennett, J. Dunkley, B. Gold, M. R. Greason, N. Jarosik, E. Komatsu, M. R. Nolta, L. Page, D. N. Spergel, E. Wollack, M. Halpern, A. Kogut, M. Limon, S. S. Meyer, G. S. Tucker and E. L. Wright. Five-year Wilkinson microwave anisotropy probe observations: Data processing, sky maps, and basic results. *Astrophys. J. Suppl.*, 180:225–245, 2009.

[127] J. Müller, J. G. Williams and S. G. Turyshev. Lunar Laser Ranging Contributions to Relativity and Geodesy. In H. Dittus, C. Lammerzahl and S. G. Turyshev, editors, *Lasers, Clocks and Drag-Free Control: Exploration of Relativistic Gravity in Space*, Vol. 349 of *Astrophysics and Space Science Library*, pages 457–472, 2008.

[128] K. Nordtvedt. Probing gravity to the second post-Newtonian order and to one part in 10^7 using the spin axis of the sun. *Astrophys. J.*, 320:871–874, 1987.

[129] T. Damour and G. Esposito-Farèse. Testing local Lorentz invariance of gravity with binary-pulsar data. *Phys. Rev. D*, 46:4128–4132, 1992.

[130] L. Shao and N. Wex. New tests of local Lorentz invariance of gravity with small-eccentricity binary pulsars. *Class. Quantum Grav.*, 29:215018, 2012.

[131] P. J. Callanan, P. M. Garnavich and D. Koester. The mass of the neutron star in the binary millisecond pulsar PSR J1012+5307. *Mon. Not. R. Astron. Soc.*, 298:207–211, 1998.

[132] K. Lazaridis, N. Wex, A. Jessner, M. Kramer, B. W. Stappers, G. H. Janssen, G. Desvignes, M. B. Purver, I. Cognard, G. Theureau, A. G. Lyne, C. A. Jordan and J. A. Zensus. Generic tests of the existence of the gravitational dipole radiation and the variation of the gravitational constant. *Mon. Not. R. Astron. Soc.*, 400:805–814, 2009.

[133] D. C. Backer, S. R. Kulkarni, C. Heiles, M. M. Davis and W. M. Goss. A millisecond pulsar. *Nature*, 300:615–618, 1982.

[134] L. Shao, R. N. Caballero, M. Kramer, N. Wex, D. Champion and A. Jessner. A new limit on local lorentz invariance violation of gravity from solitary pulsars. *Class. Quantum Grav.*, 30:165019, 2013.

[135] N. Wex and M. Kramer. A characteristic observable signature of preferred-frame effects in relativistic binary pulsars. *Mon. Not. R. Astron. Soc.*, 380:455–465, 2007.

[136] J. F. Bell and T. Damour. A new test of conservation laws and Lorentz invariance in relativistic gravity. *Class. Quantum Grav.*, 13:3121–3127, 1996.

[137] L. Shao, N. Wex and M. Kramer. New tests of local Lorentz invariance and local position invariance of gravity with pulsars. *ArXiv e-prints*, 1211.6558, 2012.

[138] L. Shao and N. Wex. New limits on the violation of local position invariance of gravity. *Class. Quantum Grav.*, 30:165020, 2013.

[139] J.-P. Uzan. Varying Constants, Gravitation and Cosmology. *Living Rev. Relativity*, 14, 2011. www.livingreviews.org/lrr-2011-2.

[140] T. Damour, G. W. Gibbons and J. H. Taylor. Limits on the variability of G using binary-pulsar data. *Phys. Rev. Lett.*, 61:1151–1154, 1988.

[141] F. Hofmann, J. Müller and L. Biskupek. Lunar laser ranging test of the Nordtvedt parameter and a possible variation in the gravitational constant. *Astron. Astrophys.*, 522:L5, 2010.

[142] Planck Collaboration, P. A. R. Ade, N. Aghanim, C. Armitage-Caplan, M. Arnaud, M. Ashdown, F. Atrio-Barandela, J. Aumont, C. Baccigalupi, A. J. Banday, and et al. Planck 2013 results. I. Overview of products and scientific results. *ArXiv e-prints*, 1303.5062, 2013.

[143] K. Nordtvedt. $\dot{G}/G$ and a cosmological acceleration of gravitationally compact bodies. *Phys. Rev. Lett.*, 65:953–956, 1990.

[144] J. M. Lattimer and M. Prakash. Neutron Star Structure and the Equation of State. *Astrophys. J.*, 550:426–442, 2001.

[145] A. T. Deller, J. P. W. Verbiest, S. J. Tingay and M. Bailes. Extremely high precision VLBI astrometry of PSR J0437-4715 and implications for theories of gravity. *Astrophys. J.*, 685:L67–L70, 2008.

[146] S. Weinreb, J. Bardin, H. Mani and G. Jones. Matched wideband low-noise amplifiers for radio astronomy. *Review of Scientific Instruments*, 80:044702, 2009.

[147] L. Blanchet and G. Schäfer. Higher order gravitational radiation losses in binary systems. *Mon. Not. R. Astron. Soc.*, 239:845–867, 1989.

[148] L. Blanchet and G. Schäfer. Higher order gravitational radiation losses in binary systems: Erratum. *Mon. Not. R. Astron. Soc.*, 242:704, 1990.

[149] R. Nan, D. Li, C. Jin, Q. Wang, L. Zhu, W. Zhu, H. Zhang, Y. Yue, and L. Qian. The Five–Hundred-Meter Aperture Spherical Radio Telescope (FAST) Project. *International Journal of Modern Physics D*, 20:989–1024, 2011.

[150] A. R. Taylor. The square kilometre array. *Proceedings of the International Astronomical Union*, 8:337–341, 2012.

[151] R. Smits, M. Kramer, B. Stappers, D. R. Lorimer, J. Cordes and A. Faulkner. Pulsar searches and timing with the Square Kilometre Array. *Astron. Astrophys.*, 493:1161–1170, 2009.

[152] R. Smits, S. J. Tingay, N. Wex, M. Kramer and B. Stappers. Prospects for accurate distance measurements of pulsars with the Square Kilometre Array: Enabling fundamental physics. *Astron. Astrophys.*, 528:A108, 2011.

[153] M. A. C. Perryman, K. S. de Boer, G. Gilmore, E. Høg, M. G. Lattanzi, L. Lindegren, X. Luri, F. Mignard, O. Pace and P. T. de Zeeuw. GAIA: Composition, formation and evolution of the Galaxy. *Astron. Astrophys.*, 369:339–363, 2001.

[154] S. M. Ransom, J. M. Cordes and S. S. Eikenberry. A new search technique for short orbital period binary pulsars. *Astrophys. J.*, 589:911–920, 2003.

[155] B. Allen, B. Knispel, J. M. Cordes, J. S. Deneva, J. W. T. Hessels, D. Anderson, C. Aulbert, O. Bock, A. Brazier, S. Chatterjee, P. B. Demorest, H. B. Eggenstein, H. Fehrmann, E. V. Gotthelf, D. Hammer, V. M. Kaspi, M. Kramer, A. G. Lyne, B. Machenschalk, M. A. McLaughlin, C. Messenger, H. J. Pletsch, S. M. Ransom, I. H. Stairs, B. W. Stappers, N. D. R. Bhat, S. Bogdanov, F. Camilo, D. J. Champion, F. Crawford, G. Desvignes, P. C. C. Freire, G. Heald, F. A. Jenet, P. Lazarus, K. J. Lee, J. van Leeuwen, R. Lynch, M. A. Papa, R. Prix, R. Rosen,

P. Scholz, X. Siemens, K. Stovall, A. Venkataraman and W. Zhu. The Einstein@Home Search for Radio Pulsars and PSR J2007+2722 Discovery. *Astrophys. J.*, 773:91, 2013.

[156] N. Wex and S. M. Kopeikin. Frame dragging and other precessional effects in black hole pulsar binaries. *Astrophys. J.*, 514:388–401, 1999.

[157] K. Liu. *Exploring Pulsar-Black Hole Binaries Using The Next Generation of Radio Telescopes.* PhD thesis, University of Manchester, UK, 2012.

[158] N. Wex, K. Liu, R. P. Eatough, M. Kramer, J. M. Cordes and T. J. W. Lazio. Prospects for probing strong gravity with a pulsar-black hole system. *Proceedings of the International Astronomical Union*, 8:171–176, 2012.

[159] K. Liu, N. Wex, M. Kramer, J. M. Cordes and T. J. W. Lazio. Prospects for probing the spacetime of Sgr A* with pulsars. *Astrophys. J.*, 747:1, 2012.

[160] R. P. Eatough, H. Falcke, R. Karuppusamy, K. J. Lee, D. J. Champion, E. F. Keane, G. Desvignes, D. H. F. M. Schnitzeler, L. G. Spitler, M. Kramer, B. Klein, C. Bassa, G. C. Bower, A. Brunthaler, I. Cognard, A. T. Deller, P. B. Demorest, P. C. C. Freire, A. Kraus, A. G. Lyne, A. Noutsos, B. Stappers and N. Wex. A strong magnetic field around the supermassive black hole at the centre of the Galaxy. *Nature*, 501: 391–394, 2013.

[161] G. Hobbs, A. Archibald, Z. Arzoumanian, D. Backer, M. Bailes, N. D. R. Bhat, M. Burgay, S. Burke-Spolaor, D. Champion, I. Cognard, W. Coles, J. Cordes, P. Demorest, G. Desvignes, R. D. Ferdman, L. Finn, P. Freire, M. Gonzalez, J. Hessels, A. Hotan, G. Janssen, F. Jenet, A. Jessner, C. Jordan, V. Kaspi, M. Kramer, V. Kondratiev, J. Lazio, K. Lazaridis, K. J. Lee, Y. Levin, A. Lommen, D. Lorimer, R. Lynch, A. Lyne, R. Manchester, M. McLaughlin, D. Nice, S. Oslowski, M. Pilia, A. Possenti, M. Purver, S. Ransom, J. Reynolds, S. Sanidas, J. Sarkissian, A. Sesana, R. Shannon, X. Siemens, I. Stairs, B. Stappers, D. Stinebring, G. Theureau, R. van Haasteren, W. van Straten, J. P. W. Verbiest, D. R. B. Yardley and X. P. You. The International Pulsar Timing Array project: using pulsars as a gravitational wave detector. *Class. Quantum Grav.*, 27:084013, 2010.

[162] B. Bertotti, B. J. Carr and M. J. Rees. Limits from the timing of pulsars on the cosmic gravitational wave background. *Mon. Not. R. Astron. Soc.*, 203:945–954, 1983.

[163] S. M. Kopeikin. Binary pulsars as detectors of ultralow-frequency gravitational waves. *Phys. Rev. D*, 56:4455–4469, 1997.

[164] F. B. Estabrook and H. D. Wahlquist. Response of Doppler spacecraft tracking to gravitational radiation. *General Relativity and Gravitation*, 6:439–447, 1975.

[165] S. Detweiler. Pulsar timing measurements and the search for gravitational waves. *Astrophys. J.*, 234:1100–1104, 1979.

[166] A. Sesana, A. Vecchio and C. N. Colacino. The stochastic gravitational-wave background from massive black hole binary systems: implications for observations with Pulsar Timing Arrays. *Mon. Not. R. Astron. Soc.*, 390:192–209, 2008.

[167] A. Sesana, A. Vecchio and M. Volonteri. Gravitational waves from resolvable massive black hole binary systems and observations with Pulsar Timing Arrays. *Mon. Not. R. Astron. Soc.*, 394:2255–2265, 2009.

[168] K. J. Lee, F. A. Jenet and R. H. Price. Pulsar Timing as a Probe of Non-Einsteinian Polarizations of Gravitational Waves. *Astrophys. J.*, 685:1304–1319, 2008.

[169] K. J. Lee, F. A. Jenet, R. H. Price, N. Wex and M. Kramer. Detecting massive gravitons using pulsar timing arrays. *Astrophys. J.*, 722:1589–1597, 2010.

[170] F. A. Jenet, A. Lommen, S. L. Larson and L. Wen. Constraining the properties of supermassive black hole systems using pulsar timing: Application to 3C 66B. *Astrophys. J.*, 606:799–803, 2004.

Jürgen Müller, Liliane Biskupek, Franz Hofmann, and Enrico Mai

Lunar laser ranging and relativity

1 Introduction

In 1969, a new era for studying Earth–Moon dynamics has started, when Apollo 11 astronauts deployed the first retroreflector on the lunar surface. With Apollo 14 and Apollo 15 in the early Seventies, two further reflector arrays were brought onto the Moon by US missions. Unmanned Soviet missions, Luna 17 and Luna 21, completed the lunar network by 1973 (see Figure 1, upper panel). Since the first returns of short laser pulses sent from observatories on Earth to reflector arrays on the Moon, a new space geodetic technique – Lunar Laser Ranging (LLR) – provides an ongoing time series of highly accurate Earth–Moon distance measurements. Over the years, the range precision has improved from the meter level of accuracy to the few millimeter level. The latter is possible, thanks to the big telescope at Apache Point (Apache Point Observatory Lunar Laser-ranging Operation, APOLLO), USA: [51]). Currently (in the year 2013), four active LLR sites track the Moon routinely: the McDonald Observatory in Texas, USA, the Observatoire de la Côte d'Azur, France, the APOLLO site in New Mexico, USA, and the Matera Laser Ranging station in Italy. In addition, the Haleakala station provided ranges from 1984 to 1990. The German Geodetic Observatory at Wettzell is still working on its system hoping to soon join the LLR tracking network (Figure 1, lower panel).

LLR tracking is quite similar to Satellite Laser Ranging (SLR) tracking. A full description of the latest implementation of an LLR apparatus can, e.g. be found in [51]. In brief, a laser (wavelength 532 nm) generates short pulses with a length of about 70 to 200 ps. One pulse carries approximately 10^{19} photons. The laser is transmitted from the telescopes (where APOLLO has a diameter of 3.5 m, that of Wettzell 0.7 m). The full aperture is utilized for beam transmission. A small portion of the outgoing beam is used to determine the exact departure time. After reflection on the passive retroreflectors on the Moon, a few photons again reach the telescope on the Earth after about 2.5 s. For the smaller telescopes, normally less than one photon is statistically received. The core of the receiver (detector) often is an avalanche photodiode array capable of high-precision timing of single photons. Using an ultrastable oscillator, i.e.

Jürgen Müller, Liliane Biskupek, Franz Hofmann, Enrico Mai: Institut für Erdmessung (IfE), Leibniz Universität Hannover, Schneiderberg 50, 30167 Hannover, Germany
Jürgen Müller: QUEST – Centre for Quantum Engineering and Space–Time Research, Leibniz Universität Hannover, Welfengarten 1, 30167 Hannover, Germany

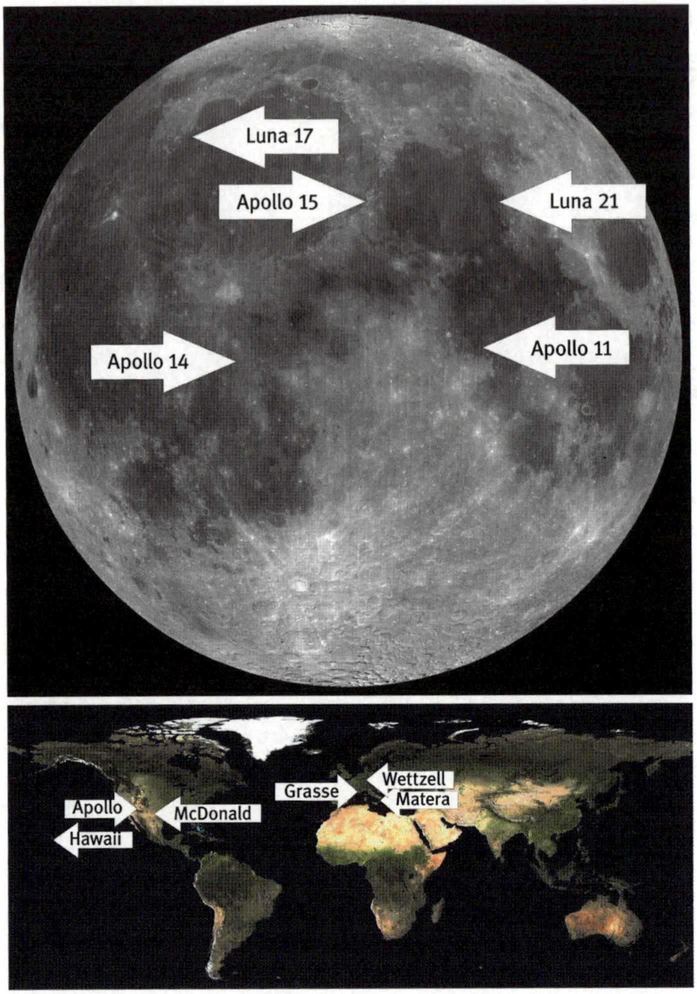

Fig. 1. Distribution of LLR observatories on the Moon (upper panel) and reflector arrays on Earth (lower panel).

an accurate atomic clock, the round-trip travel time is measured achieving mm range precision with the APOLLO telescope.

The ranges to be used in the analysis are so-called normal points (NP), i.e. a collection of round-trip travel times of single photons (this can be a few single returns up to a few thousands) obtained at one observatory to one reflector during a certain time span between 5 and 15 min. The bigger the telescope the more lunar returns are received. Special filter techniques (spatial, spectral, and temporal) are required to collect the right returns from the lunar reflectors and to separate the noise from the real lunar returns in the normal point. More details on this can be found in [50] and [42].

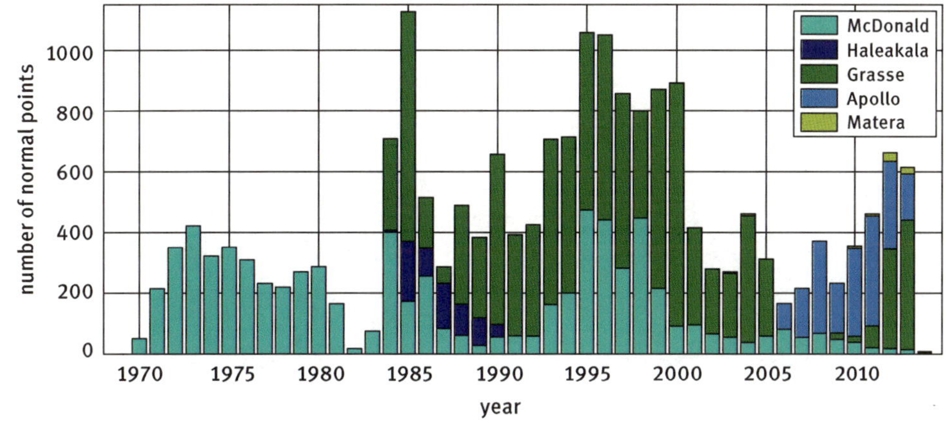

Fig. 2. Distribution of 20 061 LLR normal points taken by the major observatories over the years.

Figure 2 illustrates the amount of normal points obtained over the years 1970 to 2013, where the colours indicate the amount of data collected by each of the active LLR sites in each year. It is 20 061 normal points in total. Obviously, over long periods often only one observatory carried out LLR measurements, which is due to the difficulty to get successful lunar returns, if the equipment is not especially dedicated and optimized to lunar laser ranging. In addition, the size of the reflector array plays a central role. Also signal loss because of the big distance of about 380 000 km, beam divergence and atmospheric extinction lead to a poor tracking budget. The illuminated footprint at the Moon is about 20 km^2 where the hit reflector area is only 0.25 to 0.5 m^2. Therefore the biggest reflector of Apollo 15 is the one with – by far – the most returns. The measurement statistics of the observatories and reflectors for the whole time span is given in Figure 3. In the past years, the situation with respect to tracked reflectors has improved: Thanks to APOLLO and the upgraded French system, we see a steady increase of LLR NP, and also a much better coverage of all reflectors could be achieved than in the previous years.

The IfE LLR analysis model consists of a collection of sophisticated software modules at the cm level of accuracy. The whole measurement process is modeled at appropriate post-Newtonian approximation, i.e. the orbits of the major bodies of the Solar System, the rotation of Earth and Moon, signal propagation, but also the involved reference and time systems as well as the time-variable positions of the observatories and reflectors.

In the next section, we will show, where relativity enters the LLR analysis and what the major classical (Newtonian) effects are such as gravity field of Earth and Moon, tidal effects, ocean loading, lunar tidal acceleration (that causes the increase of the Earth–Moon distance by about 3.8 cm/year), etc.

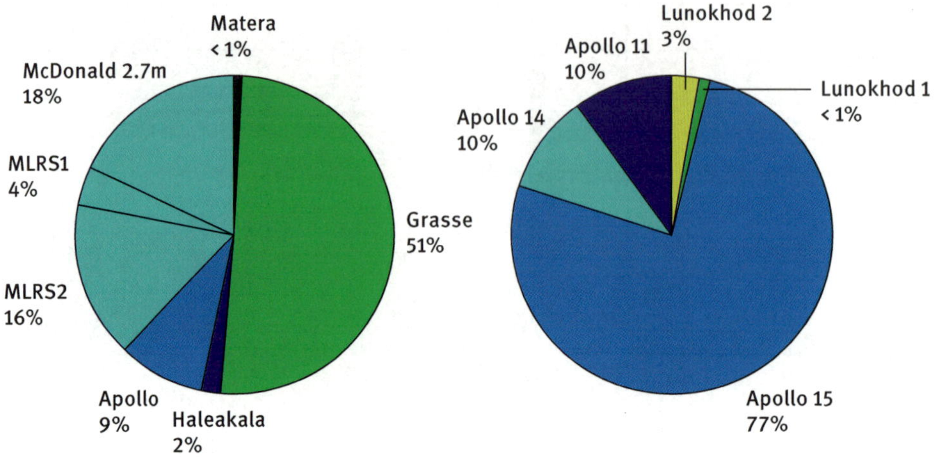

Fig. 3. Measurement statistics of observatories (left) and reflectors (right).

By analyzing the 44-year record of range data, LLR is able to provide, among others, a dynamical realization of the International Celestial Reference System, parameters related to the selenocentric and terrestrial reference frames, or (long-periodic) Earth Orientation Parameters, see Section 3.2.

Most prominently, LLR is also one of the best tools to test general relativity in the Solar System. It allows for constraining gravitational physics parameters related to the strong equivalence principle, Yukawa-like perturbations, preferred-frame effects, or the time variability of the gravitational constant. In Section 4, we will discuss recent results for the various relativistic parameters.

Former and current LLR data are electronically accessible through the CDDIS in Greenbelt, Maryland. LLR data analysis is mainly carried out by five major LLR analysis centers: Jet Propulsion Laboratory (JPL), Pasadena, USA; Center for Astrophysics (CfA), Cambridge, USA; Paris Observatory Lunar Analysis Center (POLAC), Paris, France; Institut de mécanique céleste et de calcul des éphémérides (IMCCE), Paris, France; Institute of Geodesy (IfE), Leibniz University of Hannover (LUH), Germany (see also [22]).

2 Model

At IfE the analysis model for the LLR data is based on Einstein's theory of gravity. It is fully relativistic and complete up to the first post-Newtonian ($1/c^2$) level, e.g. see [45, 49]. In Section 2.1, the general model is described and it is shown, where relativity enters the LLR analysis. A detailed discussion of the calculation of the ephemeris is given in Section 2.2.

2.1 Overview

The observation equation for the distance d between an observatory on Earth and a retroreflector on the Moon is one of the basic equations for the analysis of LLR data. It is implemented in the software of different analysis centers as outlined in [33, 41, 84]. At IfE it is given as

$$d = \frac{c}{2}\,\tau_{\mathrm{mes}} = \frac{c}{2}\,(\tau_{12} + \tau_{23} + \Delta\tau_{\mathrm{rel}} + \Delta\tau_{\mathrm{atmo}} + \Delta\tau_{\mathrm{syn}} + \Delta\tau_{\mathrm{syst}})\,, \tag{2.1}$$

where c is the speed of light and τ_{mes} the measured round trip travel time of laser pulses from the observatory to the Moon and back. In the analysis of LLR data three time epochs are considered: transmission time t_1 of the laser beam, reflection time t_2 at the retroreflector, and detection time of the returned signal at the observatory t_3. From the NP information of the LLR data only t_1 in UTC and τ_{mes} are given, see Section 1. t_2 and t_3 are obtained iteratively during the analysis. In that process transformations between different time scales (UT1, UTC, TAI, TT, TDB) must be realized as given, e.g. in [19, 35, 60, 73]. With all the determined time epochs, the pulse travel time $\tau_{12} = t_2 - t_1$ between observatory and retroreflector can be calculated as well as $\tau_{23} = t_3 - t_2$ for the way back. The separation is needed because of the motion of Earth and Moon during the measurement. $\Delta\tau_{\mathrm{rel}}$ describes corrections of the pulse travel time according to the relativistic Shapiro time delay due to the gravity fields of Earth and Sun [39] and according to the relativistic time transformation from terrestrial time TT to dynamical barycentric time TDB [19]. $\Delta\tau_{\mathrm{atmo}}$ corrects for the atmospheric time delay [37]. $\Delta\tau_{\mathrm{syn}}$ corrects the synodic oscillation of the geocentric lunar motion caused by solar radiation and thermal force effects [76], and $\Delta\tau_{\mathrm{syst}}$ considers systematic errors in the measurement equipment.

As shown in Figure 4, the distance

$$d = \left|\mathbf{r}_{\mathrm{EM}}(t_2) - \mathbf{r}_{\mathrm{obs}}(t_2) + \mathbf{r}_{\mathrm{ref}}(t_2)\right| \tag{2.2}$$

can also approximately be expressed by the vectors $\mathbf{r}_{\mathrm{EM}}$, which connects the geocenter and selenocenter, $\mathbf{r}_{\mathrm{obs}}$ representing the geocentric position of the observatory and $\mathbf{r}_{\mathrm{ref}}$ the selenocentric position of the retroreflector. All vectors are given in the relativistically defined Barycentric Celestial Reference System (BCRS) which is the reference system for the analysis of LLR data. Equation (2.2) is analyzed for the reflection time t_2 in the timescale TDB. As described in Section 3.1, it serves as the basic equation to calculate the partial derivatives for the analysis.

The Earth–Moon vector $\mathbf{r}_{\mathrm{EM}}$ in (2.2) is obtained by numerical integration of the relativistic Einstein–Infeld–Hoffmann (EIH) equations of motion, which are described in detail in Section 2.2. There the effects of the Sun, Earth, Moon, the other planets of the Solar System and the asteroids Ceres, Vesta, and Pallas are included in the standard solution. It is possible to account for further 13 asteroids in the integration. After several tests related to modeling the gravity field of Earth and Moon [42], additional Newtonian accelerations due to the gravity field up to degree and order 4 for the Earth

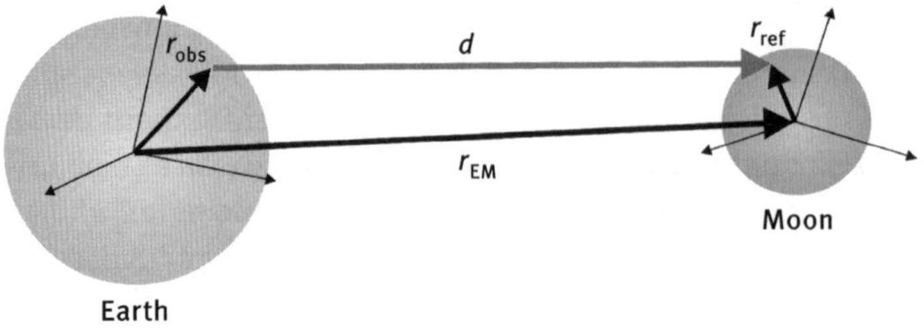

Fig. 4. LLR measurement setup with basis vectors.

and degree and order 5 for the Moon are included as well as the effect from the secular tidal acceleration [3].

The vectors $\mathbf{r}_{obs}$ of the observatory and $\mathbf{r}_{ref}$ of the retroreflector in equation (2.2) are originally given in their respective body-fixed reference systems, the International Terrestrial Reference System (ITRS) and selenocentric principal axis system PAS. To be used in equation (2.2), they have to be transformed into the space-fixed reference systems, the Geocentric Celestial Reference System (GCRS) for the Earth and the Selenocentric Celestial Reference System (SCRS) for the Moon with

$$\begin{aligned} \mathbf{r}_{obs}^{GCRS} &= \mathbf{R}_e(t)\, \mathbf{r}_{obs}^{ITRS} \\ \mathbf{r}_{ref}^{SCRS} &= \mathbf{R}_m(t)\, \mathbf{r}_{ref}^{PAS} . \end{aligned} \tag{2.3}$$

The SCRS is a reference system in the LLR analysis, which is defined analogously to the GCRS as space-fixed system with origin in the selenocenter and orientation of the axis parallel to the BCRS axis as recommended by the IAU [21].

For that first step of the transformation, the Earth orientation parameters (EOP) as described in [34] are used to build the rotation matrix $\mathbf{R}_e(t)$, including precession, nutation, Earth rotation, and polar motion. The angles of the rotation matrix $\mathbf{R}_m(t)$ are obtained by numerically integrating the Euler–Liouville equations for the Moon simultaneously with the EIH equations. Here, effects of the gravity field up to degree and order 4 for the Earth and degree and order 5 for the Moon are included, furthermore the relativistic effects of the geodetic and Lense–Thirring precession [45]. The Moon is modeled as elastic dissipative body [3, 8] with a liquid core [80].

Regarding vector $\mathbf{r}_{obs}^{ITRS}$, several effects induce time-varying displacements of the observatory. For the exact position vector of the observatory, following effects given in [60] are taken into account:

- solid Earth tides,
- ocean loading,
- atmospheric pressure loading,
- rotation deformation due to polar motion,
- motion according due to plate tectonic,
- and variations in latitude in the years 1970–1985 [11].

Analog to the time-varying displacements of the observatory also for the reflector coordinates $\mathbf{r}_{\mathrm{ref}}^{\mathrm{PAS}}$, the tidal displacement due to the Earth and Sun are considered [60].

After the first step of the transformation as shown in equation (2.3), in a second step the Lorentz–Einstein contraction of the vectors $\mathbf{r}_{\mathrm{obs}}^{\mathrm{GCRS}}$ and $\mathbf{r}_{\mathrm{ref}}^{\mathrm{SCRS}}$ is applied [60]. That second step transforms the vectors from the GCRS and SCRS into the BCRS.

After all vectors are given in the right barycentric time and space coordinates, the analysis of the LLR data is carried out, see Section 3.

2.2 Ephemerides

Construction steps

Today, there are three major renowned ephemerides available, namely various versions of the American (JPL) DE [74], the Russian (IAA RAS) EPM [61], and French (IMCCE) INPOP [14]. At IfE LUH, we currently make an effort to implement INPOP-like routines for ephemeris computations in the course of our LLR analysis. In order to do so, we seek to follow individual construction steps that were discussed in [33]. Basically, the four necessary steps are: numerical integration of equations of motion (EOM) (resulting in a discrete/tabulated solution), approximation by Chebychev polynomials (resulting in a continuous/interpolated solution), reduction of observational data (performing a comparison of the model solution with real measurements), adjustment (leading to an improved/final solution), see Figure 5. We will indicate at every step, how close our recent IfE ephemeris already is to this standard case.

First, one models the trajectories of all bodies of interest, i.e. the systems' dynamics. Those trajectories are obtained by solving first-order ordinary differential equations (ODEs) for the state vectors which describe the system. Starting with initial values one determines the mutual interactions between the bodies, based on fundamental dynamical principles, e.g. Newtons axioms and his law of gravitation, or the theorem on the angular momentum. Despite the fact that relativistic effects are taken into account in the modeling of the dynamics and in the reduction of observations, it remains a classical (Newtonian) framework. Instead of exploiting the general relativistic concept of the deformation of spacetime, one introduces supplementary forces and moments within the EOM and applies correction terms for the propagation of light when reducing observational data.

Having defined the dynamical model, one numerically integrates the EOM, e.g. by means of an Adams method [17, 18] with a chosen step size h. This results in a tabulated solution, i.e. time series for the system's state vectors. Configuration files are employed that define the dynamical model (which asteroids are taken into account, what method is used to estimate orbital phase shifts due to relativity, etc.), or containing all nominal parameter values that enter the equations (the system's initial conditions, masses of bodies, Love numbers, phase lag times due to tides, etc.).

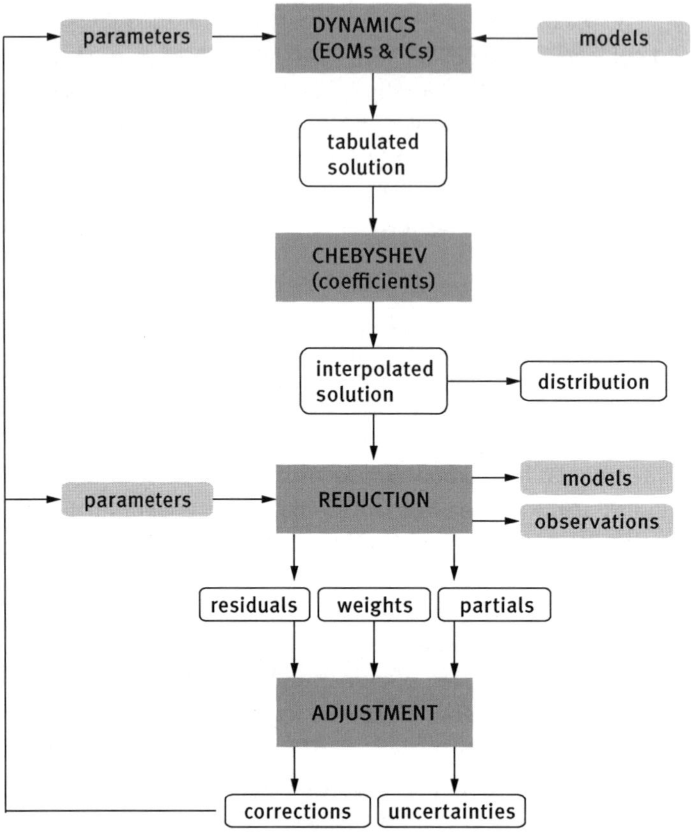

Fig. 5. Construction steps for an ephemeris.

In a next step, Chebychev polynomial coefficients are calculated to allow for a subsequent interpolation of the tabulated solution for any requested epoch of time. Another configuration file is used to specify body-dependent polynomial degrees and time-spans to which these coefficients are adjusted. The defining parameters are chosen such that the resulting interpolation errors (regarding the bodies' positions) do not exceed certain thresholds: e.g. 10 mm for the inner planets, 10 cm for the outer planets, 1 mm for the Earth, and 0.1 mm for the Moon. Variables that represent relatively short periodic motions (e.g. geocentric coordinates of the Moon) are approximated by using higher degree polynomials (e.g. degree 18), and reduced time-spans (e.g. 8 years). On the opposite, variables representing relatively long-periodic motions (e.g. heliocentric coordinates of Pluto) can sufficiently be approximated by using lower degree polynomials (e.g. degree 6), and extended time-spans (e.g. 32 years or even longer). The individual polynomial coefficients are well suited for the efficient distribution of an ephemeris.

The remaining major steps (reduction and adjustment/estimation) are addressed in the analysis section.

Force model

Within an N-body system, the acceleration of body A due to the gravitational attraction of the remaining bodies B is given by

$$\ddot{\mathbf{r}}_A = \sum_{B \neq A} \frac{\mu_B}{r_{AB}^3} (\mathbf{r}_B - \mathbf{r}_A) + \sum_\alpha \Delta\ddot{\mathbf{r}}_A^\alpha , \tag{2.4}$$

where the position vectors $\mathbf{r}$ are related to the Solar System barycenter (SSB), and the first summation term on the right-hand side denotes the classical mutual Newtonian attraction between point masses. The second term comprises all higher order effects (relativistic corrections, non-point mass effects due to mass inhomogeneities, figure–figure effects, etc.) that can be regarded as perturbations to the idealized Newtonian point mass modeling, i.e. $\sum_\alpha \Delta\ddot{\mathbf{r}}_A^\alpha = \Delta\ddot{\mathbf{r}}_A^{\text{rel}} + \Delta\ddot{\mathbf{r}}_A^{\text{fig}} + \cdots$.

Relativistic corrections are accounted for by parametrized post-Newtonian (PPN) modeling. Retaining only terms up to order $1/c^2$, one gets correction terms for the acceleration of any body A [78]:

$$\begin{aligned}
\Delta\ddot{\mathbf{r}}_A^{\text{rel}} = {} & \frac{1}{c^2} \sum_{B \neq A} \frac{\mu_B}{r_{AB}^3} (\mathbf{r}_B - \mathbf{r}_A) \Bigg(-2(\beta + \gamma) \sum_{C \neq A} \frac{\mu_C}{r_{AC}} - (2\beta - 1) \sum_{C \neq B} \frac{\mu_C}{r_{BC}} + \gamma \|\mathbf{v}_A\|^2 \\
& + (1+\gamma)\|\mathbf{v}_B\|^2 - 2(1+\gamma)\mathbf{v}_A \cdot \mathbf{v}_B - \frac{3}{2}\left(\frac{(\mathbf{r}_A - \mathbf{r}_B) \cdot \mathbf{v}_B}{r_{AB}} \right)^2 + \frac{1}{2} (\mathbf{r}_B - \mathbf{r}_A) \cdot \ddot{\mathbf{r}}_B \Bigg) \\
& + \frac{1}{c^2} \sum_{B \neq A} \frac{\mu_B}{r_{AB}^3} (\mathbf{v}_A - \mathbf{v}_B) \Big((\mathbf{r}_A - \mathbf{r}_B) \cdot \big((2 + 2\gamma)\mathbf{v}_A - (1 + 2\gamma)\mathbf{v}_B \big) \Big) \\
& + \frac{3 + 4\gamma}{2c^2} \sum_{B \neq A} \frac{\mu_B}{r_{AB}} \ddot{\mathbf{r}}_B .
\end{aligned} \tag{2.5}$$

Applying Einsteins theory of general relativity, i.e. setting $\beta, \gamma = 1$, leads to the usual Einstein–Infeld–Hoffmann (EIH) EOM. Remark: the parameters β and γ can be determined from LLR analysis.

The position and velocity vectors $\mathbf{r}$ and $\mathbf{v}$ are related to the system's barycenter in a relativistic sense. They can differ from the positions and velocities w.r.t. the fixed reference system of the integration if its origin is not connected to the barycenter. The only constraint on the origin is that it has to be inertial.

Regarding the positions, the relativistic correction terms make use of relative positions only. In the case of velocities, it is advisable to distinguish between barycentric velocities $\mathbf{v}_A$ and velocities with respect to the reference system's origin $\dot{\mathbf{r}}_A$. A special choice of the reference system will make this distinction expendable. The presence of $\ddot{\mathbf{r}}_B$ on the right-hand side of equation (2.5) implies an implicit definition of $\ddot{\mathbf{r}}_A$. As long as one is willing to neglect terms of higher order than $1/c^2$, one can avoid an iterative procedure by simply applying the Newtonian terms for the provision of $\ddot{\mathbf{r}}_B$. Knowing

$\mathbf{r}_B$ and $\dot{\mathbf{r}}_B$ then allows for the explicit and thus noniterative determination of $\ddot{\mathbf{r}}_A$:

$$\begin{aligned}\Delta\ddot{\mathbf{r}}_A^{\text{rel}} = &\frac{1}{c^2}\sum_{B\neq A}\frac{\mu_B}{r_{AB}^3}(\mathbf{r}_B-\mathbf{r}_A)\Bigg(-2(\beta+\gamma)\sum_{C\neq A}\frac{\mu_C}{r_{AC}}-(2\beta-1)\sum_{C\neq B}\frac{\mu_C}{r_{BC}}+\gamma\|\mathbf{v}_A\|^2\\&+(1+\gamma)\|\mathbf{v}_B\|^2-2(1+\gamma)\mathbf{v}_A\cdot\mathbf{v}_B-\frac{3}{2}\left(\frac{(\mathbf{r}_A-\mathbf{r}_B)\cdot\mathbf{v}_B}{r_{AB}}\right)^2\\&+\frac{1}{2}\sum_{C\neq B}\frac{\mu_C}{r_{BC}^3}(\mathbf{r}_C-\mathbf{r}_B)\cdot(\mathbf{r}_B-\mathbf{r}_A)\Bigg)\\&+\frac{1}{c^2}\sum_{B\neq A}\frac{\mu_B}{r_{AB}^3}(\mathbf{v}_A-\mathbf{v}_B)\Big((\mathbf{r}_A-\mathbf{r}_B)\cdot\big((2+2\gamma)\mathbf{v}_A-(1+2\gamma)\mathbf{v}_B\big)\Big)\\&+\frac{3+4\gamma}{2c^2}\sum_{B\neq A}\sum_{C\neq B}\frac{\mu_B}{r_{AB}}\frac{\mu_C}{r_{BC}^3}(\mathbf{r}_C-\mathbf{r}_B)\,.\end{aligned}\tag{2.6}$$

At IfE, this way of relativity modeling is implemented in the LLR analysis software.

As [33] points out, especially for the numerically precise calculation of long-term Solar System ephemerides, it is of importance to make a suitable choice of a reference for the expression of the bodies' position and velocity vectors. Translation of all position vectors by a common vector $\delta\mathbf{r}$ and of all velocity vectors by another common vector $\delta\mathbf{v}$ is equivalent to a modification of the initial conditions and thus a change of the reference, but the overall geometry of the system, i.e. the relative positions and velocities, will be conserved.

In the case of Newtonian interactions the position vector of the barycenter for a system of point masses is defined as

$$\mathbf{r}_{BC}=\sum_A\mu_A\mathbf{r}_A\,,\tag{2.7}$$

such that its second time derivative, assuming μ_A to be invariant in time and retaining only Newtonian terms, leads to the classical result of rectilinear uniform motion of the barycenter

$$\ddot{\mathbf{r}}_{BC}=\mathbf{0}\quad\rightarrow\quad\mathbf{r}_{BC}(t)=\mathbf{r}_{BC}(0)+\dot{\mathbf{r}}_{BC}(0)t\,,\tag{2.8}$$

where $t=0$ represents the initial epoch for the integration. Remark: the conservation of the system's total momentum represents 6 out of 10 well-known primary integrals of motion for N-body systems [65].

A theoretically defined reference system (RS) becomes physically realized in form of a reference frame (RF) by fixing the location of bodies (the system's geometry) for a given epoch. Let us denote the fixed distribution of bodies at certain instants of time t^* and $t_0=0$ as RF^* and RF_0, respectively.

If we choose $\mathbf{r}_{BC}(0)=\mathbf{0}$ and $\dot{\mathbf{r}}_{BC}(0)=\mathbf{0}$, then for the initial epoch of integration, the origin of RF_0 coincides with the system's barycenter.

Newtonian physics employs an absolute time which elapses uniformly and independent of the state of the bodies' masses. If the bodies move in time, the geometry of the system and thus its barycenter, in general, changes w.r.t. any reference frame RF^*. In order to ensure $\mathbf{r}_{BC}(t) = \mathbf{0}$ valid for all times w.r.t. a marked reference frame (e.g. RF_0), which is in fact of advantage (in terms of simplicity) for the description of the evolution of the system (and especially the calculation of relativistic corrections), one had to introduce a continuous transformation of the reference frame.

The idea of a changing reference is somewhat counter-intuitively in Newtonian physics and thus one can equivalently think of it as modifying the initial conditions of the bodies by the above-mentioned translation vectors $\delta\mathbf{r}$ and $\delta\mathbf{v}$, specifically

$$\delta\mathbf{r} = -\frac{\sum_A \mu_A \mathbf{r}_A}{\sum_A \mu_A}\,, \qquad \delta\mathbf{v} = -\frac{\sum_A \mu_A \dot{\mathbf{r}}_A}{\sum_A \mu_A}\,. \tag{2.9}$$

In doing so, one ensures that the barycenter stays at rest w.r.t. RF_0, whereas it remains in rectilinear uniform motion with respect to any general RF^*. From a computational point of view, it is also of advantage to avoid a secularly moving barycenter because otherwise, especially in long-term integrations, one may end up with body coordinates the numerical values of which by far exceed the values of its mutual distances. This could lead to numerical difficulties.

Switching to the relativistic case, one may use the same basic Newtonian expressions and additional relativistic corrections (2.6) but with the replacement of $\mathbf{v}_A$ by $\dot{\mathbf{r}}_A$. The latter velocities are valid for a general reference frame RF^* although, at this stage, RF^* is not yet consistent with the theory of general relativity.

The position vector of the relativistic barycenter for a system of point masses is defined as [33]

$$\mathbf{r}^*_{BC} = \sum_A \mu^*_A \mathbf{r}_A\,, \tag{2.10}$$

with

$$\mu^*_A = \mu_A \left(1 + \frac{\dot{\mathbf{r}}_A \cdot \dot{\mathbf{r}}_A}{2c^2} - \frac{1}{c^2}\sum_{B \neq A} \frac{\mu_B}{\|\mathbf{r}_A - \mathbf{r}_B\|}\right)\,. \tag{2.11}$$

In comparison to the Newtonian case, the relativistic barycenter position now additionally depends on the velocity vectors of the point masses, and the factor terms μ^*_A are itself time dependent. A more general discussion on the scaling of masses, spatial coordinates and time scales in relativistic modeling is presented in [26].

To study the EOM of the relativistic barycenter, we take the second derivative

$$\ddot{\mathbf{r}}^*_{BC} = \sum_A \left(\mu^*_A \ddot{\mathbf{r}}_A + 2\dot{\mu}^*_A \dot{\mathbf{r}}_A + \ddot{\mu}^*_A \mathbf{r}_A\right)\,. \tag{2.12}$$

Following [33] at least all terms of order $1/c^2$ in equation (2.12) cancel, and thus $\ddot{\mathbf{r}}^*_{BC} = \mathbf{0} + O(1/c^4)$. Neglecting the higher order terms, we again find the position vector of the (now relativistic) barycenter as an affine function of time:

$$\ddot{\mathbf{r}}^*_{BC} \approx \mathbf{0} \quad \rightarrow \quad \mathbf{r}^*_{BC}(t) \approx \mathbf{r}^*_{BC}(0) + \dot{\mathbf{r}}^*_{BC}(0)t\,. \tag{2.13}$$

As before, we want to choose the origin of a special reference frame RF_0 such that $\mathbf{r}^*_{BC}(0) = \mathbf{0}$ and $\dot{\mathbf{r}}^*_{BC}(0) = \mathbf{0}$. For the latter condition one has to take the first derivatives of equations (2.10) and (2.11). As in the case of the relativistic corrections itself, the right-hand side of the resulting equation, i.e. $\dot{\mu}^*_A$, contains terms with $\ddot{\mathbf{r}}_A$. As for the following expressions this would lead to terms of order $1/c^4$, we neglect them. Remark: a possibly exact cancellation of these terms requires a higher order development of the relativistic corrections and masses μ^*_A in advance.

To ensure a lasting coincidence of the reference frame's (RF_0) origin with the relativistic barycenter, one finally translates the bodies' initial conditions by using the vectors $-\delta\mathbf{r}$ (barycenter's position) and $-\delta\mathbf{v}$ (barycenter's velocity)

$$\delta\mathbf{r} = -\frac{\sum_A \mu^*_A \mathbf{r}_A}{\sum_A \mu^*_A}, \quad \delta\mathbf{v} = -\frac{\sum_A \mu^*_A \dot{\mathbf{r}}_A + \sum_A \dot{\mu}^*_A \mathbf{r}_A}{\sum_A \mu^*_A}, \tag{2.14}$$

where

$$\dot{\mu}^*_A \approx \frac{\mu_A}{2c^2} \sum_{B \neq A} \mu_B \frac{(\mathbf{r}_B - \mathbf{r}_A) \cdot (\dot{\mathbf{r}}_B + \dot{\mathbf{r}}_A)}{r^3_{AB}}. \tag{2.15}$$

Due to the dependency of μ^*_A and $\dot{\mu}^*_A$ on $\mathbf{r}_A$ and $\dot{\mathbf{r}}_A$, one has to solve this implicit formulation iteratively, which means that it takes successive transformations of the reference frame (or initial conditions of the bodies) until it eventually converges at RF_0. Regarding the number of necessary iterations, it turned out that 3 to 4 iterations are sufficient in the case of precise calculations with no more than 35 significant digits. In the end, we will again have a reference frame whose origin keeps coincident with the (relativistic) barycenter of the system of point masses, apart from minor numerical errors owing to the limited precision in computation. Employing RF_0 spares us from the determination of the barycenter's velocity.

As a remark: if the quantity $\dot{\mu}^*_A$ is neglected, as in case of some other ephemerides like DE405 or in some preliminary works on the construction of the first INPOP version (INPOP05), one obtains a different RF_0. On the other hand, the resulting effect of a moving barycenter w.r.t. the frame of integration is negligible for the geometry of the system. Barycentric positions may differ significantly, but heliocentric and geocentric (relative) positions remain very close. Regarding the LLR analysis at IfE, this effect is considered negligible for the Earth–Moon system, and thus it is not accounted for in the current operational software package.

The inaccuracy of actual observations by far exceeds any of the before-mentioned differences. In order to seek coherency of equations and a gain in precision, within later versions of INPOP these terms were included, especially since there is only a negligible additional expense. Furthermore, these calculations have to be performed only once, namely at the beginning of the integration for the determination of RF_0.

There exist two ways for the determination of the evolution of a N-body system. One either integrates the EOM for all N bodies (which implies the calculation of $(N-1)N$ interactions) with the barycenter left free to evolve within the reference frame (choice made at IMCCE for INPOP), or the EOM are integrated for $N-1$ bodies (which implies the calculation of $(N-1)^2$ interactions) such that the position of the remaining body ensues from a function of the other bodies' positions by forcing the barycenter to remain fixed at the reference frame's origin (choice made at JPL for DE). In the latter case, the position of the Sun can be determined as a function of the other Solar System bodies' coordinates, e.g. in the case of relativistically corrected interactions one may stipulate

$$\mathbf{r}_\odot = -\frac{\sum_{A\neq\odot}\mu_A^*\mathbf{r}_A}{\mu_\odot^*}, \quad \dot{\mathbf{r}}_\odot = -\frac{\sum_{A\neq\odot}\mu_A^*\dot{\mathbf{r}}_A + \sum_A \dot{\mu}_A^*\mathbf{r}_A}{\mu_\odot^*}. \tag{2.16}$$

Otherwise, these two vectors add six supplemental components to the system's overall state vector. The first method (INPOP case) requires an integration of the Sun's motion, whereas the second method (DE case) places the Sun according to equations (2.16). JPL/DE neglected the $\dot{\mu}_A^*$ terms [33], and thus their coordinates of the Sun result from incomplete expressions. At IfE we also integrate the Sun's EOM [3].

Besides the fundamental mutual interactions (Newtonian plus relativistic corrections), several minor effects $\Delta\ddot{\mathbf{r}}_A^\alpha$ have to be accounted for, which will now be shortly addressed without an extensive explicit presentation of corresponding formulas.

The preceding formulas are equally valid for point masses and spherically symmetric extended bodies. In some cases, this approximation is not sufficient. For example, especially in LLR analysis, the small distance between the Earth and the Moon (in comparison to the extension of the Solar System as a whole) does not allow us to neglect so-called figure–figure effects between these two bodies, or the lunar secular tidal acceleration.

In a first step of better approximation, one considers a non-spherical but somehow symmetrical body under perturbation by a point mass (or spherical extended body), where the term "spherical" here relates to the shape of the equipotential surfaces of the bodies' gravity fields, not to their geometrical figures.

To describe the gravity field of a body A, it is appropriate to introduce a body-fixed reference frame RF_A, the origin of which (O_A) coincides with the center of mass of that body. In general, an extended body will rotate and thus RF_A is mobile w.r.t. an inertial frame RF_0, which is chosen for integration.

The classical development of the extended body's potential U of the gravitational force $\mathbf{f} = \nabla U$ in spherical coordinates (r, θ, λ) (w.r.t. the body-fixed reference frame) depends on the total mass on the body M, its mean equatorial radius R, the spherical harmonic coefficients c_{nm}, s_{nm} of degree n and order m, the associated Legendre functions $P_{nm}(\theta)$, as well as $\cos m\lambda$ and $\sin m\lambda$.

The spherical harmonics depend on the body's internal mass distribution, i.e. on the description of the location of individual mass elements and thus on the actually chosen body-fixed reference frame RF_A. There exist various gravity models (sets of gravity field coefficients) for selected astronomical bodies, cf. NASA's Planetary Data

System [53]. For instance, gravity field models for the Moon were derived from tracking data of the Lunar Prospector spacecraft [27] or from the recent GRAIL mission [28].

The point mass expression μ/r is associated with $n, m = 0$. Furthermore, with our special choice of the coordinate system (located at the center of mass), all degree 1 terms vanish, i.e. $x_{cm}, y_{cm}, z_{cm} = 0$ or $c_{10}, c_{11}, s_{11} = 0$. The degree 2 spherical harmonics can physically be assigned to the elements of the body's real-symmetric inertia matrix (moment of inertia tensor)

$$\mathcal{I} = \int_V \begin{pmatrix} y'^2 + z'^2 & x'y' & x'z' \\ y'x' & x'^2 + z'^2 & y'z' \\ z'x' & z'y' & x'^2 + y'^2 \end{pmatrix} dm = \begin{pmatrix} I_{11} & I_{12} & I_{13} \\ I_{21} & I_{22} & I_{23} \\ I_{31} & I_{32} & I_{33} \end{pmatrix} = \begin{pmatrix} A & F & E \\ F & B & D \\ E & D & C \end{pmatrix}, \tag{2.17}$$

where primed quantities (also w.r.t. RF_A) relate to an infinitesimal mass element dm of the extended body. In equation (2.17), we applied the geodetic sign convention for the moments (A, B, C) and products (D, E, F) of inertia. The products of inertia are also called deviation moments. These would identically be zero if the body is rigid, i.e. if we assume that the body is free of any deformations (mass redistributions). All degree 2 spherical harmonics are directly related to the components of the inertia matrix. In case one can neglect polar motion, the mean axis of rotation coincides with the principal axis for the maximum moment of inertia, such that $D, E = 0$ (i.e. $c_{21}, s_{21} = 0$) and C becomes a principal moment of inertia. The dynamic form factor $c_{20} = -J_2$ represents the dynamical flattening of the body, whereas c_{22} and s_{22} characterize the deviation of its mass distribution from rotational symmetry, i.e. the ellipticity of the body's equator. In case of the Earth, c_{20} is by far the most dominant coefficient, whereas due to the synchronous rotation (tidal locking), the lunar c_{22} is more pronounced and of the same order of magnitude as the lunar dynamical flattening coefficient.

Conversely, the determination of the inertia matrix requires knowledge of degree 2 spherical harmonic coefficients and the principal moment of inertia (about the z-axis, axis of rotation of the body). Theoretically, one may always choose a reference frame (orthonormal basis for a coordinate system) where the inertia matrix becomes a diagonal matrix, i.e. where the choice of the axes directions leads to vanishing deviation moments, or $c_{21}, s_{21}, s_{22} = 0$, respectively. This implies a rigid body assumption.

In practice, the calculation of U and its partial derivatives can be performed based on representations in various sets of coordinates (Cartesian, spherical, etc.). To ease the computation as well as for a better identification and physical interpretation of the most significant terms, it is of advantage to separate the total potential into a sum of degree-(and possibly order-)dependent potential portions, e.g. $U = U_0 + U_1 + U_2 + \cdots$.

This distinction becomes important when looking at the most significant mutual interactions of bodies due to their extended figures (figure–figure effects).

Regarding the forces and moments resulting from the interaction between extended bodies and perturbing point masses, the non-point mass terms of the potential of an extended body give rise to perturbations in the translational and rotational motion of another bodies. Of course, the latter effect only applies to extended bodies.

Forces are expressed via the gradient of the potential. Thus, the force stemming from an extended body's gravitational potential U_A acting upon a point mass m_p can be written as $\mathbf{f}_{A\to p} = -m_p \nabla U_A(\mathbf{r}_p)$. To account for the figure effects in translational motion by adding a corresponding perturbation term to the EOM's right-hand side, the perturbing force has to be transformed from a body-fixed frame RF_A into the inertial frame RF_0 used for integration. Following Newton's axioms one gets the resulting effects on the perturbing body (point mass of mass m_p) and the extended body A (of total mass M_A), respectively:

$$\ddot{\mathbf{r}}_p = \frac{1}{m_p}\,\mathbf{f}_{A\to p}\,, \qquad \ddot{\mathbf{r}}_A = -\frac{1}{M_A}\,\mathbf{f}_{A\to p} = \frac{1}{M_A}\,\mathbf{f}_{p\to A}\,. \tag{2.18}$$

Besides the translational effect, there exists an effect on the momentum $\mathbf{M}_A$ of the extended body A due to the perturbing body's gravitational force $\mathbf{f}_{p\to A}$, because the individual forces $d\mathbf{f}_{p\to dm}$ regarding each mass element dm of A are, in general, not parallel to the direction joining the centers of mass of both bodies. Thus, in sum, it creates a resulting momentum $\mathbf{M}_{p\to A}$ on A. In the special case of a spherically symmetric body A, the resulting momentum would be zero.

Integrating $d\mathbf{M}_{p\to dm} = \mathbf{r}_p \times d\mathbf{f}_{p\to dm}$ over the volume of the extended body A yields $\mathbf{M}_{p\to A} = \mathbf{r}_p \times \mathbf{f}_{p\to A}$. Exemplarily looking at the effect of degree-2 potential terms on a momentum change one gets

$$\mathbf{M}_A^{n=2} = \mathbf{r}_p \times \mathbf{f}_{p\to A}^{n=2} = \mathbf{r}_p \times m_p \nabla U_2 = m_p \mathbf{r}_p \times \nabla U_2\Big|_{\mathbf{r}_p} = \frac{3Gm_p}{r_p^5}\,\mathbf{r}_p \times \mathfrak{I}\mathbf{r}_p\,. \tag{2.19}$$

We incorporated both, body-fixed as well as inertial reference frames. Their common use requires to relate them with respect to each other, either by introduction of orientation models (considering proper rotation, polar motion, precession nutation, etc.), or by simultaneous integration of the bodies' orientation and EOM. The latter option applies for recent INPOP ephemerides. At IfE, the actual treatment depends on the body. The Earth's orientation is modeled, whereas the lunar Euler angles' EOMs are integrated, we will come back to this issue in a later paragraph.

The orientation of an extended body, i.e. of its body-fixed reference frame RF_A w.r.t. the inertial frame RF_0 (used for EOM integration), can be expressed by the 3 Euler angles Φ (precession angle), Θ (nutation angle), and Ψ (proper rotation angle). In analogy to satellite orbit calculation one might identify the 1-axis of the body-fixed frame with the line of apsides in direction of the periapsis; then the angles Φ, Θ, Ψ correspond to the classical Keplerian elements Ω (right ascension of the orbit's ascending node), i (inclination of the orbit), and ω (argument of periapsis), respectively. In case, we identify the 1-axis with the satellite's radius vector, Ψ would correspond to u (argument of latitude). Various names for the Euler angles are in use.

The transformation matrices between RF_A and RF_0 (i.e. the orientation of one frame versus the other) can be obtained by multiplication of three successive elementary rotations [67], e.g. $T_{A\leftarrow 0} = R_3(\Psi)R_1(\Theta)R_3(\Phi)$, such that $\mathbf{b}_A = T_{A\leftarrow 0}\mathbf{b}_0$, where $\mathbf{b}_A$ and $\mathbf{b}_0$ denote the bases of the body-fixed and inertial reference frames, respectively. Here, we restrict ourselves to orthonormal coordinate systems and make use of Cartesian coordinates, therefore introducing the notation $\mathbf{b} = (\mathbf{i}_1, \mathbf{i}_2, \mathbf{i}_3) = (\mathbf{i}_x, \mathbf{i}_y, \mathbf{i}_z)$ with $\mathbf{i}_\alpha$ representing a unit vector for the indexed direction. In order to avoid an additional subindex (for A or 0), in the following we will use capital letters for quantities related to body-fixed frames, whereas small letters will imply an inertial frame: $\mathbf{b}_A = (\mathbf{I}_x, \mathbf{I}_y, \mathbf{I}_z)$ and $\mathbf{b}_0 = (\mathbf{i}_x, \mathbf{i}_y, \mathbf{i}_z)$.

In addition to the positions and velocities of the system's individual bodies one incorporates into the system's state vector all respective Euler angles of those bodies, whose orientation one wants to integrate (instead of modeling), e.g. the Moon.

In general, the Euler angles are subject to change in the course of time ($\dot{\Phi}, \dot{\Theta}, \dot{\Psi} \neq 0$), because the bodies change their orientation, most notably via a major rotation about the principal axis (with maximum moment of inertia).

Therefore, besides the translational EOM, one has to introduce additional second order differential equations for the Euler angles, that have to be integrated simultaneously. The equations for $\ddot{\Phi}$, $\ddot{\Theta}$ and $\ddot{\Psi}$ can be derived by introducing the instantaneous rotational vector $\mathbf{W}_A$ of a body A, and its coefficients will be related to the corresponding Euler equation $d(\mathcal{J}\mathbf{W}_A)/dt = \mathbf{M}_A - \mathbf{W}_A \times \mathcal{J}\mathbf{W}_A$, where

$$\mathbf{W} = \begin{pmatrix} W_x \\ W_y \\ W_z \end{pmatrix} = \begin{pmatrix} \dot{\Phi} \sin\Theta \sin\Psi + \dot{\Theta} \cos\Psi \\ \dot{\Phi} \sin\Theta \cos\Psi - \dot{\Theta} \sin\Psi \\ \dot{\Phi} \cos\Theta + \dot{\Psi} \end{pmatrix} \tag{2.20}$$

and consequently

$$\dot{\mathbf{W}} = \begin{pmatrix} \ddot{\Phi} \sin\Theta \sin\Psi + \dot{\Phi}\dot{\Theta} \cos\Theta \sin\Psi + \dot{\Phi}\dot{\Psi} \sin\Theta \sin\Psi + \ddot{\Theta} \cos\Psi - \dot{\Theta}\dot{\Psi} \sin\Psi \\ \ddot{\Phi} \sin\Theta \cos\Psi + \dot{\Phi}\dot{\Theta} \cos\Theta \cos\Psi - \dot{\Phi}\dot{\Psi} \sin\Theta \sin\Psi - \ddot{\Theta} \sin\Psi - \dot{\Theta}\dot{\Psi} \cos\Psi \\ \ddot{\Phi} \cos\Theta - \dot{\Phi}\dot{\Theta} \sin\Theta + \ddot{\Psi} \end{pmatrix}. \tag{2.21}$$

The instantaneous rotational vector can be used to ease the transformation from the inertial to the body-fixed state vector via $\mathbf{R}_A = T_{A\leftarrow 0}\, \mathbf{r}_A$ and $\dot{\mathbf{R}}_A = T_{A\leftarrow 0}\, \dot{\mathbf{r}}_A + T\, \mathbf{R}_A$ with

$$T = \begin{pmatrix} 0 & +W_z & -W_y \\ -W_z & 0 & +W_x \\ +W_y & -W_x & 0 \end{pmatrix}, \tag{2.22}$$

such that $\dot{\mathbf{R}}_A = T_{A\leftarrow 0}\dot{\mathbf{r}}_A - \mathbf{W}_A \times \mathbf{R}_A$. The subindex A specifies a certain body but not a certain body-fixed reference frame. We did not yet fix the reference frame's origin. The only indication of relating to an (arbitrary) body-fixed reference frame or to an inertial frame is given by the use of upper or lower case letters, respectively.

The second-order ODEs for the Euler angles can easily be retrieved by solving the linear system (2.21) for $\ddot{\Phi}$, $\ddot{\Theta}$, and $\ddot{\Psi}$. One gets a coupled equation system [33]

$$\begin{aligned} \ddot{\Phi} &= \frac{\dot{W}_x \sin\Psi + \dot{W}_y \cos\Psi + \dot{\Theta}(\dot{\Psi} - \dot{\Phi}\cos\Theta)}{\sin\Theta}, \\ \ddot{\Theta} &= \dot{W}_x \cos\Psi - \dot{W}_y \sin\Psi - \dot{\Psi}\dot{\Phi}\sin\Theta, \\ \ddot{\Psi} &= \dot{W}_z + \dot{\Theta}\dot{\Phi}\sin\Theta - \ddot{\Phi}\cos\Theta. \end{aligned} \tag{2.23}$$

The treatment of the coefficients $\dot{W}_x$, $\dot{W}_y$, and $\dot{W}_z$ depends on our assumptions on the rigidity of the bodies. In the absence of any deformations the inertia matrix is constant and real-symmetric, and thus can be diagonalized as mentioned before by choosing a suitable body-fixed frame RF_A, such that $c_{21}, s_{21}, s_{22} = 0$ and the inertia matrix takes the form $\mathcal{J} = \mathrm{diag}(A, B, C) = \mathrm{const.}$, such that $\dot{\mathcal{J}} = \mathbf{0}$, and the constant principal moments of inertia satisfy the inequality $A \leq B \leq C$, which implies that the z-axis is the principal axis of RF_A. For a rigid body we have $d(\mathcal{J}\mathbf{W})/dt = \mathcal{J}\dot{\mathbf{W}}$, and thus $\mathcal{J}\dot{\mathbf{W}} = \mathbf{M} - \mathbf{W} \times \mathcal{J}\mathbf{W}$. Solving for $\dot{W}_x$, $\dot{W}_y$, and $\dot{W}_z$ yields

$$\begin{aligned} \dot{W}_x &= \frac{M_x}{A} - \frac{C - B}{A} W_y W_z, \\ \dot{W}_y &= \frac{M_y}{B} - \frac{A - C}{B} W_x W_z, \\ \dot{W}_z &= \frac{M_z}{C} - \frac{B - A}{C} W_x W_y, \end{aligned} \tag{2.24}$$

which has to be introduced into equations (2.23).

In the case of non-rigid bodies, the inertia matrices are no longer constant and, in general, one cannot find favorable diagonalizations (i.e. suitable choices for RF_A) as before. For instance, in case of the Moon, at IfE one considers the body's elasticity and inner structure (core modeling), cf. [41, 80].

So far, we assumed minor changes of orientation of a body due to the perturbing gravitational attraction of a remote point mass. More generally, we have to account for an additional (most often comparatively fast) change of orientation of a body due to proper rotation (about its principal axis). This leads to a major change of the Euler angle Ψ. To a first approximation this effect is a linear function of time, which may require a special treatment [33] to avoid an accumulation of large numerical errors during the computation.

Instead of integrating the orientation one can enforce a kinematical model, where the orientation parameters (e.g. Euler angles) are given as explicit functions of time. Actually, the bodies are treated differently. In case of the Moon, its orientation angles are regarded as main solve-for parameters in the operational IfE LLR software package and they result numerically from a simultaneous integration of Euler's gyroscopic equations (as stated above) with the bodies' EOMs (as in INPOP). The consideration of the Earth's orientation parameters is analytical (via models) with a twofold handling of possible modeling errors [3]. Regarding these errors, the ones of proper rotation and polar motion are separated from the other (perturbing) rotations, especially the errors in precession and nutation. The latter rotations are adjusted afterwards, whereas in the ephemeris part they were originally neglected. In later software versions, at least the nutation corrections were included as well as in the ephemeris calculation [45]. In opposition to INPOP, at IfE the precession–nutation model directly enters the EOMs via the figure–figure effects, where the spherical harmonic coefficients of the secondary body (constant in that body's principal axes system) have to be expressed w.r.t. the primary body's principal axes system and thus become time dependent, in general. This transformation step requires the bodies' mutual orientation, which makes use of models. Details on the differing INPOP approach are given below.

The transformation matrix between body-fixed and inertial frames is composed of several steps. Its construction makes use of a precession–nutation model and a bias to account for differences between a true-of-date inertial frame and an inertial reference frame referring to a mean equator instead.

The origin and axes of the International Celestial Reference Frame (ICRF) refer to the Solar System's barycenter and directions toward remote extragalactic sources of negligible proper motion, according to IERS standards. The integration of the Solar System's state is performed w.r.t. this kinematically defined non-rotating reference frame. We also used body-fixed frames in order to apply (and determine) body-related parameters, e.g. gravity field coefficients, that otherwise would be hard to physically interpret. Depending on the actual parameters in question, various body-fixed frames might be incorporated where any of these can be related to the commonly used International Terrestrial Reference Frame (ITRF) via given ICRS/ITRS transformation formulas [34].

Regarding Earth's proper rotation, INPOP only considers zonal terms in the development of Earth's gravitational potential and thus assumes rotational symmetry. Therefore, in the modeling of any gravitationally induced perturbations within the transition between reference frames, its terms will be independent of the perturbing bodies' longitudes which is equivalent to neglecting the Earth's rotation about itself. At IfE, refinements of the force model have been studied, regarding the significance of higher degree terms as well as of non-zonal terms [41].

The above-mentioned transformation comprises several models with parameter values which are only valid for a limited time span. Therefore, especially important to long-term numerical integrations of the Solar System's state, INPOP does not make use of a modeling for the bodies' orientation but instead integrates their orientation simultaneously with the translational EOM. Remark: at IfE this is done solely in case of the Moon, i.e. for the lunar Euler angles. Regarding the Earth, only the first (sub-)version(s) of INPOP incorporated certain precession–nutation models in order to perform studies on the consequences of varying nominal values therein.

For the next level of approximation, we consider the deformation of extended bodies. Forces acting upon an extended body's mass elements can vary in both strength and direction and thus generate internal stresses that may deform the body depending on its rigidity. Potential coefficient values depend on the chosen (body-fixed) reference frame and the body's internal distribution of mass. Deformations lead to time-variable coefficients and tensor of inertia, respectively. Corresponding expressions for any changes of the coefficients due to solid body tides as well as rotation variations can be derived, which eventually modifies the EOMs.

Tidal effects between two bodies or two mass elements of a common body arise due to the gravitational forces that are generated by an external body. Massive generating bodies induce deformations of extended bodies, the effect of which we call solid tides.

Using a (deformable-)body-fixed RF, and assuming the generating body being a point mass with geographical coordinates $P_g(r_g, \theta_g, \lambda_g)$, the tide-generating potential at a point $P(r, \theta, \lambda)$ of the deformable body due to a mass M_g (or $\mu_g = GM_g$) at P_g reads [2, 33, 67]

$$V(r, \theta, \lambda) = \sum_{n=0}^{\infty} V_n \tag{2.25}$$

with

$$V_n = -\frac{\mu_g}{r_g}\left(\frac{r}{r_g}\right)^n \sum_{m=0}^{n} (2 - \delta_{0m}) \frac{(n-m)!}{(n+m)!} P_{nm}(\cos\theta) P_{nm}(\cos\theta_g) \cos m(\lambda - \lambda_g) \,. \tag{2.26}$$

The first two harmonics do not lead to any deformations of the extended body at all, and one can start the development (series expansion) for the effective tide-generating potential at $n = 2$. Even the higher degree harmonics do not necessarily give raise to tidal effects. For instance, in case of a homogeneous extended body, all exerted forces due to the perturbing body finally remain independent on the location of the mass elements. Likewise, a uniform gravity field does not create any constraints on the distances between free-falling masses within it.

Assuming an elastic and non-dissipative body, each harmonic in the variation of its gravitational potential, evaluated at the surface, is proportional to the same harmonic in the tide-generating potential ($\Delta U_n(r = R, \theta, \lambda) \propto V_n$), and one can introduce degree-dependent Love numbers k_n as proportionality factors, such that in total [2, 68]

$$\Delta U(R, \theta, \lambda) = \sum_{n=2}^{\infty} k_n V_n \,. \tag{2.27}$$

Similarly, proportionality factors for the radial and tangential spatial displacement of surface points can be introduced, namely the Love- and Shida-numbers h_n and l_n, respectively [36].

After solving a Dirichlet-type problem one can finally express the variation $\Delta U(r,\theta,\lambda)$ via modifications of the spherical harmonic coefficients as

$$\Delta U(r,\theta,\lambda) = -\frac{GM}{r}\sum_{n=2}^{\infty}\left(\frac{R}{r}\right)^n \sum_{m=0}^{n} P_{nm}(\cos\theta)(\Delta c_{nm}\cos m\lambda + \Delta s_{nm}\sin m\lambda) \quad (2.28)$$

with [32–34]

$$\begin{Bmatrix}\Delta c_{nm}\\ \Delta s_{nm}\end{Bmatrix} = (2-\delta_{0m})k_{nm}\frac{M_g}{M}\left(\frac{R}{r_g}\right)^{n+1}\frac{(n-m)!}{(n+m)!}P_{nm}(\cos\theta_g)\begin{Bmatrix}\cos m\lambda_g\\ \sin m\lambda_g\end{Bmatrix}, \quad (2.29)$$

where the Love numbers were generalized to be also order-dependent which is necessary for inelastic bodies. In case of elasticity, one replaces k_{nm} by k_n.

In reality, the reaction of a body to an external excitation does not happen instantaneously. Rather, due to frictional forces in the (Earth's) viscous mantle and at the ocean floors, there will be a timely retardation τ_{nm}, the magnitude of which depends on the individual harmonics' degree and order. Thus, any deformation at epoch t is due to a perturbing body's position at epoch $t-\tau_{nm}$, i.e. one has to apply $\mathbf{r}_g(t-\tau_{nm}) = \tilde{\mathbf{r}}_g = (\tilde{x}_g, \tilde{y}_g, \tilde{z}_g)^T$ instead of $\mathbf{r}_g(t)$. Any difference between these two vectors can be caused by the motion of the perturbing body within its orbit and/or by the proper rotation of the extended body about its principal axis. In general, the numerical integration refers to epoch t, therefore it is difficult to get an estimation for $\tilde{\mathbf{r}}_g$, whereas it is easy to obtain $\mathbf{r}_g$.

The outlined method of incorporating tidal effects by modification of the spherical harmonic coefficients is more flexible than an alternative method which introduces the phase lag angle by additional perturbation terms in the EOM. The former approach is applied within INPOP, whereas the latter approach was being used for various DE versions. Also at IfE, currently a (single) lag angle is introduced which gives rise to a secular tidal acceleration term for the Moon [3].

The advantage of the INPOP method is the fact that it formally distinguishes between tide generating bodies and perturbing bodies in general; the (Earth's) potential coefficients at each instant are determined by a rigid part as well as a part stemming from (solid) tides. With modified coefficients one then calculates the interactions with any perturbing bodies. This approach, in comparison to the other one, enables the inclusion of more perturbing bodies at basically the same computational costs. Thus, it seems to be more efficient and flexible.

Further modifications of a deformable extended body's potential coefficients originate from proper rotation. The associated disturbing potential at point $P(\mathbf{r})$ of that body, in dependence on its instantaneous rotation vector $\mathbf{W}$, has to be expressed in spherical coordinates and finally reads

$$V(\mathbf{r}) = -\frac{1}{2}\left(r^2W^2 - (\mathbf{r}\cdot\mathbf{W})^2\right)$$
$$= -r^2\sum_{n=0}^{2}\sum_{m=0}^{n} P_{nm}(\cos\theta)\left(\alpha_{nm}\cos m\lambda + \beta_{nm}\sin m\lambda\right), \tag{2.30}$$

where the non-zero coefficients are [32, 33]

$$4\alpha_{00} = \tfrac{1}{3}W^2,\quad \alpha_{20} = \tfrac{1}{6}(W^2 - 3W_z^2),\quad \alpha_{21} = -\tfrac{1}{3}W_xW_z,\quad \alpha_{22} = \tfrac{1}{12}(W_y^2 - W_x^2),$$
$$\beta_{21} = -\tfrac{1}{3}W_yW_z,\quad \beta_{22} = -\tfrac{1}{6}W_xW_y. \tag{2.31}$$

Looking at individual degree-dependent terms, one finds resulting modifications Δc_{nm} and Δs_{nm}. Only terms of degree 0 and 2 need to be considered. For $n = 0$ one can conclude that each volume element is accelerated in a radial direction and thus the deformation of the body is spherically symmetric. The concept of mass conservation induces a corresponding change in the inertia matrix. For the degree-2 terms one applies the proportionality hypothesis as before ($\Delta U_{nm}(r = R, \theta, \lambda) \propto V_{nm}$), and takes degree- and order-dependent Love numbers k_{nm} as proportionality factors similar to equation (2.27), thus

$$\Delta U(R,\theta,\lambda) = \sum_{n=0}^{2}\sum_{m=0}^{n} k_{nm}V_{nm}. \tag{2.32}$$

Comparable to equation (2.28), after solving a Dirichlet-type problem, one expresses the variation $\Delta U(r,\theta,\lambda)$ via modifications of the spherical harmonic coefficients as

$$\Delta U(r,\theta,\lambda) = -\frac{GM}{r}\sum_{n=0}^{2}\left(\frac{R}{r}\right)^n\sum_{m=0}^{n} P_{nm}(\cos\theta)(\Delta c_{nm}\cos m\lambda + \Delta s_{nm}\sin m\lambda) \tag{2.33}$$

with

$$\Delta c_{nm} = \frac{R^3}{GM}k_{nm}\alpha_{nm},\qquad \Delta s_{nm} = \frac{R^3}{GM}k_{nm}\beta_{nm}. \tag{2.34}$$

Regarding the actual integration, it is advisable to consider the permanent deformation that is induced by a mean proper rotation $\overline{W}$ about the principle axis of inertia. As a consequence, the mean modification of c_{20} vanishes.

As in the tidal case, the reaction of the body is retarded by some time τ, i.e. accounting for dissipation requires us to use $\mathbf{W}(t-\tau) = \tilde{\mathbf{W}} = (\tilde{W}_x, \tilde{W}_y, \tilde{W}_z)^T$ instead of $\mathbf{W}(t)$. Similarly to the treatment of tides, the vector $\tilde{\mathbf{W}}$ is not accessible in a direct way. The Euler angles in the state vector (to be integrated numerically) refer to epoch t. One has to estimate phase shifts of the Euler angles, which then allow for the derivation of the phase-shifted instantaneous proper rotation vector.

At IfE, the phase shift is accounted for by the introduction of a single lag angle, and (as in case of the other perturbation effects like figure–figure interactions) it leads to an additional term on the EOMs' right-hand sides. For details we refer to [3].

The above-mentioned (INPOP) approach via the modification of the potential coefficients and inertia matrix eventually affects the EOMs too, of course, but indirectly. It avoids the introduction of additional right-hand side terms. Instead, all calculations keep formally the same, i.e. one can apply the already existing formulas. This equally pertains

- to the force (generated by an extended body) exerted on a point mass, and vice versa
- to the (reaction) force and moment (generated by a point mass) exerted on an extended body.

One merely has to modify the values of the potential coefficients as indicated, whereas the (total) masses remain constant. By doing so, one can apply the fundamental relations of dynamics as usual and determine the accelerations of the bodies as if they were all rigid.

Regarding Euler's equation, in the present case of a non-rigid body, now a time-variable inertia matrix has to be taken into account. One considers the additional term $\dot{\mathcal{I}}\mathbf{W}$ such that $d(\mathcal{I}\mathbf{W})/dt = \dot{\mathcal{I}}\mathbf{W} + \mathcal{I}\dot{\mathbf{W}} = \mathbf{M} - \mathbf{W} \times \mathcal{I}\mathbf{W}$.

In the following, we will distinguish the two cases (rigid vs. deformable body) by using a tilde to indicate a deformable body, i.e. its corresponding inertia matrix is denoted by $\widetilde{\mathcal{I}}$, and the corresponding potential coefficients are $\tilde{c}_{nm}$ and $\tilde{s}_{nm}$, respectively. The latter are additively composed of the associated rigid coefficients (c_{nm}, s_{nm}) and its modifications (Δc_{nm}, Δs_{nm}) due to any deformations.

Tidal effects lead to a redistribution of surface masses which conserves the trace of the inertia matrix, i.e. $\mathrm{tr}(\mathcal{I}) = \mathrm{tr}(\widetilde{\mathcal{I}})$. Therefore, we get a constraint and finally

$$\widetilde{\mathcal{I}} = \mathcal{I} + MR^2 \begin{pmatrix} \frac{1}{3}\Delta c_{20} - 2\Delta c_{22} & -2\Delta s_{22} & -\Delta c_{21} \\ -2\Delta s_{22} & \frac{1}{3}\Delta c_{20} + 2\Delta c_{22} & -\Delta s_{21} \\ -\Delta c_{21} & -\Delta s_{21} & -\frac{2}{3}\Delta c_{20} \end{pmatrix} . \tag{2.35}$$

Formally, at each instant in time, one can decompose the time-dependent inertia matrix $\mathcal{I}(t)$ into three additive portions, namely $\mathcal{I}_{\mathrm{rigid}}$ (fixed rigid body part), $\mathcal{I}_{\mathrm{tide}}(t)$ (time variable part due to solid tides), and $\mathcal{I}_{\mathrm{spin}}(t)$ (time variable part due to proper rotation/spinning), such that

$$\mathcal{I}(t) = \mathcal{I}_{\mathrm{rigid}} + \mathcal{I}_{\mathrm{tide}}(t) + \mathcal{I}_{\mathrm{spin}}(t) \quad \Rightarrow \quad \dot{\mathcal{I}}(t) = \dot{\mathcal{I}}_{\mathrm{tide}}(t) + \dot{\mathcal{I}}_{\mathrm{spin}}(t) , \tag{2.36}$$

and after separation of $\dot{\mathbf{W}}$ from $\mathbf{W}$ one obtains

$$\mathcal{I}(t)\dot{\mathbf{W}} = \mathbf{M} - \mathbf{W} \times \mathcal{I}(t)\mathbf{W} - \dot{\mathcal{I}}_{\mathrm{tide}}(t)\mathbf{W} - \dot{\mathcal{I}}_{\mathrm{spin}}(t)\mathbf{W} . \tag{2.37}$$

Retaining only potential terms up to degree 2, and applying the average mean motion of the body and furthermore dropping the harmonics' indicating subindex (which is obsolete in a simplified approach where one only considers a single Love number k and a single retardation or displacement time $t - \tau$), [33] explicitly provides

$$\mathcal{J}_{\text{tide}}(t) = M_g k \left(\frac{R}{r_g}\right)^5 \begin{pmatrix} \frac{1}{3}(-2x_g^2 + y_g^2 + z_g^2) & -x_g y_g & -x_g z_g \\ -x_g y_g & \frac{1}{3}(x_g^2 - 2y_g^2 + z_g^2) & -y_g z_g \\ -x_g z_g & -y_g z_g & \frac{1}{3}(x_g^2 + y_g^2 - 2z_g^2) \end{pmatrix} \tag{2.38}$$

and

$$\mathcal{J}_{\text{spin}}(t) = \frac{kR^5}{3G} \begin{pmatrix} W_x^2 - \frac{1}{3}(W^2 - n^2) & W_x W_y & W_x W_z \\ W_x W_y & W_y^2 - \frac{1}{3}(W^2 - n^2) & W_y W_z \\ W_x W_z & W_y W_z & W_z^2 - \frac{1}{3}(W^2 + 2n^2) \end{pmatrix}. \tag{2.39}$$

The final expressions, to be used in equation (2.37), are then

$$\begin{aligned} \dot{\mathcal{J}}_{\text{tide}} &= 5M_g k \left(\tfrac{R}{r_g}\right)^5 \tfrac{\mathbf{r}_g \cdot \dot{\mathbf{r}}_g}{r_g^2} \begin{pmatrix} x_g^2 - \frac{1}{3} r_g^2 & x_g y_g & x_g z_g \\ x_g y_g & y_g^2 - \frac{1}{3} r_g^2 & y_g z_g \\ x_g z_g & y_g z_g & z_g^2 - \frac{1}{3} r_g^2 \end{pmatrix} \\ &\quad - M_g k \left(\tfrac{R}{r_g}\right)^5 \begin{pmatrix} 2(x_g \dot{x}_g - \frac{1}{3} \mathbf{r}_g \cdot \dot{\mathbf{r}}_g) & \dot{x}_g y_g + x_g \dot{y}_g & \dot{x}_g z_g + x_g \dot{z}_g \\ \dot{x}_g y_g + x_g \dot{y}_g & 2(y_g \dot{y}_g - \frac{1}{3} \mathbf{r}_g \cdot \dot{\mathbf{r}}_g) & \dot{y}_g z_g + y_g \dot{z}_g \\ \dot{x}_g z_g + x_g \dot{z}_g & \dot{y}_g z_g + y_g \dot{z}_g & 2(z_g \dot{z}_g - \frac{1}{3} \mathbf{r}_g \cdot \dot{\mathbf{r}}_g) \end{pmatrix} \end{aligned} \tag{2.40}$$

and

$$\dot{\mathcal{J}}_{\text{spin}} = \frac{kR^5}{3G} \begin{pmatrix} 2(W_x \dot{W}_x - \frac{1}{3} \mathbf{W} \cdot \dot{\mathbf{W}}) & \dot{W}_x W_y + W_x \dot{W}_y & \dot{W}_x W_z + W_x \dot{W}_z \\ \dot{W}_x W_y + W_x \dot{W}_y & 2(W_y \dot{W}_y - \frac{1}{3} \mathbf{W} \cdot \dot{\mathbf{W}}) & \dot{W}_y W_z + W_y \dot{W}_z \\ \dot{W}_x W_z + W_x \dot{W}_z & \dot{W}_y W_z + W_y \dot{W}_z & 2(W_z \dot{W}_z - \frac{1}{3} \mathbf{W} \cdot \dot{\mathbf{W}}) \end{pmatrix}. \tag{2.41}$$

At this stage, we know how to calculate forces and moments (torques) due to the interactions between an extended body of arbitrary shape and a perturbing point mass. For a next level of approximation one considers the interactions with a rotationally symmetric body in fast rotation.

So far, we made use of a body-fixed RF for the coordinates of the perturbing body and, by knowing the transition matrix, of a space-fixed (inertial) RF. In case of a rotationally symmetric body (w.r.t. the principal (z-)axis of inertia), the formulas can be simplified, because one of the Euler angles, namely Ψ, can be eliminated. If non-zonal terms are taken into account [41] one has to apply the full transformation matrix. At IfE, rotational symmetry is applicable in case of the Earth.

The stepwise execution of the 3 elementary rotations of the transition $T_{A\leftarrow 0}$ leads to the intermediary bases

$$\mathbf{b}_1 = T_{1\leftarrow 0}\mathbf{b}_0 \quad \text{with } T_{1\leftarrow 0} = R_3(\Phi) \qquad \text{and} \qquad \mathbf{b}_2 = T_{2\leftarrow 1}\mathbf{b}_1 \quad \text{with } T_{2\leftarrow 1} = R_1(\Theta)\,, \tag{2.42}$$

where $\mathbf{b}_1 = (\mathbf{i}_z \times \mathbf{I}_z, \mathbf{i}_z \times (\mathbf{i}_z \times \mathbf{I}_z), \mathbf{i}_z)$ and $\mathbf{b}_2 = (\mathbf{i}_z \times \mathbf{I}_z, \mathbf{I}_z \times (\mathbf{i}_z \times \mathbf{I}_z), \mathbf{I}_z)$. Basis $\mathbf{b}_2$ refers to a body-fixed (mobile, i.e. non-inertial) RF, the principal (z-)axis of which is perpendicular to the body's equator. The Ψ-independent portion of $T_{A\leftarrow 0}$ constitutes the transition between $\mathbf{b}_0$ and $\mathbf{b}_2$, e.g.

$$\mathbf{b}_0 = T_{0\leftarrow 2}\mathbf{b}_2 \quad \text{with} \quad T_{0\leftarrow 2} = T_{0\leftarrow 1}T_{1\leftarrow 2} = R_3(-\Phi)R_1(-\Theta)\,. \tag{2.43}$$

Denoting the base vectors of the body-fixed RF (basis $\mathbf{b}_A$) $\mathbf{I}_x$, $\mathbf{I}_y$, and $\mathbf{I}_z$ by $\mathbf{I}$, $\mathbf{J}$, and $\mathbf{K}$, respectively, if they are expressed in coordinates w.r.t. the inertial RF (basis $\mathbf{b}_0$), the transition matrix $T_{0\leftarrow 2}$ can be formulated solely in dependence on $\mathbf{K} = (K_x, K_y, K_z)^T$.

In case of a rotationally symmetric body, the calculation of forces and moments does not necessarily require a transition to $\mathbf{b}_A$, but instead it can be performed within $\mathbf{b}_2$ or even directly within $\mathbf{b}_0$.

First we neglect any tide-raising portions (due to non-zonal terms) within the interactions and thus retain only the longitude-independent (rotationally symmetric) zonal terms of various degree.

Noting that $\cos\theta = (\mathbf{K}\cdot\mathbf{r})/r$, one can simplify the calculations by directly working in the inertial frame (being used for our prospective numerical integration of the EOM). Based on $\mathbf{b}_0$ one obtains expressions for the individual U_{n0}. Taking the respective gradients yields expressions for the perturbing accelerations $\ddot{\mathbf{r}}_{A_{n0}}$ of the extended body A (mass M_A) due to a zonal harmonic of degree n as well as for associated moments (torques), all of which are induced by the gravitational attraction of a perturbing point mass (mass m_P), namely

$$\begin{aligned}
\ddot{\mathbf{r}}_{A_{20}} &= -\frac{3Gm_P R^2 c_{20}}{2r^7}\Big(\big(r^2 - 5(\mathbf{K}\cdot\mathbf{r})^2\big)\mathbf{r} + 2r^2(\mathbf{K}\cdot\mathbf{r})\,\mathbf{K}\Big)\,,\\
\ddot{\mathbf{r}}_{A_{30}} &= -\frac{Gm_P R^3 c_{30}}{2r^9}\Big(5(\mathbf{K}\cdot\mathbf{r})\big(3r^2 - 7(\mathbf{K}\cdot\mathbf{r})^2\big)\mathbf{r} + 3r^2\big(5(\mathbf{K}\cdot\mathbf{r})^2 - r^2\big)\mathbf{K}\Big)\,,\\
\ddot{\mathbf{r}}_{A_{40}} &= -\frac{5Gm_P R^4 c_{40}}{8r^{11}}\Big(\big(42r^2(\mathbf{K}\cdot\mathbf{r})^2 - 63(\mathbf{K}\cdot\mathbf{r})^4 - 3r^4\big)\mathbf{r}\\
&\quad + 4r^2(\mathbf{K}\cdot\mathbf{r})\big(7(\mathbf{K}\cdot\mathbf{r})^2 - 3r^2\big)\mathbf{K}\Big)
\end{aligned} \tag{2.44}$$

and

$$
\begin{aligned}
\mathbf{M}_{A_{20}} &= \frac{3GM_A m_P R^2 c_{20}}{r^5}\,(\mathbf{K}\cdot\mathbf{r})\,\mathbf{K}\times\mathbf{r}\,,\\
\mathbf{M}_{A_{30}} &= \frac{3GM_A m_P R^3 c_{30}}{2r^7}\Big(5(\mathbf{K}\cdot\mathbf{r})^2 - r^2\Big)\mathbf{K}\times\mathbf{r}\,,\\
\mathbf{M}_{A_{40}} &= \frac{5GM_A m_P R^4 c_{40}}{2r^9}\,(\mathbf{K}\cdot\mathbf{r})\Big(7(\mathbf{K}\cdot\mathbf{r})^2 - 3r^2\Big)\mathbf{K}\times\mathbf{r}\,.
\end{aligned}
\tag{2.45}
$$

The corresponding perturbing accelerations $\ddot{\mathbf{r}}_{P_{n0}}$ of the perturbing body (point mass) P follow immediately from $m_P\ddot{\mathbf{r}}_{P_{n0}} = -M_A\ddot{\mathbf{r}}_{A_{n0}}$ (actio = reactio axiom). Relations (2.44) and (2.45) represent the rigid part only and are directly applicable w.r.t. basis $\mathbf{b}_0$.

In [3] the force model of the current IfE implementation for LLR analysis is given in detail.

In a next step, one considers the forces and moments that result from a deformation, which corresponds to a potential variation of form (2.28). Again, in a first approximation, we focus on the most significant effects, i.e. the degree-2 harmonics. Tidally induced variations Δc_{2m}, Δs_{2m} lead to a variation of the inertia matrix, where $\Delta\mathfrak{J} = \tilde{\mathfrak{J}} - \mathfrak{J}$ was already given by equation (2.35). This expression remains valid for any chosen basis and it is used in $\Delta U_2 = (3G/2r^5)\mathbf{r}\cdot\Delta\mathfrak{J}\mathbf{r}$, from which one derives the perturbing acceleration $\ddot{\mathbf{r}}_{P_{2m}}$ of the point mass P and the moment $\mathbf{M}_{A_{2m}}$ exerted upon the extended body A, both being due to the deformation of the latter body by tide-generating bodies:

$$
\ddot{\mathbf{r}}_{P_{2m}} = \frac{15G}{2r^7}(\mathbf{r}\cdot\Delta\mathfrak{J}\mathbf{r})\,\mathbf{r} - \frac{3G}{r^5}\,\Delta\mathfrak{J}\mathbf{r}\,, \qquad \mathbf{M}_{A_{2m}} = \frac{3Gm_P}{r^5}\,\mathbf{r}\times\Delta\mathfrak{J}\mathbf{r}\,. \tag{2.46}
$$

The realistic evaluation of the equations for Δc_{nm}, Δs_{nm} requires the use of $\tilde{\mathbf{r}}_g$, and because they relate to $\mathbf{b}_A$, we must consider $\tilde{\mathbf{R}}_g = \tilde{T}_{A\leftarrow 0}\tilde{\mathbf{r}}_g$ with $\tilde{T}_{A\leftarrow 0} = \tilde{T}_{A\leftarrow 2}\tilde{T}_{2\leftarrow 0} = R_3(\tilde{\Psi})\,R_1(\tilde{\Theta})R_3(\tilde{\Phi})$. Here, any quantity $\tilde{\alpha}$ implies a retardation, i.e. $\tilde{\alpha} = \alpha(t-\tau)$. In case of a fast rotating extended body one can assume a linear variation of the fast angle Ψ between epochs t and $t-\tau$. The body's rotational velocity ω_A (e.g. $\omega_\oplus$ in case of Earth) itself is not necessarily a constant. Furthermore, in comparison to Ψ and especially for short phase shift times τ, we can treat Θ and Φ as slow angles. Eventually, one applies the assumptions $\tilde{\Psi} \approx \Psi - \omega_A\tau$, $\tilde{\Phi} \approx \Phi$, and $\tilde{\Theta} \approx \Theta$, such that $\tilde{T}_{A\leftarrow 0} = R_3(-\omega_A\tau)R_3(\Psi)R_1(\Theta)R_3(\Phi)$.

A combination of before-mentioned equations yields the variation of the inertia matrix w.r.t. basis $\mathbf{b}_A$, i.e. $\Delta\mathfrak{J}_{\mathbf{b}_A}$, but in practice it is more convenient to use $\Delta\mathfrak{J}_{\mathbf{b}_0}$. In an intermediate step, we can calculate $\Delta\mathfrak{J}_{\mathbf{b}_2}$ in two different ways, either by $\Delta\mathfrak{J}_{\mathbf{b}_2} = R_3(-\Psi)\Delta\mathfrak{J}_{\mathbf{b}_A}R_3(\Psi)$, or by introducing phase-shifted coordinates (of the tide-generating body) w.r.t. basis $\mathbf{b}_2$, which could be obtained from $\tilde{\mathbf{r}}_g$ by applying the transformation matrix $R_3(-\omega_A\tau)\,T_{2\leftarrow 0}$. The rotational symmetry of the body is not of importance for the calculation of $\Delta\mathfrak{I}$ but for the calculation of the total matrix of inertia $\mathfrak{I}$. Without this symmetry, the rigid part $\mathfrak{I}_{\text{rigid}}$ would not be invariant against rotations of vector $\mathbf{K}$.

From the intermediate step we deduce likewise $\Delta\mathfrak{I}_{\mathbf{b}_0} = T_{0\leftarrow 2}\Delta\mathfrak{I}_{\mathbf{b}_2}T_{2\leftarrow 0}$. The expressions for Δc_{nm} and Δs_{nm} w.r.t. basis $\mathbf{b}_0$, based on the phase shift vector $\tilde{\mathbf{r}}_g$, are very complex. In the following we will assume that Δc_{2m} and Δs_{2m} remain given w.r.t. a body-fixed basis. Following formula (2.35) for $\Delta\mathfrak{I}$, one can express $\Delta\mathfrak{I}_{\mathbf{b}_2}$ due to the individual harmonics directly in dependence on the base vectors $\mathbf{I}$, $\mathbf{J}$, $\mathbf{K}$ [33]:

$$
\begin{aligned}
2\Delta\mathfrak{I}_{\mathbf{b}_2} &= \frac{1}{3}MR^2\Delta c_{20}(\mathbf{I}_3 - 3\mathbf{K}^T\mathbf{K})\,,\\
\Delta\mathfrak{I}_{\mathbf{b}_2} &= -MR^2\Delta c_{21}(\mathbf{I}^T\mathbf{K} + \mathbf{K}^T\mathbf{I})\,,\\
\Delta\mathfrak{I}_{\mathbf{b}_2} &= -MR^2\Delta s_{21}(\mathbf{J}^T\mathbf{K} + \mathbf{K}^T\mathbf{J})\,,\\
\Delta\mathfrak{I}_{\mathbf{b}_2} &= 2MR^2\Delta c_{22}(\mathbf{J}^T\mathbf{J} - \mathbf{I}^T\mathbf{I})\,,\\
\Delta\mathfrak{I}_{\mathbf{b}_2} &= -2MR^2\Delta s_{22}(\mathbf{I}^T\mathbf{J} + \mathbf{J}^T\mathbf{I})\,.
\end{aligned}
\tag{2.47}
$$

The matrices within the brackets of relations (2.47) are fully occupied when expressed in $\mathbf{b}_0$, such that, in essence, every single potential coefficient with respect to this basis will depend on all individual Love numbers k_{nm} and phase shift times τ_{nm}.

To summarize, we did not need to apply basis $\mathbf{b}_A$ but could use basis $\mathbf{b}_2$ to calculate any forces and moments, and thus we got rid of the fast-changing angle Ψ. We assumed period(s) of ω_A being longer than the phase shift time τ, i.e. Ψ could be treated as linearly changing with time between epochs t and $t-\tau$. This assumption is not in conflict with a secular slowdown of Earth's proper rotation. Secondly, we approximated the true-of-date equator by the equator at $t-\tau$. Eventually, the easiest way to calculate $\Delta\mathfrak{I}_{\mathbf{b}_0}$ is to calculate $\Delta\mathfrak{I}_{\mathbf{b}_2}$ and subsequently apply a change of basis from $\mathbf{b}_2$ to $\mathbf{b}_0$.

The orientation of a rotating body can be described by a differential equation, too. Retaining only moments of degree 2 one gets [33], cf. equations (2.45),

$$
\dot{\bar{\mathbf{H}}}{}^u_A = \frac{3(2-3\sin^2 J)(C-A)}{2H_A}\sum_{B\neq A}\frac{GM_B}{r_B^5}\,(\mathbf{r}_B\cdot\bar{\mathbf{H}}{}^u_A)\,\mathbf{r}_B\times\bar{\mathbf{H}}{}^u_A\,,
\tag{2.48}
$$

where $\mathbf{H}_A$ denotes a body's angular momentum vector, $\bar{\mathbf{H}}_A$ the mean (with respect to the fast angle Ψ) of the unit vector $\mathbf{H}^u_A = \mathbf{H}_A/H_A$, and J represents the (mean) angle between $\mathbf{K}$ and $\mathbf{H}_A$. The index B denotes any perturbing bodies, whereas C and A are again the principal moments of inertia of the extended body A. In the case of a non-symmetric body, one replaces $(C-A)$ by the more general factor $C-(A+B)/2$. The vector $\mathbf{K}$ precesses around the unit vector $\bar{\mathbf{H}}_A$. As we assume fast rotation, angle J becomes small and its corresponding term can be neglected.

Applying the theorem of angular momentum in an inertial RF, $\dot{\mathbf{H}}$ equals the sum of all external moments that are exerted upon our extended body, i.e. $\dot{\mathbf{H}} = \sum \mathbf{M}_{\text{external}}$. $\mathbf{M}_{\text{external}}$ may comprise any moments $\mathbf{M}_{A_{nm}}$ as given in equations (2.45) and (2.46).

In general, including higher degree moments and those due to deformation, with equations (2.45) as well as the above-mentioned transition $\Delta\mathfrak{J}_{\mathbf{b}_0} \leftarrow \Delta\mathfrak{J}_{\mathbf{b}_2}$, one obtains an expression for $\dot{\mathbf{H}}_A$ comprising $T_{0\leftarrow 2}$ and $T_{2\leftarrow 0} = T^T_{0\leftarrow 2}$, which solely depend on $\mathbf{K}$. Provided that we can apply the gyroscopic approximation and thus replace $\mathbf{K}$ by $\mathbf{H}^u_A$:

$$\begin{aligned}\dot{\mathbf{H}}_A &= 3M_A R^2 c_{20} \sum_{B\neq A} \frac{GM_B}{r_B^5} (\mathbf{H}^u_A \cdot \mathbf{r}_B)\, \mathbf{H}^u_A \times \mathbf{r}_B \\ &+ \frac{3}{2} M_A R^3 c_{30} \sum_{B\neq A} \frac{GM_B}{r_B^7} \left(5(\mathbf{H}^u_A \cdot \mathbf{r}_B)^2 - r_B^2\right) \mathbf{H}^u_A \times \mathbf{r}_B \\ &+ \frac{5}{2} M_A R^4 c_{40} \sum_{B\neq A} \frac{GM_B}{r_B^9} (\mathbf{H}^u_A \cdot \mathbf{r}_B)\left(7(\mathbf{H}^u_A \cdot \mathbf{r}_B)^2 - 3r_B^2\right) \mathbf{H}^u_A \times \mathbf{r}_B \\ &+ 3 \sum_{B\neq A} \frac{GM_B}{r_B^5}\, r_B \times \widetilde{T}_{0\leftarrow 2} \Delta\widetilde{\mathfrak{J}}_{\mathbf{b}_2} \widetilde{T}_{2\leftarrow 0}\end{aligned} \tag{2.49}$$

with a transformation matrix that now solely depends on $\mathbf{H}_A = (H_x, H_y, H_z)^T$, the components of which again relate to basis $\mathbf{b}_0$ via ($H_{xy}^2 = H_x^2 + H_y^2$)

$$\widetilde{T}_{0\leftarrow 2} = \frac{1}{H_A} \begin{pmatrix} -H_A H_y / H_{xy} & -H_x H_z / H_{xy} & H_x \\ H_A H_x / H_{xy} & -H_y H_z / H_{xy} & H_y \\ 0 & H_{xy} & H_z \end{pmatrix}. \tag{2.50}$$

Within the calculation of $\Delta\widetilde{\mathfrak{J}}_{\mathbf{b}_2}$ one also applies $\widetilde{T}_{0\leftarrow 2}$ instead of $T_{0\leftarrow 2}$ for the determination of the phase-shifted coordinates of the tide-generating bodies.

In general, due to retardation time(s) τ_n, the shape of an extended and deformable body at epoch t depends on the system's state at epoch $t-\tau$. Here, phase shifts as such are only necessary for the calculation of Δc_{nm}, Δs_{nm} or $\Delta\mathfrak{J}$ and thus do not need to be determined with very high precision. The variations of the potential coefficients are small in comparison to its nominal values and play just a role within the interaction of the bodies' shapes, but those effects are inferior to the principal Newtonian attraction.

Within INPOP as well as in the IfE software, only two bodies are regarded as being non-rigid: the Earth is deformed, e.g. by solid tides which are generated by Moon and Sun, and the Moon is deformed by solid tides which are generated by Earth (and optionally also by the Sun) and its rotational fluctuations. Thus only a few (phase shift) components of the state vector enter the calculations of the deformations (Δc_{nm}, Δs_{nm}), namely the relative positions between Earth and Moon, Earth and Sun, Moon and Sun, the lunar Euler angles (called lunar libration angles), as well as the time derivatives of all these quantities.

With respect to LLR analysis, phase shift estimation is relevant to both the relative positions of the Solar System bodies within the ICRF at epoch $t - \tau_\alpha$ and the orientation of the selenocentric RF. The latter issue requires the calculation of the lunar Euler angles (libration angles), which are also involved in the determination of the Moon's deformation due to its variational spin, or of the instantaneous lunar spin vector, respectively.

In a last step of approximation, one generally considers non-spherical bodies. The derivation of mutual interactions (forces and moments/torques) between two extended solid bodies S_1 and S_2 of the arbitrary shape is made under two simplifying assumptions: both bodies are rigid, and its potential developments are limited to degree-2 terms. Nonetheless, the results will be extendable to deformable bodies and higher degree expansions. Furthermore, the base vectors $(\mathbf{I}^{(i)}, \mathbf{J}^{(i)}, \mathbf{K}^{(i)})$ $(i = 1, 2)$ of the respective body-fixed RFs coincide with the bodies' principal axes of inertia such that their corresponding inertia matrices $\mathcal{I}_i$ are diagonal matrices and their second-degree potential coefficients are limited to $c_{20}^{(i)}$ and $c_{22}^{(i)}$ only.

Regarding the resulting forces, the derivation can be done based on energy considerations. The potential energy $\mathcal{E}_{\text{pot}}$ of the interaction between S_1 and S_2 results from an integration of the relation $d\mathcal{E}_{\text{pot}} = U dm$, where U denotes the gravitational potential of S_1 at the location of a single mass element dm of S_2.

Again, the main idea here is to keep the structure of the EOMs unchanged, but rather to incorporate the resulting effect of the mutual interaction via a change of the potential coefficients of the involved bodies.

The potential coefficients of body S_2 naturally relate to S_2's body-fixed RF_2, i.e. the basis vectors $(\mathbf{I}^{(2)}, \mathbf{J}^{(2)}, \mathbf{K}^{(2)})$. If they are related to the body-fixed RF_1 of S_1 by using basis vectors $(\mathbf{I}^{(1)}, \mathbf{J}^{(1)}, \mathbf{K}^{(1)})$, we will indicate the basis by a second upper index in square brackets (the first one in round brackets still indicating the body), i.e. $c_{nm}^{(2)[1]}$, $s_{nm}^{(2)[1]}$. In case one only considers a single body and a single (body-fixed) RF, the upper indices are identical and thus expendable, i.e. $c_{nm} = c_{nm}^{(i)[i]}$ and $s_{nm} = s_{nm}^{(i)[i]}$. Likewise, we can apply this notation to express a change of RF $(\mathbf{I}^{(2)}, \mathbf{J}^{(2)}, \mathbf{K}^{(2)})^T = T^{(2)\leftarrow(1)}(\mathbf{I}^{(1)}, \mathbf{J}^{(1)}, \mathbf{K}^{(1)})^T$ via the transformation matrix

$$T^{(2)\leftarrow(1)} = \left(\mathbf{I}^{(2)[1]}, \mathbf{J}^{(2)[1]}, \mathbf{K}^{(2)[1]}\right) = \begin{pmatrix} I_x^{(2)[1]} & J_x^{(2)[1]} & K_x^{(2)[1]} \\ I_y^{(2)[1]} & J_y^{(2)[1]} & K_y^{(2)[1]} \\ I_z^{(2)[1]} & J_z^{(2)[1]} & K_z^{(2)[1]} \end{pmatrix}. \tag{2.51}$$

Introducing the position vector $\mathbf{d}$ (distance) of S_2 w.r.t. S_1 (regarding the origins of their corresponding body-fixed RFs), as well as functions $X_{nm}(\mathbf{r}) = P_{nm}(\cos\theta)\cos m\lambda$ and $Y_{nm}(\mathbf{r}) = P_{nm}(\cos\theta)\sin m\lambda$, one can derive a final formula for the potential energy, depending on the degree-2 coefficients, as follows [33]:

$$
\begin{aligned}
\mathcal{E}^{n=2}_{\text{pot}} &= -\frac{GM_1 M_2 R_1^2 R_2^2}{d^5} \\
&\times \Big(c^{(1)}_{20} \big(6X_{40}(\mathbf{d})\, c^{(2)}_{20} + 3X_{41}(\mathbf{d})\, c^{(2)}_{21} + 3\,Y_{41}(\mathbf{d})\, s^{(2)}_{21} + X_{42}(\mathbf{d})\, c^{(2)}_{22} + Y_{42}(\mathbf{d})\, s^{(2)}_{22} \big) \\
&+ \frac{1}{2}\, c^{(1)}_{22} \big(24X_{40}(\mathbf{d})\, c^{(2)}_{22} - 6X_{41}(\mathbf{d})\, c^{(2)}_{21} + 6\,Y_{41}(\mathbf{d})\, s^{(2)}_{21} + X_{42}(\mathbf{d})\, c^{(2)}_{20} \\
&\qquad + X_{43}(\mathbf{d})\, c^{(2)}_{21} + Y_{43}(\mathbf{d})\, s^{(2)}_{21} + X_{44}(\mathbf{d})\, c^{(2)}_{22} + Y_{44}(\mathbf{d})\, s^{(2)}_{22} \big) \Big),
\end{aligned}
\tag{2.52}
$$

where all terms are consistently expressed w.r.t. the same basis, that means RF_1 here. Within this final formula, we could thus drop the second upper index (square brackets indicating the RF) at the potential coefficients' symbols. Before that, one had to express the coefficients of S_2 w.r.t. RF_1, which is possible via

$$
\begin{pmatrix} c^{(2)[1]}_{20} \\ c^{(2)[1]}_{21} \\ s^{(2)[1]}_{21} \\ c^{(2)[1]}_{22} \\ s^{(2)[1]}_{22} \end{pmatrix}
= \frac{1}{3}\, c^{(2)[2]}_{22}
\begin{pmatrix} 12X_{20}\big(\mathbf{I}^{(2)[1]}\big) \\ 4X_{21}\big(\mathbf{I}^{(2)[1]}\big) \\ 4\,Y_{21}\big(\mathbf{I}^{(2)[1]}\big) \\ X_{22}\big(\mathbf{I}^{(2)[1]}\big) \\ Y_{22}\big(\mathbf{I}^{(2)[1]}\big) \end{pmatrix}
+ \frac{1}{12}\Big(c^{(2)[2]}_{20} + 2c^{(2)[2]}_{22}\Big)
\begin{pmatrix} 12X_{20}\big(\mathbf{K}^{(2)[1]}\big) \\ 4X_{21}\big(\mathbf{K}^{(2)[1]}\big) \\ 4\,Y_{21}\big(\mathbf{K}^{(2)[1]}\big) \\ X_{22}\big(\mathbf{K}^{(2)[1]}\big) \\ Y_{22}\big(\mathbf{K}^{(2)[1]}\big) \end{pmatrix} .
\tag{2.53}
$$

Obviously, the potential coefficients of the (rigid) body S_2, even though being constant w.r.t. RF_2, are time-varying w.r.t. RF_1, i.e.

$$
\begin{aligned}
c^{(2)[2]}_{nm} &= c_{nm} = \text{const.}\,, & & & c^{(2)[1]}_{nm} &\neq \text{const.}\,, \\
& & &\text{but} & & \\
s^{(2)[2]}_{nm} &= s_{nm} = \text{const.}\,, & & & s^{(2)[1]}_{nm} &\neq \text{const.}
\end{aligned}
\tag{2.54}
$$

[33] lists explicit formulas for transitions up to degree 4, i.e. expressions for $c^{(2)[1]}_{3m}$ and $s^{(2)[1]}_{3m}$ in dependence on $c^{(2)[2]}_{3m}$ and $s^{(2)[2]}_{3m}$, as well as expressions for $c^{(2)[1]}_{4m}$ and $s^{(2)[1]}_{4m}$ in dependence on $c^{(2)[2]}_{4m}$ and $s^{(2)[2]}_{4m}$. In the IfE software, by default, the Earth's gravity field is considered up to degree and order 4, whereas the lunar field also comprises degree and order-5 terms. Regarding LLR analysis, optionally higher degree/order terms can be included, which may influence the Earth–Moon distance [41].

Taking the gradient of (2.52) with respect to $\mathbf{d}$ then provides the requested forces, acting upon bodies S_1 and S_2 due to their mutual figure–figure interactions. Following Newton's axioms, the forces are equal but of opposite sign.

Finally, it remains to obtain the moment exerted on S_1 due to its figure–figure effect with S_2, and the moment exerted on S_2 due to the existence of S_1, respectively. In opposition to the calculation of the forces, one cannot simply apply the opposite sign but instead one has to develop the other expression in a likewise manner. Here we explicitly focus on the moment on S_1 due to S_2 $\mathbf{M}^{S_1 \leftarrow S_2}$.

The derivation actually follows the preceding annotations regarding the lower level approximations, based on the consideration of the inertia matrix.

Introducing a unit vector $\mathbf{u} = \mathbf{d}/d$ pointing from $\mathcal{O}_1$ toward $\mathcal{O}_2$, one can finally express the moment exerted upon S_1 due to S_2 (figure–figure effect) by degree-2 terms:

$$\begin{aligned}\mathbf{M}^{S_1 \leftarrow S_2} = \frac{3GM_2R_2^2}{2d^5} \Bigg\{ & 4c_{22}^{(2)}\Big(2\mathbf{I}^{(2)} \times \mathcal{J}_1\mathbf{I}^{(2)} + 5\Big(7\big(\mathbf{u}\cdot\mathbf{I}^{(2)}\big)^2 - 1\Big)\mathbf{u} \times \mathcal{J}_1\mathbf{u} \\ & - 10\big(\mathbf{u}\cdot\mathbf{I}^{(2)}\big)\big(\mathbf{u} \times \mathcal{J}_1\mathbf{I}^{(2)} + \mathbf{I}^{(2)} \times \mathcal{J}_1\mathbf{u}\big)\Big) \\ & + \big(c_{20}^{(2)} + 2c_{22}^{(2)}\big)\Big(2\mathbf{K}^{(2)} \times \mathcal{J}_1\mathbf{K}^{(2)} + 5\Big(7\big(\mathbf{u}\cdot\mathbf{K}^{(2)}\big)^2 - 1\Big)\mathbf{u} \times \mathcal{J}_1\mathbf{u} \\ & - 10\big(\mathbf{u}\cdot\mathbf{K}^{(2)}\big)\big(\mathbf{u} \times \mathcal{J}_1\mathbf{K}^{(2)} + \mathbf{K}^{(2)} \times \mathcal{J}_1\mathbf{u}\big)\Big)\Bigg\}\,. \end{aligned} \tag{2.55}$$

All vectors $\mathbf{u}$, $\mathbf{I}^{(2)}$, $\mathbf{K}^{(2)}$ are expressed in the reference frame RF_1 of the solid body S_1. If the solid S_2 is assumed to be rotationally symmetric, one only retains the zonal coefficient $c_{20}^{(2)} = -J_2^{(2)}$.

Apart from the Earth's Moon, the other Solar System planets' natural satellites are accounted for in an indirect way. The motion of those minor bodies is not integrated individually. Instead, their masses are supposed to influence the motion of the barycenter of a planetary subsystem only. Thus, the EOMs of Mars, Jupiter, Saturn, etc. relate to the barycenter motion of respective secondary multi-body systems (planet plus satellites).

On the other hand, the treatment of further minor bodies, namely the Solar System's asteroids, is twofold. Their number (several hundreds of thousands) is by far too large to integrate all of them individually. At least, within INPOP, a selection of about 300 asteroids is treated individually. The actual number differs between the various INPOP versions and it is the result of statistical investigations on the most significant influences of the asteroids, e.g. on the Earth–Mars distance [31]. Remark: only a selection of main belt asteroids is considered.

Besides the individual integration of a subset, the combined effect of the remaining asteroids is modeled via a ring, the perturbing force of which can be expressed via hypergeometric functions [30]. These asteroids are supposed to be uniformly distributed (at any instant in time) on a circular ring which is centered at SSB (former INPOP versions) or at the Sun's center (recent versions), respectively. One task in practice is then to determine the ring parameters, i.e. its radius R_r and total mass M_r. The orbits of the asteroids itself are not necessarily circular.

The asteroidal ring is regarded as a perturber and one is interested in the calculation of its cumulative gravitational effect on single perturbed bodies (the individual planets of the Solar System). For the perturbing effect of the ring, we distinguish different cases. The affected bodies, which are all to be numerically integrated via their respective EOM, are either located outside or within the ring's plane. In the latter case, in addition, we distinguish between bodies that are located inside or outside the ring.

In general, the perturbed body is located out of the ring's plane. We introduce a ring-related reference frame RF_r by choosing its origin $\mathcal{O}_r$ to be coincident with the SSB (as mentioned above, recent INPOP versions make use of the Sun's center instead), and its orthonormal base vectors $\mathbf{I}^{(r)}$, $\mathbf{J}^{(r)}$, $\mathbf{K}^{(r)}$ are defined in dependence on the location of the perturbed body as follows: vector $\mathbf{K}^{(r)}$ is normal to the ring's plane, vector $\mathbf{I}^{(r)}$ thus lies within the orbital plane and it shall result from a projection of the perturbed body's barycentric position vector $\mathbf{r}$ such that these three vectors are coplanar, or its scalar triple product is zero, i.e. $(\mathbf{I}^{(r)} \times \mathbf{K}^{(r)}) \cdot \mathbf{r} = 0$, respectively. Finally, vector $\mathbf{J}^{(r)}$ consequently lies within the ring's plane and completes a right-handed coordinate system via $\mathbf{J}^{(r)} = \mathbf{K}^{(r)} \times \mathbf{I}^{(r)}$.

By introducing a longitude- or right ascension-like angle α for the localization of any of the ring's mass elements, and a latitude-, declination- or elevation-like angle δ for the localization of the perturbed body, its respective coordinates w.r.t. $\mathrm{RF}_r = (\mathcal{O}_r, \mathbf{b}_r)$ with $\mathbf{b}_r = (\mathbf{I}^{(r)}, \mathbf{J}^{(r)}, \mathbf{K}^{(r)})$ are given by the position vectors $\mathbf{S} = (R_r \cos\alpha, R_r \sin\alpha, 0)^T$ and $\mathbf{R} = (R\cos\delta, 0, R\sin\delta)^T$.

The Newtonian acceleration of the (perturbed) body due to a single mass element (of the ring) is given by

$$d\ddot{\mathbf{R}} = G\frac{\mathbf{S}-\mathbf{R}}{\|\mathbf{S}-\mathbf{R}\|^3}\,dm = \frac{GM_r}{2\pi(R_r^2+R^2)^{3/2}}(1-z\cos\alpha)^{-3/2}\begin{pmatrix} R_r\cos\alpha - R\cos\delta \\ R_r\sin\alpha \\ -R\ \sin\delta \end{pmatrix} d\alpha \tag{2.56}$$

with $R = \|\mathbf{R}\|$, $R_r = \|\mathbf{S}\|$, and $z = 2R_rR\cos\delta/(R_r^2+R^2)$. If the body is not located on the ring (either due to $R \neq R_r$ or $\delta \neq 0$) then $|z\cos\alpha| < 1$ is always satisfied. So, a binomial series expansion of $(1-z\cos\alpha)^{-3/2}$ can be applied.

With $R_r\mathbf{I}^{(r)} = (R_r, 0, 0)^T$ and the use of the Pochhammer symbol $(q)_n$, defined as

$$(q)_n = q(q+1)\cdots(q+n-1) = \Gamma(q+n)/\Gamma(q)\,, \tag{2.57}$$

one gets

$$\ddot{\mathbf{R}} = \frac{GM_r}{(R_r^2+R^2)^{3/2}}\left(R_r\sum_{n=1}^{\infty}\frac{(\frac{3}{2})_{2n-1}(\frac{1}{2})_n}{(1)_{2n-1}(1)_n}z^{2n-1}\mathbf{I}^{(r)} - \sum_{n=0}^{\infty}\frac{(\frac{3}{2})_{2n}(\frac{1}{2})_n}{(1)_{2n}(1)_n}z^{2n}\mathbf{R}\right) \tag{2.58}$$

as a final expression [33] for a body's perturbation because of the ring's gravitational attraction.

For the special case (perturbed body located within the ring's plane) we have $\delta = 0$ or $\cos\delta = 1$ and $\sin\delta = 0$, such that $\mathbf{R} = (R, 0, 0)^T = R\mathbf{I}^{(r)}$. Consequently, the perturbation equation now reads

$$d\ddot{\mathbf{R}} = G\frac{\mathbf{S}-\mathbf{R}}{\|\mathbf{S}-\mathbf{R}\|^3}\,dm = \frac{GM_r}{2\pi}\frac{(R_r\cos\alpha - R)\mathbf{I}^{(r)} + R_r\sin\alpha\,\mathbf{J}^{(r)}}{(R_r^2 + R^2 - 2R_rR\cos\alpha)^{3/2}}\,d\alpha\,, \tag{2.59}$$

and the expression to be evaluated eventually simplifies to

$$\ddot{\mathbf{R}} = \frac{GM_r}{2\pi}\int\limits_0^{2\pi}\frac{(R_r\cos\alpha - R)\mathbf{I}^{(r)}}{(R_r^2 + R^2 - 2R_rR\cos\alpha)^{3/2}}\,d\alpha\,. \tag{2.60}$$

For a body *inside* the asteroidal ring, one finally gets

$$\ddot{\mathbf{R}} = \frac{GM_r}{2R_r^3}\left(\sum_{n=0}^{\infty}\frac{(\frac{3}{2})_n(\frac{3}{2})_n}{(2)_n(1)_n}\left(\frac{R}{R_r}\right)^{2n}\right)\mathbf{R} \qquad (R < R_r)\,, \tag{2.61}$$

whereas for a body *outside* the asteroidal ring, we find

$$\ddot{\mathbf{R}} = -\frac{GM_r}{R^3}\left(1 + \sum_{n=1}^{\infty}\frac{1}{2n+1}\cdot\frac{(\frac{3}{2})_n(\frac{3}{2})_n}{(1)_n(1)_n}\left(\frac{R_r}{R}\right)^{2n}\right)\mathbf{R} \qquad (R > R_r)\,. \tag{2.62}$$

Consideration of observational data

Despite the use of an elaborated force model, any ephemeris calculation will quickly deteriorate in time with respect to available highly precise observational planetary data. In order to optimize the planetary (and lunar) solution, a variety of observational data, comprising historic planetary meridional transit observations as well as current ranging to orbiting planetary spacecraft, is exploited. The consistent use of those data, e.g. LLR measurements, requires different reduction and adjustment procedures.

In the case of LLR, for instance, one has to consider a correct localization of the point of emission and reception of the laser signal. This task comprises corrected coordinates for the terrestrial stations and lunar reflectors. The former require (geophysical) models of the Earth's plate tectonics, solid tides, atmospheric loading, ocean loading, pole tides, and so on [60]. Similarly, reflector sites on the Moon undergo displacements due to lunar solid tides, or lunar deformations because of the Moon's rotation.

In the case of both bodies, site coordinates are usually given in body-fixed frames, whereas the numerical integration of the EOMs is performed in a (barycentric) inertial frame. This implies necessary reduction steps regarding coordinate system transformations, e.g. ITRS<>GCRS and GCRS<>BCRS, and likewise PAS<>SCRS and SCRS<>BCRS, see Section 2.1. These transformations also comprise different time scales (TT, TCG, TCB, TDB). Actually, TCG and TCB are only intermediary time scales in the calculation. In the end, one needs to be able to relate TT (connected to UTC and TAI via the dating of the observations) with TDB (time scale for the EOM of the planetary solution). Because the chain of time scale transformations also depends on planetary ephemeris, one option is to integrate a respective time-related differential equation, e.g. the one for TT-TDB, simultaneously with the EOMs of the Solar System bodies. This approach was actually chosen for the recent INPOP versions. Alternatively, one may apply a time ephemeris like TE405 [24] which provides TT-TDB based on the DE405 solution, or implement SOFA routines [13] for the TT-TDB calculation.

The observation signals itself have to be reduced because of time delays due to relativistic or atmospheric effects. Again, certain models are involved, e.g. following the IERS conventions 2003 [34]. Furthermore, possible measurement biases have to be considered.

Eventually, a classical least-squares adjustment is applied, where a suitable weighting scheme is required when observations of different precision are incorporated. Several methods can be used to get an appropriate selection of solve-for parameters (selection of unknowns), e.g. via the elimination of the least eigenvalues, the elimination of parameters (from the list of potential unknowns) that degrade the residuals the least, the elimination of strongly correlated parameters, or based on the ratio between their formal errors and values itself. Last but not the least, the whole estimation procedure is accompanied by various tests on the stability of parameters [33, 41].

Current parameter estimation approaches at IfE are discussed in detail in [45]. The main goal is to estimate parameters of the Earth–Moon system, which requires an iterative procedure. It is performed in three major steps, in order to achieve fast convergence. First, one estimates main parameters (station and reflector coordinates, lunar orbit and rotation, potential coefficients as well as elasticity and dissipation parameters of the Moon, etc.), all of which are very well suited for LLR analysis. In a second step, approximative initial values for the Earth's orientation parameters get corrected. Finally, these parameters are fixed, and one estimates individually or simultaneously the parameters of less-significant physical effects (PPN parameters, Nordtvedt parameter, time rate of change of Newton's gravitational constant, plate tectonics parameters, etc.).

Eventually, as in INPOP, the IfE results are not judged solely based on formal errors from the adjustment. Instead, one tries to get a more realistic estimate on the achieved accuracy level by several means: comparison with independent analyses from other observation techniques, analysis of dependencies/correlations among individual parameters, spectral analysis of normalized residuals (which should be normally distributed), modified worst case analysis, and so on [49]. In practice, one combines all of these methods.

Differences between INPOP and the current IfE approach

The current IfE software package portion on ephemeris calculation differs from the before-mentioned INPOP approach in certain respects. It is primarily used for LLR analysis and thus tailored to this task. Minor differences exist in the overall procedure regarding the creation of a planetary/lunar solution. In view of the stated construction steps, cf. Figure 5, one major difference is to waive the computation of Chebyshev polynomials/coefficients, because it was not intended, in the past, to distribute the IfE solution to the public.

Regarding the numerical integration technique, one makes use of the initial value solver of Shampine and Gordon, which is a variable order variable step-size Adams–Bashforth–Moulton method (output every 8 h) and supports dense output by interpolation. The drift of the Solar System barycenter is not accounted for by a transformation of the system's initial state. No ODE for a time scale difference is set up, nor does the IfE software integrate the Earth's orientation. Regarding the latter issue, one rather models the orientation based on recommended precession–nutation models (IERS conventions 2003 [34]). A fixed orientation is also implemented in INPOP, but only for the reduction of observations step. At IfE, time scale transformations TT<>TDB are performed according to [19].

Asteroids are treated similar to the planets, i.e. individual equations of motion are integrated simultaneously with the major Solar System bodies. So far, only up to 16 asteroids can be included, but the actual number in routine operations is usually less than this maximum. A ring model has not been implemented yet.

Details on the force model can be found in [3, 16, 45], all of which can be regarded as primary references for the actual IfE software coding.

The reduction and adjustment steps are quite similar to the INPOP approach, but the (LLR station) bias handling and the lunar modeling show significant differences. For instance, a different lunar gravity field model is applied, and one solves for a single lag angle. On the other hand, different dissipation/tidal models for the description of the Moon's orientation can be used: either a constant time delay (delayed variation of the lunar tensor of inertia) in combination with a single Love number (for a degree-independent response to the perturbing potential), or a fluid core model.

Regarding the estimation step, the partial derivatives can be obtained in two ways, either by numerical difference methods (as in INPOP), or via the integration of variational equations. The weighting scheme is solely based on the (LLR) normal point uncertainty of every single observation, whereas INPOP incorporates not only LLR data.

Additionally, at IfE, much effort has been spent on the treatment of various relativistic effects and their testing. The ephemeris program comprised different extensions for including and testing various relativistic effects like:
- time variable gravitational constant [20, 46],
- geodetic precession of the lunar orbit in addition to EIH [71],
- violation of equivalence principle [43, 48],

- acceleration due to dark matter in the galactic center, [48, 57],
- Yukawa term for modifying Newton's $1/r^2$ law of gravity [1, 52],
- preferred-frame effects and metric parameters [44, 78],
- gravitomagnetic effects [72],
- optional spin-orbit coupling (Brumberg/Kopeikin, [45]).

See also Section 4 for the results of the relativistic tests.

3 Analysis

3.1 Software package LUNAR

The analysis model of LLR at IfE is implemented in the software package LUNAR. Initiated by M. Schneider, the development of this package started at the Forschungseinrichtung Satellitengeodäsie (FESG) in Munich within the scope of the Sonderforschungsbereich (SFB) 78 [12, 66] and is mainly based on the work of [3, 16, 64] and [45]. It is performed in a weighted least-squares adjustment where equations (2.1) and (2.2) are used. To build the entries of the Jacobi matrix, equation (2.2) is differentiated with respect to the unknown parameters as

$$\frac{\partial d}{\partial p} = \frac{\mathbf{d}}{d}\left(\frac{\partial \mathbf{r}_{\mathrm{EM}}}{\partial p} - \frac{\partial \mathbf{r}_{\mathrm{obs}}}{\partial p} + \frac{\partial \mathbf{r}_{\mathrm{ref}}}{\partial p}\right) . \tag{3.1}$$

In the following, the analysis package and the strategy will be described in more detail. The package consists of three major parts:

Calculation of the ephemeris of the Solar System bodies by the numerical integration of the Einstein–Infeld–Hoffmann equation of motion and the simultaneous calculation of the Euler angles and angular velocities of the Moon (see Section 2.2) by means of a Adams–Bashford–Moulton method with variable step size. The result is a tabulated solution of the ephemeris of the Solar System bodies and the Euler angles for the Moon, with one data set for every 8 h of integrated time. The dynamical partial derivatives with respect to the relativistic parameters which affect $\mathbf{r}_{\mathrm{EM}}$ are optionally calculated by numerical differentiation as given in [45].

Calculation of the dynamical partial derivatives of the lunar and solar orbit as well as the lunar rotation with respect to the Newtonian parameters (which influence $\mathbf{r}_{\mathrm{EM}}$ and the Euler angles) by a numerical integration of the ephemeris. This is based on the Newtonian equation of motion, including the effects of the gravitational field of Earth and Moon up to degree and order 3 and the tidal acceleration. Here, the same integration method is used as in the ephemeris part. The result is a file with the respective partial derivatives for every 8 h of integrated time.

Parameter estimation using a weighted least-squares adjustment where in a first step, equation (2.1) is solved. In a second step, the geometrical partial derivatives $\partial \mathbf{r}_{obs}/\partial p$ and $\partial \mathbf{r}_{ref}/\partial p$ are calculated and the entries for the Jacobi matrix according to (3.1) are computed by the chain rule combining the partial derivatives computed in all three parts of the analysis package. So the functional model of the adjustment is given. The input NP are weighted according to the accuracy of the measured travel time of the laser pulse as provided by the observatories and are regarded as uncorrelated. This information represents the stochastic model of the adjustment. The result of the calculation are values for the unknown parameters and their standard deviation.

The strategy of the analysis is to calculate the ephemeris and partial derivatives on the basis of initial values. With this input, the parameter estimation is carried out. Using the improved fitted parameters, the calculation starts again and iterates until the solution converges.

The residuals of a standard solution are shown in Figure 6. In the early years of the LLR measurements, the weighted annual residuals reach about 30 cm. Between 1980 and 1984 the residuals are about 25 cm. From the mid-80s on, when more observatories start tracking the Moon (see Figure 2), the residuals become better, down to 6 cm. From the middle 90s on, the residuals continue to reduce until 2006 to 3.5 cm. Between 2006 and 2010 they are about 2–3 cm and increase again up to 4 cm.

If new NP from the observatories are available, they are included in the consisting LLR data set after some preprocessing. In the preprocessing, the recent data set from a standard solution is extended by the new NP and the parameter estimation is performed. The residuals of this solution are analyzed regarding the ratio between the residual value and its formal error. If the ratio exceeds the threshold of 3.0, i.e. the residual is three times bigger than the formal error, the observation is treated as an outlier and removed from the analysis process. Afterwards this preprocessing step is repeated until all observations remain below the threshold. A smaller threshold leads to smaller residuals for the whole analysis result, but with the disadvantage that more NP have to be removed. So here it has to be decided carefully about the threshold.

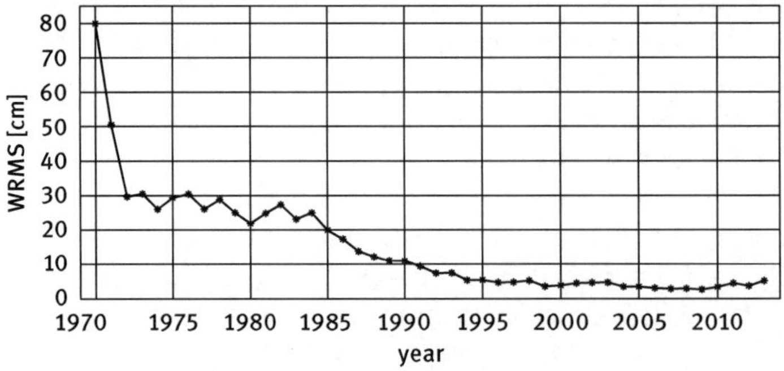

Fig. 6. Weighted annual residuals of the LLR analysis of 20 061 NP from 1970 to 2013.

With the longest observation time series of all space geodetic techniques, LLR allows to determine various parameters of the Earth–Moon system [49]. They can be divided into two parts, the first part comprises the non-relativistic Newtonian parameters, the second part contains parameters related to general relativity [38]. The next section gives an overview of some Newtonian parameters which can be estimated with the IfE LLR analysis package. The relativistic parameters are addressed in Section 2.2, recent results and a detailed discussion are given in Section 4.

3.2 Newtonian parameters

Only non-relativistic Newtonian parameters are estimated when analyzing the LLR data for a typical solution. The relativistic parameters are held fixed to their Einsteinian values. The result of this first step is called standard solution, with the following possible parameters estimated:

- *Lunar orbit and rotation:* Within the Solar System ephemeris computation (see Section 2.2), the lunar ephemeris is adjusted by estimating the corresponding set of initial values for the integration process. For the translational motion of the Moon, the initial position and velocity are estimated. For the rotational motion, the lunar orientation with respect to the used inertial frame via Euler angles and the corresponding angular velocities is estimated. The result is a highly accurate lunar ephemeris, see also [14, 86].
- *Lunar gravity field:* The complete lunar gravity field up to degree and order 5 is used for the ephemeris computation. It is also possible to estimate single gravity field coefficients up to degree and order 4. In general, all coefficients of degree and order 4 and degree and order 5 are held fixed with their model values.
- *Lunar tidal parameters:* The Moon is, as the Earth, an elastic, deformable body. At the moment it is possible to estimate the Love number k_2, which is a value for the tidal changes of the tensor of inertia, and a dissipation parameter D to account for a temporal delay of the lunar tidal reaction, see also [3, 85].
- *Lag angle:* The tidal bulge lag angle of the Earth is a measure for the lunar tidal acceleration. The interaction between the Earth's body and Moon with the Earth's tidal bulge leads to an deceleration of the Earth rotation and, due to the conservation of angular momentum, to an acceleration of the Moon. This effect leads to a secular increase of the Earth–Moon distance of about 3.8 cm per year [3, 86].
- *Coordinates:* The coordinates and velocities of LLR observatories on Earth and retroreflectors on the Moon can be estimated [3, 47, 86].
- *Mass of the Earth–Moon system:* The product of the gravitational constant times the combined mass of the Earth and Moon can be estimated [86].
- *Earth orientation:* Due to the sparse distribution of measurements, LLR is not sensitive to short-term Earth orientation parameters (EOP). But on longer timescales, it is possible to estimate the precession rate and nutation parameters with periods

of 18.6 years, 9.3 years, 365.26 days and 182.6 days [7]. From nights with dense LLR data, it is also possible to get informations about UT0 and polar motion/variation of latitude [5, 6].
- *Lunar interior:* Parameters of the lunar interior due to a fluid core can also be derived from LLR measurements [81, 82].

4 Results for relativistic parameters

In this section, we discuss a few relativistic tests, which can be carried out with LLR data.

4.1 Gravitational constant

One basic assumption of Einstein's theory of gravity is a constant gravitational constant G. Some alternative theories predict temporal variations in G in the order of $\dot{G}/G = 10^{-11}\ \mathrm{yr}^{-1}$ to $\dot{G}/G = 10^{-14}\ \mathrm{yr}^{-1}$ [63, 75]. Recent LLR analysis at JPL gives an upper bound for a variation of G as $\dot{G}/G = 3 \times 10^{-13}\ \mathrm{yr}^{-1}$ [79] while the analysis of planetary motion and interplanetary spacecraft seem to constrain $\dot{G}/G$ at a level of $7 \times 10^{-14}\ \mathrm{yr}^{-1}$ [62].

In our analysis, the gravitational constant can be varied in time by introducing

$$G(t) = G_0 \left(1 + \frac{\dot{G}}{G_0} \Delta t + \frac{1}{2} \frac{\ddot{G}}{G_0} \Delta t^2 \right) \tag{4.1}$$

into the EIH equations of motion (Section 2.2), where G_0 is the gravitational constant and Δt is the time difference to a reference epoch. For testing a possible acceleration part $\ddot{G}/G_0$, we refer to [46]. Figure 7 shows the power spectrum of the differences in the Earth–Moon distances with and without a $\dot{G}$ perturbation, where a value of $\dot{G}/G = 1.5 \times 10^{-13}\ \mathrm{yr}^{-1}$ was assumed.

Prominent periods are connected to the lunar anomalistic and synodic month. The anomalistic period shows the strongest effect and represents the change in the Earth–Moon perigee distance. Due to the strong influence of the solar gravitation on the lunar orbit, a change in G also shows a strong synodic period. The increased energy at lower frequencies indicates the long-term effect on the lunar orbit due to the different evolution of the Solar System ephemeris with respect to $\dot{G}$.

In our parameter fit, we obtained a value of

$$\frac{\dot{G}}{G} = (1.4 \pm 1.5) \times 10^{-13}\ \mathrm{yr}^{-1} ,$$

which agrees with the recent result from JPL [79].

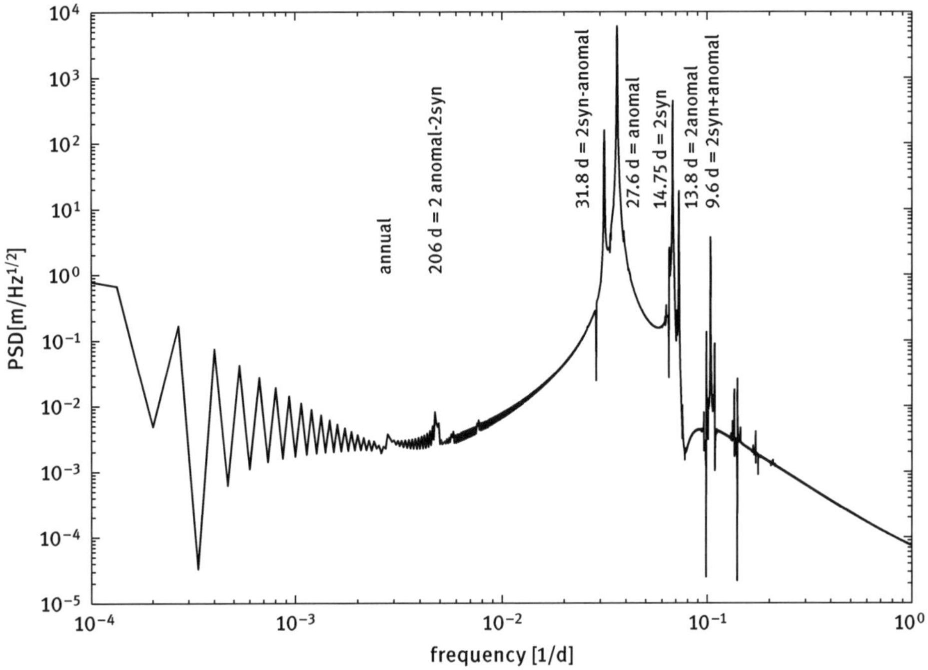

Fig. 7. Power spectrum of the difference in the Earth–Moon distance with and without a $\dot{G}$ perturbation with $\dot{G}/G = 1.5 \times 10^{-13}\ \text{yr}^{-1}$. Prominent periods and their frequencies are labeled. Here, "syn" denotes the synodic frequency and "anomal" the anomalistic frequency.

The result shows a high correlation (>0.9) with the value for the tidal bulge lag angle of the Earth. Both parameters cause a secular effect on the Earth–Moon distance. Fixing the lag angle on the value of the standard solution, which was generated before estimating the relativistic parameters, we would get a value of

$$\frac{\dot{G}}{G} = (0.8 \pm 3.0) \times 10^{-14}\ \text{yr}^{-1}\ .$$

The given accuracies are 3σ values, i.e. 3 times the standard deviation as obtained from the least-squares adjustment. It considers model inaccuracies and other possible error sources that affect the result, for more details see [46]. The latter value is highly optimistic but shows the potential of LLR.

4.2 Equivalence principle

More than 400 years ago, in the times of Galileo, it was discovered, that the acceleration of falling bodies in the same gravitational field without atmospheric friction is independent of their composition, mass, and shape. This was the birth of the so-called equivalence principle. This principle was extended by Newton, who introduced in his second axiom two types of masses, called gravitational m_g and inertial mass m_i, which must be the same to fulfill the equivalence principle $m_g = m_i$.

The equivalence principle belongs to the corner stones of Einstein's theory of gravity. Some alternative theories predict a violation of the equivalence principle at some level, so it is still an ongoing research topic to test Einstein's theory as well as other theories of gravity [15, 58].

The equivalence principle can be divided into different "types," e.g. into the weak and the strong principle. Testing the free fall acceleration of two bodies, which are made of different materials, e.g. beryllium and titanium [77], directly leads to a test of the weak equivalence principle as

$$\Delta \frac{m_g}{m_i} = \left(\frac{m_g}{m_i}\right)_1 - \left(\frac{m_g}{m_i}\right)_2 . \tag{4.2}$$

A violation of the weak principle leads to an non-zero value of $\Delta m_g/m_i$. Free-fall experiments confirm the weak equivalence principle within an accuracy of 1.3×10^{-13} for a differential acceleration toward the Earth and 4.0×10^{-13} toward the Sun [77]. LLR tests at JPL confirm the equivalence principle at the level of 1.3×10^{-13} [83].

Considering astronomically sized bodies, the strong type of the equivalence principle can be tested [54]. This is caused by the non-negligible amount of gravitational self-energy of these bodies compared to the small laboratory-sized bodies. The ratio m_g/m_i can then be expressed as

$$\frac{m_g}{m_i} = 1 + \eta \left(\frac{U}{Mc^2}\right) , \tag{4.3}$$

with the Nordtvedt parameter η (equals to zero in general relativity), the gravitational self-energy U, the mass of the body M, and the speed of light c. A possible violation of the strong equivalence principle would lead to an non-zero value of η.

Testing the equivalence principle in the Earth–Moon system means, that the Earth and the Moon get the role of test bodies falling in the gravitational field of the Sun [48, 80]. Due to the different composition and gravitational self-energy of Earth and Moon, this test is a combined test of the weak and the strong principle. A violation would cause an additional acceleration of the Moon into the direction to the Sun. This depends on the difference of the mass ratios

$$\Delta \left(\frac{m_g}{m_i}\right)_{\mathrm{EM}} = \left(\frac{m_g}{m_i}\right)_{\mathrm{Earth}} - \left(\frac{m_g}{m_i}\right)_{\mathrm{Moon}} . \tag{4.4}$$

In our LLR analysis, there are two ways to estimate the equivalence principle violation parameter:

(1) Estimation of the Nordtvedt parameter η: The additional acceleration leads to an additional range term Δr_{EM} between Earth and Moon. This term depends on the

synodic angle D and the parameter η and can be expressed in the first order as

$$\Delta r_{\mathrm{EM}} = 13.1\mathrm{m}\ \eta \cos D, \tag{4.5}$$

see [9, 57].

(2) Estimation of the ratio $\Delta(m_g/m_i)_{\mathrm{EM}}$: The corresponding numerically integrated partial derivative is computed by introducing an additional acceleration of the Moon into the direction to the Sun. Here, the ratio $(m_g/m_i)_{\mathrm{Earth}}$ is set to 1. Due to embedding it in the equations of motion, this approach keeps the correlation to all other forces of the ephemeris of Earth and Moon. Therefore this approach is preferred over the first version, even if the results are rather equivalent [48].

Figure 8, again as power spectrum, shows the effect of a possible equivalence principle violation on the Earth–Moon distance in the order of $\Delta(m_g/m_i)_{\mathrm{EM}} = 2\times10^{-13}$. Due to the additional acceleration into the direction of the Sun, a strong synodic signal appears and dominates the spectrum together with its combination with the anomalistic period. We obtain a value of

$$\Delta\left(\frac{m_g}{m_i}\right)_{\mathrm{EM}} = (0.1 \pm 1.3) \times 10^{-13}\,.$$

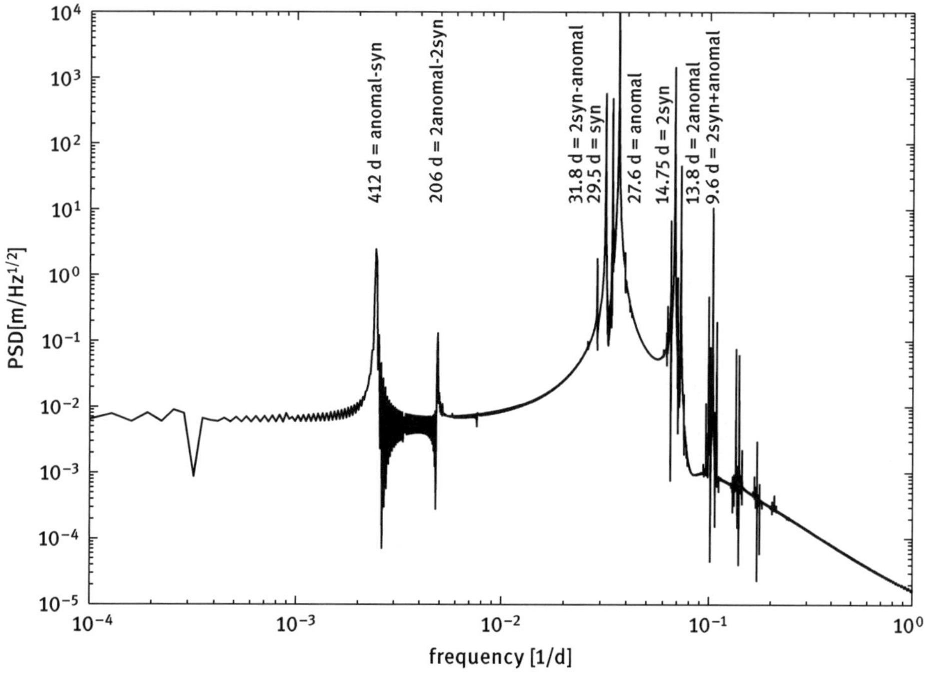

Fig. 8. Power spectrum of the difference in the Earth–Moon distance with and without a perturbation due to a violation of the equivalence principle with $\Delta(m_g/m_i)_{\mathrm{EM}} = 2\times10^{-13}$. See Figure 7 for labeling.

The result shows a strong correlation with the mass of the Earth–Moon system $GM_{\mathrm{E+M}}$. This is caused by the influence of the synodic angle D on both parameters, see [43, 56] for further discussion. Fixing $GM_{\mathrm{E+M}}$ at the value of the standard solution leads to

$$\Delta\left(\frac{m_g}{m_i}\right)_{\mathrm{EM}} = (0.1 \pm 1.0) \times 10^{-13}\,.$$

The given errors again are 3σ accuracies. For a more detailed error discussion see [48].

In combination with laboratory tests, it is possible to give an upper limit for the validity of the strong equivalence principle (SEP) from the LLR fit. Therefore, a weak equivalence principle (WEP) test with two special bodies was carried out that represented the miniaturized Earth and Moon with corresponding composition of these bodies. Combining the random and systematic errors of the result, the WEP test for Earth and Moon gives [1]

$$\Delta\left(\frac{m_g}{m_i}\right)_{\mathrm{EM_{WEP}}} = (1.0 \pm 1.4) \times 10^{-13}\,.$$

As the estimated result of the LLR equivalence principle test provides the sum of WEP and SEP, the combination with laboratory results leads to a result for the SEP alone:

$$\Delta\left(\frac{m_g}{m_i}\right)_{\mathrm{EM_{SEP}}} = (-0.9 \pm 1.9) \times 10^{-13}\,,$$

$$\Delta\left(\frac{m_g}{m_i}\right)_{\mathrm{EM_{SEP}}} = (-0.9 \pm 1.7) \times 10^{-13}$$

according to the two LLR results given above. With equations (4.5) and

$$\Delta r_{\mathrm{EM}} = S\Delta\left(\frac{m_g}{m_i}\right)_{\mathrm{EM}} \cos D\,, \tag{4.6}$$

where $S = -2.9427 \times 10^{10}$ m [9], these results can be converted into a value for the Nordtvedt parameter as

$$\eta = (2.0 \pm 4.0) \times 10^{-4}\,,$$

see also [48, 83] for further recent results.

A LLR equivalence principle test with respect to possible dark matter into the direction of the galactic center is discussed in [48].

4.3 Yukawa term

For testing Newton's inverse square law. i.e. the quadratic decrease of gravitational acceleration with increasing distance to the attracting body ($\ddot{r} \propto 1/r^2$), one can consider an additional Yukawa potential in the model:

$$V_{\mathrm{EM}}(r) = -\frac{GM_E M_M}{r}\left(1 + \alpha e^{-r/\lambda}\right)\,. \tag{4.7}$$

M_E and M_M denote the masses of Earth and Moon, respectively, λ is the interacting range and α is the coupling constant. Tests of the inverse square law were carried out at various interacting length scales, see, e.g. [25] and [23]. In our LLR analysis, λ is fixed to 380 000 km and a corresponding additional differential acceleration between Earth E and Moon M with $\Delta\ddot{\mathbf{r}}_{\text{yuk}} = \ddot{\mathbf{r}}_{\text{M yuk}} - \ddot{\mathbf{r}}_{\text{E yuk}}$

$$\Delta\ddot{\mathbf{r}}_{\text{yuk}} = -\frac{\mathbf{r}_{\text{EM}}}{r_{\text{EM}}}\frac{GM_{\text{E+M}}}{r_{\text{EM}}}\alpha e^{-r_{\text{EM}}/\lambda}\left(\frac{1}{r_{\text{EM}}} + \frac{1}{\lambda}\right) \tag{4.8}$$

is introduced in the equations of motion. Figure 9 again shows the spectrum of this effect on the Earth–Moon distance. The strongest signals are at frequencies of the anomalistic month, the half-synodic month and their combinations. Here, no pure synodic signal is present.

Our parameter estimation of the coupling constant α gives

$$\alpha = (-1.8 \pm 0.5) \times 10^{-11},$$

indicating again a 3σ error. This possible non-null result should not be misinterpreted. Due to nonmodeled effects and possible systematics, the given error might be too small. We do not consider this result as a significant deviation from general relativity which requires $\alpha = 0$, but further investigation is needed to understand this behavior.

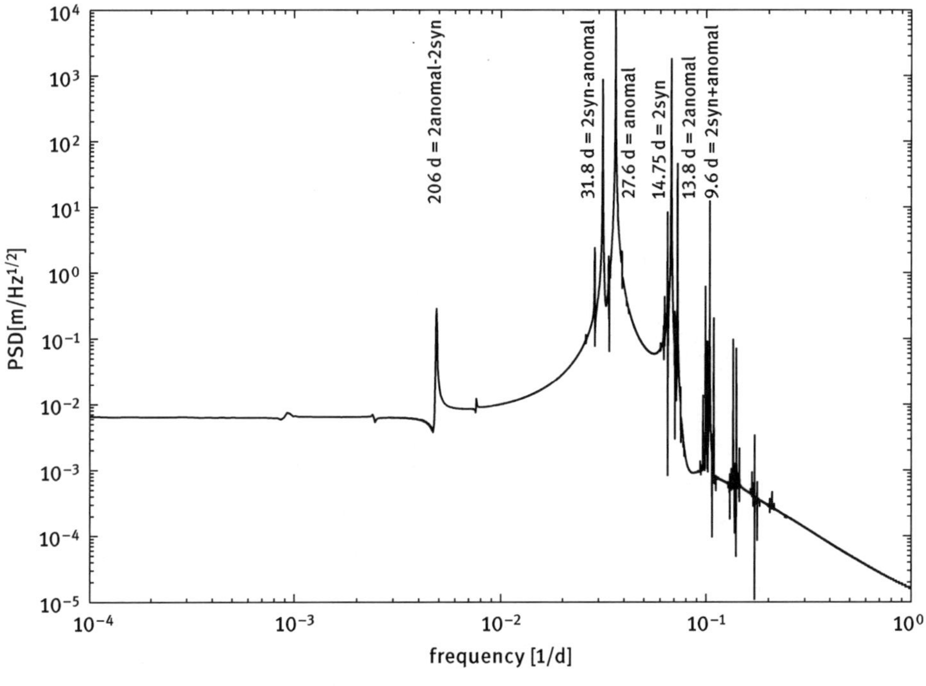

Fig. 9. Power spectrum of the difference in the Earth–Moon distance with and without a perturbation due to an additional Yukawa term with $\alpha = 1.5 \times 10^{-11}$. See Figure 7 for labeling.

A Yukawa potential would also introduce an additional precession of the lunar orbit [1]. This is also obvious from the very high correlation with the parameter for an additional geodetic precession (see below), if both are estimated simultaneously. In this case, we get for the Yukawa coupling constant

$$\alpha = (-0.6 \pm 1.8) \times 10^{-11} .$$

The latter error estimate might be the more realistic one for α.

4.4 Geodetic precession

The Moon orbits the Earth in the gravitational field of the Sun. As a consequence of the relativistic description, an acceleration acts on the Moon causing a relativistic precession of the lunar orbit, called de Sitter or geodetic precession [10]. The effect of the geodetic precession is included in the relativistic equations of motion, see Section 2.2. Following a method from Shapiro et al. [71], we introduce the geodetic precession separately a second time in the equations of motion by adding an additional acceleration

$$\mathbf{a}_{\mathrm{gp}} = 2h\Omega_{\mathrm{gp}} \times \mathbf{v}_{\mathrm{ME}} . \tag{4.9}$$

Here, Ω_{gp} denotes the angular velocity of the geodetic precession, $\mathbf{v}_{\mathrm{ME}}$ the velocity of the Earth–Moon system and h is the solve-for parameter which indicates a possible deviation of the observed from the predicted geodetic precession by Einstein's theory. The value for h indicates a relative deviation from general relativity, where $h = 1$ means a 100 % error of the geodetic precession included in the EIH equations of motion. Results from LLR analysis at JPL confirm the predicted geodetic precession ($\approx$ 1.9 as/cy) within 0.4% [79].

The effect on the Earth–Moon distance is shown as power spectrum in Figure 10. The additional precessional movement nearly excites the same frequencies as the Yukawa term.

We estimated for the parameter h

$$h = -0.005 \pm 0.002 ,$$

with the 3σ error. Analog to the estimated Yukawa coupling parameter, we do not interpret this result as a deviation from general relativity, because the given error might be too small due to nonmodeled effects or undetected systematics. Estimating the geodetic precession together with the highly correlated Yukawa coupling constant, we obtain

$$h = -0.003 \pm 0.005 .$$

The geodetic precession and its relation to the Yukawa coupling constant as well as to other model parts is subject to further studies.

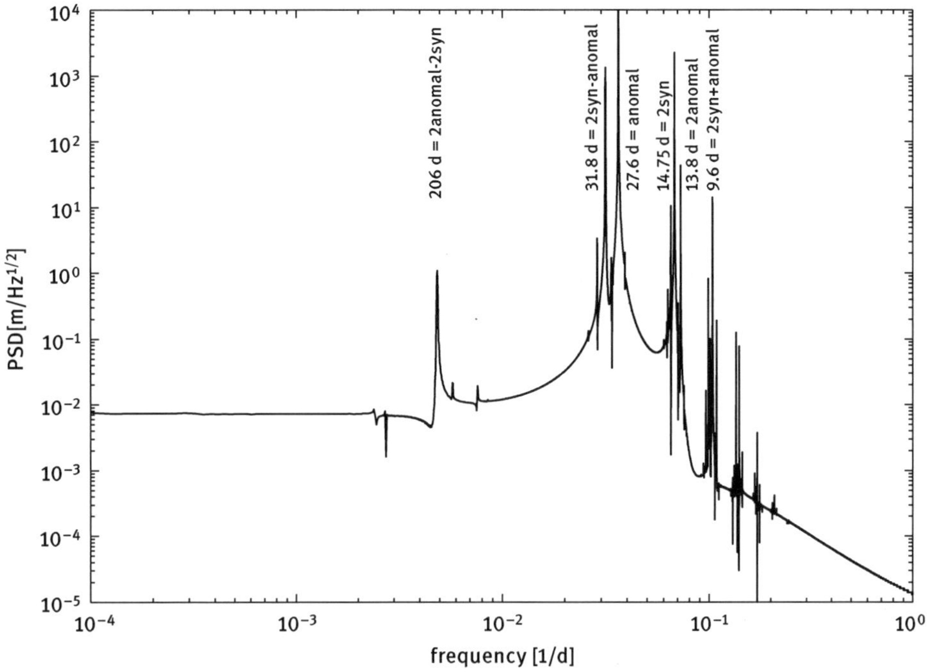

Fig. 10. Power spectrum of the difference in the Earth–Moon distance with and without a perturbation due to an additional geodetic precession with $h = 0.003$. See Figure 7 for labeling.

4.5 Metric parameter β

In the parametrized post-Newtonian (PPN) framework, the parameter β indicates the nonlinearity of gravity with a value of one in general relativity. Recent results from LLR at JPL agree with general relativity with an accuracy of 1.1×10^{-4} [79]. A determination based on planetary ephemeris leads to $\beta - 1 = -2 \pm 3 \times 10^{-5}$ [62] and $\beta - 1 = -6 \pm 8 \times 10^{-5}$ [14].

The parameter β is introduced in the EIH equations of motion, see Section 2.2. Computing the power spectrum of the difference of the Earth–Moon distance between an ephemeris with $\beta = 1$ and $\beta = 1+3\times10^{-4}$, Figure 11 is obtained. It shows dominant periods which are connected with the anomalistic and two times synodic frequencies.

In our LLR analysis, we have two possibilities to estimate β. The first is the direct estimation from the EIH equations (2.5) as a further parameter in the analysis. The corresponding partial derivatives are computed numerically. All given uncertainties are 3σ errors. In this case, we obtained

$$\beta - 1 = (3.5 \pm 1.1) \times 10^{-4} ,$$

with a moderate correlation to the also estimated initial values for orbital elements of the Earth. Fixing the Earth's orbit to the values of the standard solution, see Section 3,

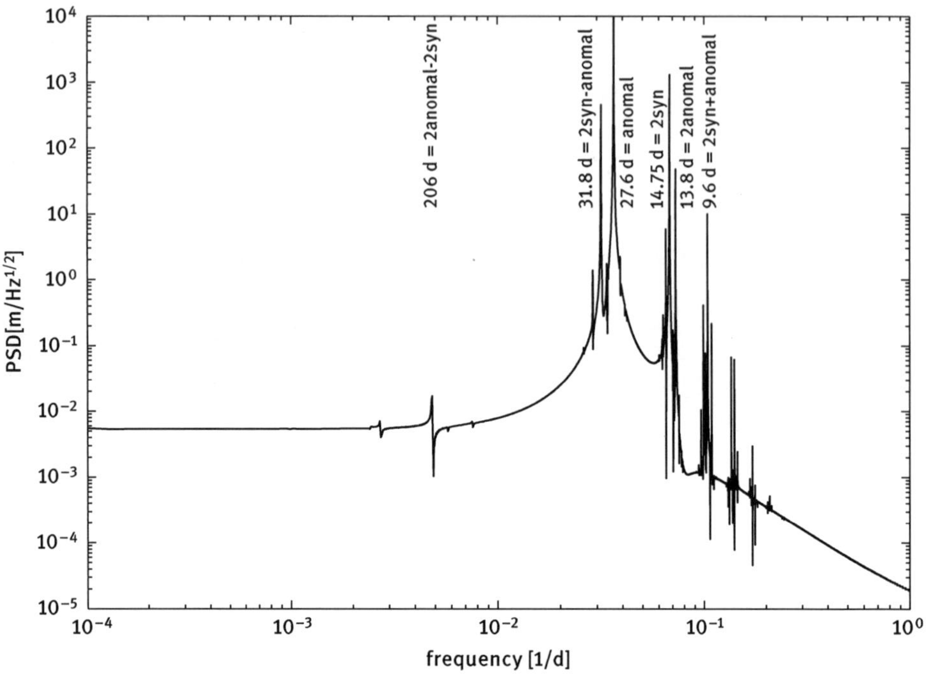

Fig. 11. Power spectrum of the difference in the Earth–Moon distance with and without a perturbation due to a nonunity value for the β PPN parameter with $\beta = 1 + 3 \times 10^{-4}$. See Figure 7 for labeling.

we get

$$\beta - 1 = (2.0 \pm 1.1) \times 10^{-4} .$$

When estimating β together with the geodetic precession, which is correlated with a factor of 0.7, we get

$$\beta - 1 = (1.7 \pm 1.5) \times 10^{-4} .$$

The second way to derive a value for β is based on a linear combination of further PPN parameters. Under the assumption of no composition-induced equivalence principle violation and no preferred-frame effects the parameter β is related to the strong equivalence principle parameter η and metric parameter γ as

$$\beta = \frac{1}{4} (\eta + \gamma + 3) . \tag{4.10}$$

The parameter γ is determined with high accuracy by data from the Cassini mission with a value of $\gamma = 1 + (2.1 \pm 2.3) \times 10^{-5}$ [4]. By taking η from Section 4.2, we obtain for the PPN parameter β

$$\beta - 1 = (0.6 \pm 1.1) \times 10^{-4} .$$

Here, the absolute value is within the error boundaries as we benefit from the high quality of the equivalence principle fit.

4.6 Preferred-frame parameters α_1, α_2

Preferred-frame effects can be described in the PPN formalism by three parameters α_1, α_2 and α_3 with corresponding counterparts $\hat{\alpha}_1$, $\hat{\alpha}_2$, and $\hat{\alpha}_3$ in a strong gravitational field generalization, e.g. by investigating pulsars [70, 78]. All parameters are zero in general relativity. The weak field parameters α_1 and α_2 can be estimated within our LLR analysis program with respect to a frame related to the cosmic microwave background, see [44]. Regarding the Earth–Moon distance, Figures 12 and 13 show the power spectra of a non-null result of the two parameters with typical annual signals for α_1. In contrast, the α_2 spectrum shows mainly signals with the anomalistic and two times the synodic frequency. A further idea is to consider the BCRS as a preferred frame. In this way, even a special variant of testing gravitomagnetism using LLR data is set up, for discussion and results see [72].

The best limit for $\hat{\alpha}_1$ is derived from binary pulsar observations with a value of $\hat{\alpha}_1 = (-0.4 \pm 3.4) \times 10^{-5}$ [70]. The second parameter in a weak field regime is derived by the alignment of the solar spin with the angular momentum of the whole Solar System and is determined as $|\alpha_2| < 2.4 \times 10^{-7}$ [55]. In the strong gravitational field regime of pulsars, the limit of $\hat{\alpha}_2$ is determined as $|\hat{\alpha}_2| < 1.6 \times 10^{-9}$ [69].

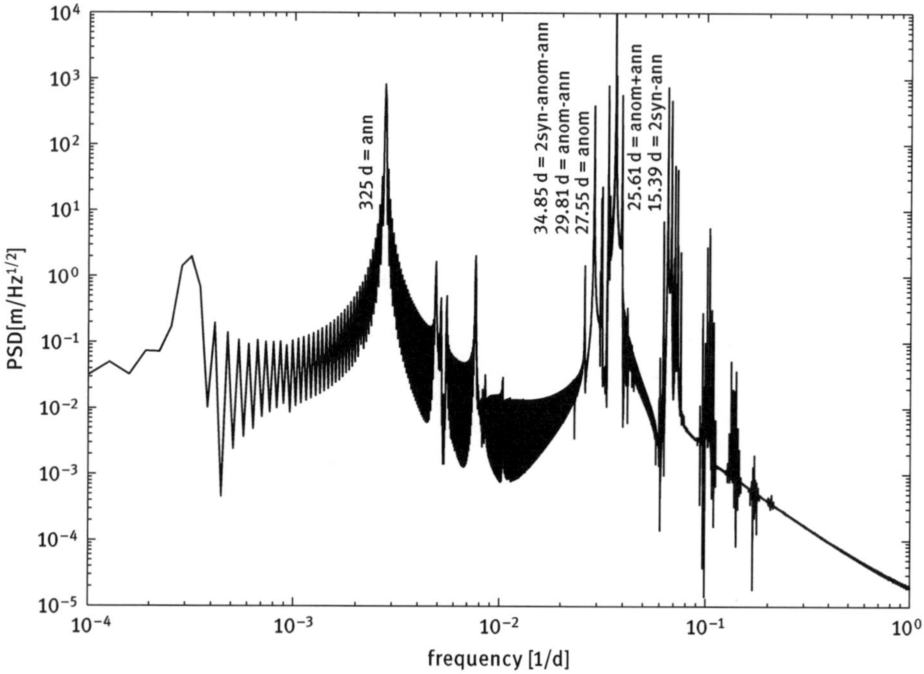

Fig. 12. Power spectrum of the difference in the Earth–Moon distance with and without a perturbation due to a non-zero value for the α_1 preferred-frame parameter with $\alpha_1 = 4 \times 10^{-5}$. "Ann" denotes the annual frequency, for further labels see Figure 7.

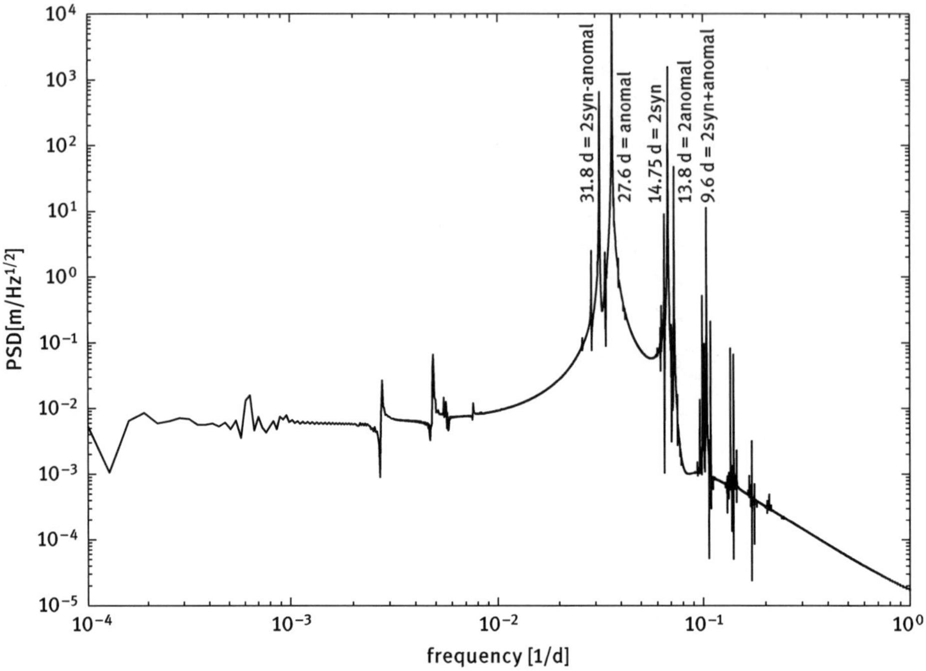

Fig. 13. Power spectrum of the difference in the Earth–Moon distance with and without a perturbation due to a non-zero value for the α_2 preferred-frame parameter with $\alpha_2 = 3 \times 10^{-5}$. See Figure 7 for labeling.

In our LLR-based solution, we obtained

$$\alpha_1 = (3.7 \pm 2.7) \times 10^{-5} \,,$$
$$\alpha_2 = (2.3 \pm 1.0) \times 10^{-5} \,,$$

with corresponding 3σ errors, see [44] for further error discussion of α_1. In contrast to α_1, our LLR solution can not compete with the α_2 solutions from other methods as mentioned above, but nevertheless it is an independent determination of these parameters.

5 Summary and outlook

Lunar Laser Ranging (LLR) by now is carried out for more than 44 years. With this space geodetic technique, parameters of the Earth–Moon system (e.g. station and reflector coordinates, lunar orbit, gravity field, tidal acceleration, and Earth orientation) could be obtained with high accuracy; we gave a few examples. The major application of LLR, however, is to test elements or predictions of Einstein's theory of gravity. Lunar

Table 1. IfE results for relativistic parameters. See Section 4 for a detailed explanation and comparison with results from other groups.

Parameter	**IfE results**
$\dot{G}/G$	$(1.4 \pm 1.5)\times 10^{-13}$ yr^{-1}
$\Delta(m_g/m_i)_{\mathrm{EM}}$	$(-0.9 \pm 1.9)\times 10^{-13}$
η	$(2.0 \pm 4.0)\times 10^{-4}$
α_{Yukawa}	$(-0.6 \pm 1.8)\times 10^{-11}$
$h_{\mathrm{Geodetic\ Precession}}$	$(-3.0 \pm 5.0)\times 10^{-3}$
PPN $\beta - 1$	$(1.7 \pm 1.5)\times 10^{-4}$
PPN $\beta - 1$ from η	$(0.6 \pm 1.1)\times 10^{-4}$
PPN α_1	$(3.7 \pm 2.7)\times 10^{-5}$
PPN α_2	$(2.3 \pm 1.0)\times 10^{-5}$

Laser Ranging analysis allows for constraining gravitational physics parameters related to the strong equivalence principle, Yukawa-like perturbations, preferred-frame effects, or the time variability of the gravitational constant. We discussed recent results for those relativistic quantities in detail where no violation of general relativity has been found so far. A summary of the results is given in Table 1.

The prerequisite for highly precise parameter estimation is a sophisticated analysis model that itself has to be consistently formulated in the context of Einstein's theory. We explicitly showed, where and how relativity enters the LLR model at IfE. Recent technical improvements (upgraded tracking system in Grasse, big APOLLO telescope) support and stimulate to improve modeling capabilities also from a relativistic point of view [29]. The future goal is to reach the 1 mm level of accuracy for our LLR model, which may allow us to touch the possible breakdowns of Einstein's theory. At IfE, we will start in improving the ephemeris calculation of which the various steps were discussed in detail in this report. This will certainly include the relevant numerical accuracy. Modeling improvements, among others, will comprise the asteroids, the lunar interior, the lunar rotation, etc.

The accuracy of lunar science parameters depends on the number and distribution of retroreflectors or transponders on the Moon as well as of the observatories on the Earth. Installing active laser transponders on the lunar surface would support both parts, as it would improve the lunar network and it would allow the extensive SLR network on Earth to engage in LLR on a routine basis. This would have essential impact in data volume, global distribution (no LLR from southern hemisphere, currently), improvements in Earth surface/atmospheric modeling, etc. Simulations at IfE [40] show the great benefit of including more observatories on Earth or more reflectors on the Moon.

Forty four years of Lunar Laser Ranging provided impressive results especially by putting Einstein's theory to test. The coming years promise even more spectacular results based on technological and modeling improvements.

Acknowledgement: Current LLR data are collected, archived, and distributed under the auspices of the International Laser Ranging Service (ILRS) ([59]). We acknowledge with thanks, that the more than 44 years of processed LLR data has been obtained under the efforts of the personnel at the Observatoire de la Côte d'Azur in France, the LURE Observatory in Maui, Hawaii, the McDonald Observatory in Texas as well as the Apache Point Observatory in New Mexico. We would also like to thank the International Space Science Institute (ISSI) in Bern for supporting this research. LLR-related research at the University of Hannover was funded by the Centre for Quantum Engineering and Space–Time Research QUEST and the DFG, the German Research Foundation, within the research unit FOR584 "Earth rotation and global dynamic processes" as well as within the research unit FOR1503 "Space–Time Reference Systems for Monitoring Global Change and for Precise Navigation in Space".

References

[1] E. G. Adelberger, New tests of Einstein's equivalence principle and Newton's inverse-square law, *Classical and Quantum Gravity* **18** (2001), 2397–2405.

[2] D. C. Agnew, *Earth Tides*, Treatise on Geophysics, Elsevier, 2007, pp. 163–195.

[3] R. Bauer, *Bestimmung von Parametern des Erde-Mond-Systems – Ein Beitrag zur Modellerweiterung und Bewertung, Ergebnisse*, Ph.D. thesis, Technische Universität München, 1989, Veröffentlichungen der Deutschen Geodätischen Kommision bei der Bayerischen Akademie der Wissenschaften, Reihe C, Nr. 353.

[4] B. Bertotti, L. Iess and P. Tortora, A test of general relativity using radio links with the Cassini spacecraft, *Nature* **425** (2003), 374–376.

[5] L. Biskupek and J. Müller, Lunar Laser Ranging and Earth Orientation, in: *Proceedings of the Journées 2008 «Systèmes de référence spatio-temporels»* (M. Soffel and N. Capitaine, eds.), pp. 182–185, Lohrmann-Observatorium and Observatoire de Paris, 2009, Les Journées 2008 «Systèmes de référence spatio-temporels» and X. Lohrmann-Kolloquium, Dresden, Germany, 22.–24.09.2008.

[6] L. Biskupek and J. Müller, Relativity and Earth Orientation Parameters from Lunar Laser Ranging, in: *Proceedings of the 16th International Workshop on Laser Ranging, October 2008, Poznan, Poland* (S. Schillak, ed.), 1, pp. 270–276, Space Research Centre, Polish Academy of Sciences, Warsaw, 2009, 16th International Workshop on Laser Ranging, Poznan, Poland, 12.–17.10.2008.

[7] L. Biskupek, J. Müller and F. Hofmann, *Determination of Nutation Coefficients from Lunar Laser Ranging*, Geodesy for Planet Earth (S. Kenyon, M. C. Pacino, U. Marti and M. G. Sideris, eds.), International Association of Geodesy Symposia 136, Springer, Berlin and Heidelberg, 2012, Proceedings of the 2009 IAG Symposium, Buenos Aires, Argentina, 31.08.–04.09.2009, pp. 521–525.

[8] R. J. Cappallo, Tidal dissipation in the moon, *Journal of Geophysical Research* **86** (1981), 7180–7184.

[9] T. Damour and D. Vokrouhlický, Equivalence principle and the Moon, *Physical Review* **53** (1996), 4177–4201.

[10] W. de Sitter, On Einstein's theory of gravitation and its astronomical consequences. Second paper, *Monthly Notices of the Royal Astronomical Society* **77** (1916), 155–184.

[11] J. O. Dickey, X. X. Newhall and J. G. Williams, Earth orientation from lunar laser ranging and an error analysis of polar motion services, *Journal of Geophysical Research* **90** (1985), 9353–9362.

[12] D. Egger, *Systemanalyse der Laserentfernungsmessung.*, Ph.D. thesis, Technische Universität München, 1985, Deutsche Geodätische Kommision bei der Bayerischen Akademie der Wissenschaften, Reihe C, Nr. 311.

[13] L. Fairhead and P. Bretagnon, An analytical formula for the time transformation TB-TT, *Astronomy and Astrophysics* **229** (1990), 240–247.

[14] A. Fienga, J. Laskar, P. Kuchynka, H. Manche, G. Desvignes, M. Gastineau, I. Cognard and G. Theureau, The INPOP10a planetary ephemeris and its applications in fundamental physics, *Celestial Mechanics and Dynamical Astronomy* **111** (2011), 363–385, 10.1007/s10569-011-9377-8.

[15] D. Giulini, Equivalence principle, quantum mechanics, and atom-interferometric tests, *ArXiv e-prints, 2011arXiv1105.0749G* (2011).

[16] H. Gleixner, *Ein Beitrag zur Ephemeridenrechnung und Parameterschätzung im Erde-Mond-System*, Ph.D. thesis, Technische Universität München, 1986, Veröffentlichungen der Deutschen Geodätischen Kommision bei der Bayerischen Akademie der Wissenschaften, Reihe C, Nr. 319.

[17] E. Hairer, S. Nørsett and G. Wanner, *Solving Ordinary Differential Equations I – Nonstiff Problems*, Springer Series in Computational Mathematics 8, Springer-Verlag, Berlin u. a., 2009.

[18] E. Hairer and G. Wanner, *Solving Ordinary Differential Equations II – Stiff and Differential-Algebraic Problems*, Springer Series in Computational Mathematics 14, Springer-Verlag, Berlin Heidelberg, 2010.

[19] T. Hirayama, H. Kinoshita, M.-K. Fujimoto and T. Fukushima, Analytical expression of TDB – TDT_0, in: *Relativistic Effects in Geodesy*, Proceedings of the International Association of Geodesy (IAG) - Symposia Tome I, pp. 91–100, Bureau Central de l'A.I.G., Paris, 1987, International Union of Geodesy and Geophysics (IUGG), XIX General Assembly, Vancouver, Canada, 10.–22.08.1987.

[20] F. Hofmann, J. Müller and L. Biskupek, Lunar laser ranging test of the Nordtvedt parameter and a possible variation in the gravitational constant, *Astronomy and Astrophysics* **522** (2010), L5+.

[21] IAU, *IAU Resolution 1991*, 1991, XXIst General Assembly, Buenos Aires, Argentina.

[22] ILRS, *International Laser Ranging Service, Lunar Laser Ranging website*, http://ilrs.gsfc.nasa.gov/science/scienceContributions/lunar.html, October 2013.

[23] L. Iorio, Constraints on the range λ of Yukawa-like modifications to the Newtonian inverse-square law of gravitation from Solar System planetary motions, *Journal of High Energy Physics* **10** (2007), 41.

[24] A. Irwin and T. Fukushima, A numerical time ephemeris of the Earth, *Astronomy and Astrophysics* **348** (1999), 642–652.

[25] D. J. Kapner, T. S. Cook, E. G. Adelberger, J. H. Gundlach, B. R. Heckel, C. D. Hoyle and H. E. Swanson, Tests of the gravitational inverse-square law below the dark-energy length scale, *Physical Review Letters* **98** (2007), 021101.

[26] S. Klioner, Relativistic scaling of astronomical quantities and the system of astronomical units, *Astronomy and Astrophysics* **478** (2008), 951–958.

[27] A. S. Konopliv, S. W. Asmar, E. Carranza, W. L. Sjogren and D. N. Yuan, Recent gravity models as a result of the lunar prospector mission, *Icarus* **150** (2001), 1–18.

[28] A. Konopliv, R. Park, D.-N. Yuan, S. Asmar, M. Watkins, J. Williams, E. Fahnestock, G. Kruizinga, M. Paik, D. Strekalov, N. Harvey, D. Smith and M. Zuber, The JPL lunar gravity field to spherical harmonic degree 660 from the GRAIL Primary Mission, *Journal of Geophysical Research: Planets* **118** (2013), 1415–1434.

[29] S. M. Kopeikin, E. Pavlis, D. Pavlis, V. A. Brumberg, A. Escapa, J. Getino, A. Gusev, J. Müller, W.-T. Ni and N. Petrova, Prospects in the orbital and rotational dynamics of the Moon with the

advent of sub-centimeter lunar laser ranging, *Advances of Space Research* **42** (2008), 1378–1390.

[30] G. Krasinsky, E. Pitjeva, M. Vasilyev and E. Yagudina, Hidden Mass in the Asteroid Belt, *Icarus* **158** (2002), 98–105.

[31] P. Kuchynka, J. Laskar, A. Fienga and H. Manche, A ring as a model of the main belt in planetary ephemerides, *Astronomy and Astrophysics* **514** (2010), A96+.

[32] K. Lambeck, *Geophysical geodesy – the slow deformations of the earth*, Oxford science publications, Clarendon Press, Oxford, 1988.

[33] H. Manche, *Élaboration de l'éphéméride INPOP: modèle dynamique et ajustements aux données de télémétrie laser Lune*, Promotion, L'Observatoire de Paris en Astronomie et Astrophysique, 2011.

[34] D. D. McCarthy and G. Petit (eds.), *IERS Conventions (2003)*, IERS Technical Note 32, Verlag des Bundesamts für Kartographie und Geodäsie, Frankfurt am Main, 2004.

[35] D. D. McCarthy and P. K. Seidelmann, *Time – From Earth Rotation to Atomic Physics*, Wiley-VCH, Berlin, 2009.

[36] P. Melchior, *The Tides of Planet Earth*, Pergamon Press, Oxford, 1983.

[37] V. B. Mendes and E. C. Pavlis, High-accuracy zenith delay prediction at optical wavelengths, *Geophysical Research Letters* **31** (2004), L14602.

[38] S. M. Merkowitz, Tests of Gravity Using Lunar Laser Ranging, *Living Reviews in Relativity* **13** (2010), 1–30.

[39] T. D. Moyer, *Mathematical formulation of the Double Precision Orbit Determination Program /DPODP/*, Jet Propulsion Laboratory, California Institute of Technology, Pasadena, California, Technical Report no. JPL-TR-32-1527, 1971.

[40] J. Müller, L. Biskupek, F. Hofmann and E. Mai, *Lunar Laser Ranging Contribution to GGOS*, http://www.iag2013.org/IAG_2013/Publication.html, 2013, talk at IAG Scientific Assembly 2013, Potsdam, 01.–06.09.2013, will be available online.

[41] J. Müller, F. Hofmann, X. Fang and L. Biskupek, Lunar Laser Ranging: Recent Results based on Refined Modelling, in: *Earth on the Edge: Science for a Sustainable Planet* (C. Rizos and P. Willis, eds.), IAG Symposia Series 139, 2013.

[42] J. Müller, T. W. Murphy, U. Schreiber, P. J. Shelus, J.-M. Torre, J. G. Williams, D. H. Boggs, S. Bouquillon and G. Francou, Lunar laser ranging – A tool for general relativity, lunar geophysics and earth science, *Journal of Geodesy* (2013), submitted.

[43] J. Müller and K. Nordtvedt, Lunar laser ranging and the equivalence principle signal, *Physical Review D* **58** (1998), 062001+.

[44] J. Müller, K. Nordtvedt and D. Vokrouhlický, Improved constraint on the α_1 PPN parameter from lunar motion, *Physical Review D* **54** (1996), 5927+.

[45] J. Müller, *Analyse von Lasermessungen zum Mond im Rahmen einer post-Newton'schen Theorie*, Ph.D. thesis, Technische Universität München, 1991, Veröffentlichungen der Deutschen Geodätischen Kommision bei der Bayerischen Akademie der Wissenschaften, Reihe C, Nr. 383.

[46] J. Müller and L. Biskupek, Variations of the gravitational constant from lunar laser ranging data, *Classical and Quantum Gravity* **24** (2007), 4533–4538.

[47] J. Müller, L. Biskupek, J. Oberst and U. Schreiber, Contribution of Lunar Laser Ranging to Realise Geodetic Reference Systems, in: *Geodetic Reference Frames* (H. Drewes, ed.), International Association of Geodesy Symposia 134, pp. 55–59, Springer, Berlin, Heidelberg, 2009.

[48] J. Müller, F. Hofmann and L. Biskupek, Testing various facets of the equivalence principle using lunar laser ranging, *Classical and Quantum Gravity* **29** (2012), 184006.

[49] J. Müller, J. G. Williams and S. G. Turyshev, Lunar Laser Ranging Contributions to Relativity and Geodesy, in: *Lasers, Clocks and Drag-Free Control: Exploration of Relativistic Gravity in Space* (H. Dittus, C. Lämmerzahl and S. G. Turyshev, eds.), Astrophysics and Space Science Library

349, pp. 457–472, Springer, Berlin, Heidelberg, 2008, ZARM, Bremen 30.05.–01.06.2005.

[50] T. W. Murphy, Lunar laser ranging: the millimeter challenge, *Reports on Progress in Physics* **76** (2013), 076901.

[51] T. W. Murphy, E. G. Adelberger, J. B. R. Battat, L. N. Carey, C. D. Hoyle, P. LeBlanc, E. L. Michelsen, K. Nordtvedt, A. E. Orin, J. D. Strasburg, C. W. Stubbs, H. E. Swanson and E. Williams, The Apache Point Observatory Lunar Laser-ranging Operation: Instrument Description and First Detections, *Publications of the Astronomical Society of the Pacific* **120** (2008), 20–37.

[52] J. Müller, M. Schneider, M. Soffel and H. Ruder, Testing Einstein's theory of gravity by lunar laser ranging, in: *Symposia Gaussiana, Proceedings of the 2nd Gauss Symposium, Conference A: Mathematics and Theoretical Physics* (M. Behara, R. Fritsch and R. G. Lintz, eds.), pp. 637–647, Walter de Gruyter, 01 1995.

[53] NASA, *Planetary Data System Geosciences Node – Gravity Models*, http://pds-geosciences.wustl.edu/dataserv/gravity models.htm, last visit: 07 Oct 2013.

[54] K. Nordtvedt, Equivalence Principle for Massive Bodies. I. Phenomenology, *Physical Review* **169** (1968), 1014–1016.

[55] K. Nordtvedt, Probing gravity to the second post-Newtonian order and to one part in 10 to the 7th using the spin axis of the sun, *Astrophysical Journal* **320** (1987), 871–874.

[56] K. Nordtvedt, Optimizing the observation schedule for tests of gravity in lunar laser ranging and similar experiments, *Classical and Quantum Gravity* **15** (1998), 3363–3381.

[57] K. Nordtvedt, J. Müller and M. Soffel, Cosmic acceleration of the Earth and Moon by dark matter., *Astronomy and Astrophysics* **293** (1995), L73–L74.

[58] J. Overduin, J. Mitcham and Z. Warecki, Expanded solar-system limits on violations of the equivalence principle, *ArXiv e-prints, 2013arXiv1307.1202O* (2013).

[59] M. R. Pearlman, J. J. Degnan and J. M. Bosworth, The International Laser Ranging Service, *Advances in Space Research* **30** (2002), 135–143.

[60] G. Petit and B. Luzum (eds.), *IERS Conventions (2010)*, IERS Technical Note 36, Verlag des Bundesamts für Kartographie und Geodäsie, Frankfurt am Main, 2010.

[61] E. V. Pitjeva, Ephemerides EPM2008: the updated model, constants, data, in: *Proceedings of the Journées 2008 «Systèmes de référence spatio-temporels»* (M. Soffel and N. Capitaine, eds.), pp. 57–59, Lohrmann-Observatorium and Observatoire de Paris, 2009, Les Journées 2008 «Systèmes de référence spatio-temporels» and X. Lohrmann-Kolloquium, Dresden, Germany, 22.–24.09.2008.

[62] E. V. Pitjeva and N. P. Pitjev, Relativistic effects and dark matter in the Solar system from observations of planets and spacecraft, *Monthly Notices of the Royal Astronomical Society* **432** (2013), 3431–3437.

[63] A. J. Sanders, G. T. Gillies and E. Schmutzer, Implications upon theory discrimination of an accurate measurement of the time rate of change of the gravitational parameter and other cosmological parameters, *arXiv 1005.3226* (2010).

[64] J. Schastok, *Drei Studien zur Dynamik des Erde–Mond–Systems*, Ph.D. thesis, Eberhard–Karls–Universität zu Tübingen, 1987.

[65] M. Schneider and C. Cui, *Theoreme über Bewegungsintegrale und ihre Anwendung in Bahntheorien*, Deutsche Geodätische Kommission bei der Bayerischen Akademie der Wissenschaften 121, Verlag der Bayerischen Akademie der Wissenschaften in Kommissionbeim Verlag C. H. Beck, 2005.

[66] M. Schneider and D. Deutsche Forschungsgemeinschaft, *Satellitengeodäsie. Ergebnisse aus dem gleichnamigen Sonderforschungsbereich der Technischen Universität München.*, 1990.

[67] M. Schneider, *Himmelsmechanik – Band I: Grundlagen, Determinierung*, B.I.-Wissenschaftsverlag, 1992.

[68] M. Schneider, *Himmelsmechanik – Band II: Systemmodelle*, B.I.-Wissenschaftsverlag, 1993.

[69] L. Shao, R. N. Caballero, M. Kramer, N. Wex, D. J. Champion and A. Jessner, A new limit on local Lorentz invariance violation of gravity from solitary pulsars, *Classical and Quantum Gravity* **30** (2013), 165019.
[70] L. Shao and N. Wex, New tests of local Lorentz invariance of gravity with small-eccentricity binary pulsars, *Classical and Quantum Gravity* **29** (2012), 215018.
[71] I. I. Shapiro, R. D. Reasenberg, J. F. Chandler and R. W. Babcock, Measurement of the de Sitter precession of the Moon: A relativistic three-body effect, *Physical Review Letters* **61** (1988), 2643–2646, American Physical Society.
[72] M. Soffel, S. Klioner, J. Müller and L. Biskupek, Gravitomagnetism and lunar laser ranging, *Physical Review D* **78** (2008), 024033.
[73] M. Soffel and R. Langhans, *Space-Time Reference Systems*, Astronomy and Astrophysics Library, Springer, Berlin, Heidelberg, 2013.
[74] E. M. Standish, Jr., *JPL Planetary and Lunar Ephemerides, DE405/LE405*, Jet Propulsion Laboratory, California Institute of Technology, JPL Interoffice Memorandum no. IOM 312.F-98-048, Passadena, Califonia, 1998.
[75] P. J. Steinhardt and D. Wesley, Exploring extra dimensions through observational tests of dark energy and varying Newton's constant, *arXiv 1003.2815* (2010).
[76] D. Vokrouhlický, A Note on the Solar Radiation Perturbations of Lunar Motion, *Icarus* **126** (1997), 293–300.
[77] T. A. Wagner, S. Schlamminger, J. H. Gundlach and E. G. Adelberger, Torsion-balance tests of the weak equivalence principle, *Classical and Quantum Gravity* **29** (2012), 184002.
[78] C. M. Will, *Theory and Experiment in Gravitational Physics*, 2nd ed, Cambridge University Press, England, 1993.
[79] J. G. Williams and W. M. Folkner, Lunar Laser Ranging: Relativistic Model and Tests of Gravitational Physics, *IAU Symposium 261,Bulletin of the American Astronomical Society* **41** (2009).
[80] J. G. Williams and D. H. Boggs, Lunar Core and Mantle. What Does LLR See?, in: *Proceedings of the 16th International Workshop on Laser Ranging, October 2008, Poznan, Poland* (S. Schillak, ed.), 1, pp. 101–120, Space Research Centre, Polish Academy of Sciences, Warsaw, 2009, 16th International Workshop on Laser Ranging, Poznan, Poland, 12.–17.10.2008.
[81] J. G. Williams, D. H. Boggs and R. L. Ricklefs, A Larger Lunar Core?, in: *Lunar and Planetary Institute Conference Abstracts*, Lunar and Planetary Inst. Technical Report 40, 1452, mar 2009, Provided by the SAO/NASA Astrophysics Data System.
[82] J. G. Williams, D. H. Boggs, C. F. Yoder, T. J. Ratcliff and J. O. Dickey, Lunar rotational dissipation in solid body and molten core, *Journal of Geophysical Research* **106 (E11)** (2001), 27933–27968.
[83] J. G. Williams, S. G. Turyshev and D. H. Boggs, Lunar laser ranging tests of the equivalence principle, *Classical and Quantum Gravity* **29** (2012), 184004.
[84] J. G. Williams, S. G. Turyshev and D. H. Boggs, Lunar Laser Ranging Tests of the Equivalence Principle with the Earth and Moon, *International Journal of Modern Physics D* **18** (2009), 1129–1175.
[85] J. G. Williams, S. G. Turyshev, D. H. Boggs and T. J. Ratcliff, Lunar laser ranging science: Gravitational physics and lunar interior and geodesy, *Advances in Space Research* **37** (2006), 67–71, Presented at 35th COSPAR Scientific Assembly, 18.–24.07.2004, Paris.
[86] J. Williams, D. Boggs and W. Folkner, *DE430 Lunar Orbit, Physical Librations and Surface Coordinates*, Jet Propulsion Laboratory, Report no. IOM 335-JW,DB,WF-20080314-001, 2013.

Ignazio Ciufolini, Antonio Paolozzi, Vahe Gurzadyan, Erricos C. Pavlis, Rolf König, John Ries, Richard Matzner, Roger Penrose, Giampiero Sindoni, and Claudio Paris

Dragging of inertial frames, fundamental physics, and satellite laser ranging

1 Introduction

The composition of most of the universe in which we live is one of the biggest riddles of science. This is the mystery of the nature of dark energy and dark matter, which have been estimated to constitute about 95% of the universe [1–5]. The laws of physics cover the four fundamental interactions: gravitational, electromagnetic, weak, and strong. The last three interactions have been encompassed in the standard model theory of gauge symmetries, and the recent verification of the existence of the Higgs boson has confirmed an essential component of the standard model. Other theories, not yet experimentally verified, such as string and brane-world theories, try to unify gravitation with the other three interactions and thus unify the two great physical theories of general relativity and quantum mechanics. The aim is the unification of the four interactions of nature in a theory that can be experimentally tested, and that can address the riddle of dark energy and dark matter.

Gravitational interaction and the evolution of the universe are currently described by Einstein's gravitational theory of general relativity [6]. General relativity is a triumph of classical thought, created by Einstein to satisfy the competing requirements of the equivalence principle (local inertial physics can show no evidence of gravity), and the large scale effects of gravity. Einstein's gravitational theory succeeded by postulating that gravitation is the curvature of spacetime: this is a fundamental compo-

Ignazio Ciufolini: Dipartimento Ingegneria dell'Innovazione, Università del Salento, Lecce, Italy
Ignazio Ciufolini: Centro Fermi, Rome, Italy
Antonio Paolozzi, Giampiero Sindoni, Claudio Paris: Scuola di Ingegneria Aerospaziale and DIAEE, Sapienza Università di Roma, Italy
Vahe Gurzadyan: Center for Cosmology and Astrophysics, Alikhanian National Laboratory, Yerevan, Armenia
Erricos C. Pavlis: Goddard Earth Science and Technology Center (GEST), University of Maryland, Baltimore County, USA
Rolf König: Helmholtz Centre Potsdam – GFZ German Research Centre for Geosciences, Germany
John Ries: Center for Space Research, University of Texas at Austin, USA
Richard Matzner: Center for Relativity, University of Texas at Austin, USA
Roger Penrose: Mathematical Institute, University of Oxford, UK

nent for understanding the universe that we observe. During the past century, general relativity achieved a cumulative experimental triumph [7–10]. A number of key predictions of Einstein's gravitational theory have been experimentally confirmed with impressive accuracy. General relativity today has practical applications in space research, geodesy, astronomy, and navigation in the Solar System [11], from the global navigation satellite systems (GNSS) to the techniques of very long baseline interferometry (VLBI) and satellite laser ranging (SLR), and is a basic ingredient for understanding astrophysical and cosmological observations such as the expanding universe and the dynamics of the binary systems of neutron stars.

Despite being a well-verified description of gravity, general relativity has encountered somewhat unexpected developments in observational cosmology and is affected by some theoretical problems. Indeed, the study of distant supernovae in 1998 led to the discovery that they accelerate away from us. Since then, what is now referred to as dark energy is at the center of attention of many theoreticians. Observational data currently support its interpretation as the cosmological constant introduced by Einstein. However its current value, comparable with the critical density, needs to be reconciled with the expectations of quantum field theory or any analogous fundamental theory (e.g. [12]). Combining gravity with quantum field theory may be expected to reveal the nature of dark energy and hence resolve the mystery of its value, and whether it might be related to dark matter.

Among its theoretical problems, general relativity predicts the occurrence of spacetime singularities [13], events in which every known physical theory ceases to be valid, the spacetime curvature diverges, and time ends. Furthermore, general relativity is a classical theory that does not include quantum mechanics and no one has succeeded in producing a quantized version of general relativity, although this is a serious ongoing effort, with both loop quantum gravity and string theory approaches. A breakdown of general relativity should occur at the quantum level, but some viable modifications of Einstein's theory already give different predictions at the classical level and might explain the riddle of dark energy. Modifications of general relativity on cosmological scales, for instance the so-called $f(R)$ theories (with higher order curvature terms in the action), have been proposed to explain the acceleration of the universe without dark energy [14].

Discovery of the dark energy conundrum has especially emphasized that every aspect of Einstein's gravitational theory should be directly tested and the accuracy of the present measurements of general relativity and of the foundations of gravitational theories should be further improved. This chapter discusses observations and tests based on precision location and navigation techniques in the Solar System (and, in principle, beyond).

2 Dragging of inertial frames

Dragging of inertial frames, or frame dragging, and gravitomagnetism are produced by currents of mass–energy, for example, by the angular momentum of a body, in a way similar to what happens in electrodynamics where magnetism is generated by electric currents [15]. The accurate measurement of frame dragging and gravitomagnetism is one of the main challenges in experimental gravitation, together with the direct detection of gravitational waves, with the improvement of the accuracy in the measurement of the post-Newtonian parameters that test general relativity versus alternative metric theories of gravity, and with the improvement in the accuracy of testing the equivalence principle. Observational tests of gravitational physics might be divided into Solar System measurements, binary pulsar observations, and intermediate- and long-range cosmological observations via gravitational radiation. The measurements in the Solar System include redshift and clock measurements, light deflection and time delay of electromagnetic waves, Lunar Laser Ranging (LLR), geodetic precession, and frame-dragging measurements.

With regard to frame dragging, the nature and origin of inertia has intrigued scientists and philosophers for centuries. What determines an inertial frame? In the Newtonian theory of gravitation, the inertial systems have absolute existence: they are not affected by the matter in the Universe. In Einstein's theory of gravitation, the local inertial systems play a key role [6, 10, 16]. The equivalence principle, at the foundation of general relativity and other metric theories, states that the gravitational field is locally "unobservable" in a sufficiently small freely falling frame and, therefore, in these local inertial frames, all the laws of physics are the laws of special relativity.

However, the local inertial frames are determined and dragged by the distribution and flow of mass–energy in the Universe. The axes of these local inertial frames are determined by freely falling, torque-free gyroscopes, i.e. "spinning tops" sufficiently small and accurate. Therefore, these gyroscopes are dragged by the motion and rotation of a mass [6, 10, 16], i.e. they change their orientation with respect to the distant stars: this is the "dragging of inertial frames" or "frame dragging," as Einstein called it in a letter to Ernst Mach [17]. If we were to rotate with respect to these gyroscopes, we would then feel centrifugal forces, even though we would not rotate with respect to the "distant stars," in contrast with our everyday intuition. In fact, a gyroscope is dragged by a rotating mass, i.e. it changes its orientation with respect to the "distant stars." Frame dragging is, in Einstein's theory, the remnant of the ideas of Mach on the origin of inertia: Mach thought that inertial and centrifugal forces were caused by the rotation and acceleration of a system with respect to all the masses in the universe, and this is known as Mach's principle [10].

Frame dragging has also an intriguing influence on the flow of time and on the propagation of electromagnetic waves around a rotating body. In fact, the synchronization of clocks over a closed path around a rotating body is not possible [18, 19] in any rigid frame which is nonrotating with respect to the 'fixed stars.' In fact, the light that travels in the same direction as the rotation of the central body takes less time to return to the starting point (fixed with respect to the "distant stars") than does light traveling in the opposite direction [18–22]. Frame-dragging influences clocks, electromagnetic waves, gyroscopes [23, 24] (for example, the gyroscopes of the space experiment Gravity Probe B (GP-B)), orbiting particles [25] (see below, the sections on the LAGEOS satellites and on the space experiment LARES), and matter orbiting and falling on a rotating body. In fact, an explanation of the constant orientation of the spectacular jets from active galactic nuclei (AGN) and quasars, emitted in the same direction for a period of time that can reach millions of years, is based on frame dragging of the accretion disk by the central rotating supermassive black hole [26, 27], which acts as a gyroscope.

In general relativity, a small torque-free gyroscope defines an axis of a local inertial frame. Also the orbital plane of a test particle is a kind of gyroscope. In fact, as with a small gyroscope, the orbital plane of a planet, moon, or satellite is a huge gyroscope that feels general relativistic effects. Indeed, frame dragging produces a change of the orbital angular momentum vector of a test particle, i.e. the Lense–Thirring effect, that is, the precession of the nodes of a satellite, i.e. the rate of change of its nodal longitude: $\dot{\Omega}_{\text{Lense–Thirring}} = \frac{2GJ}{c^2 a^3 (1-e^2)^{3/2}}$, where Ω is the longitude of the nodal line of the satellite (the intersection of the satellite's orbital plane with the equatorial plane of the central body), J is the angular momentum of the central body, a the semimajor axis of the orbiting test particle, e its orbital eccentricity, G the gravitational constant and c the speed of light. A similar formula also holds for the rate of change of the longitude of the pericentre of a test particle, that is, of the so-called Runge–Lenz vector [10, 25].

The phenomenon of frame dragging, due to mass–energy currents and to the rotation of a mass, can be usefully described by a formal analogy of general relativity with electrodynamics in the case of a weak gravitational field and slow moton (see Figure 1) [10, 27]. This formal analogy is called gravitomagnetism. Whereas an electric charge generates an electric field and a current of electric charge generates a magnetic field, in the Newtonian theory the mass of a body generates a gravitational field but a current of mass, for example the rotation of a body, does not generate any additional gravitational field. On the other hand, Einstein's theory of gravitation states that a current of mass generates a gravitomagnetic field that exerts a force on surrounding bodies, i.e. a current of mass changes the structure of spacetime by generating an additional spacetime curvature [28]. Moreover, in general relativity, a small current of mass in a loop (i.e. a gyroscope) has a behavior formally similar to that of a magnetic dipole in electrodynamics, which is made of an electric current in a loop. The gravitomagnetic field produces the frame dragging of a gyroscope, in a way similar to the magnetic field producing the change of orientation of a magnetic needle (magnetic

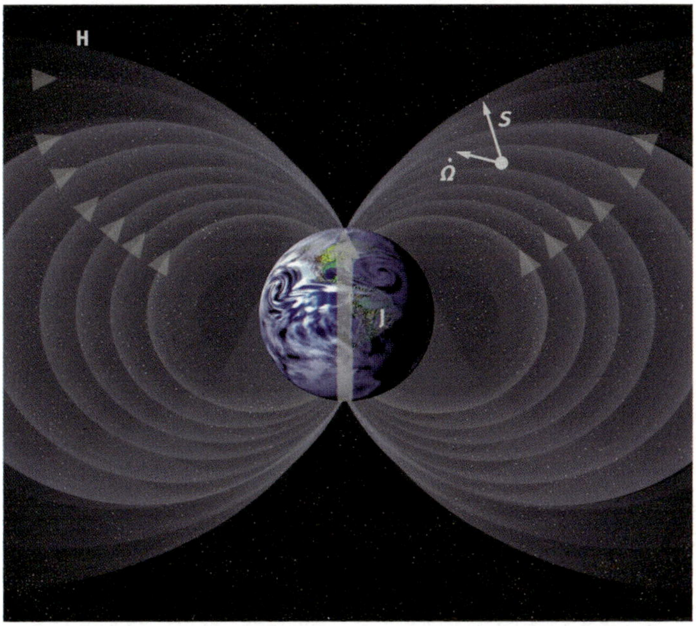

Fig. 1. The gravitomagnetic field [27] $\boldsymbol{H}$, in weak field and slow motion, generated in general relativity by the angular momentum $\boldsymbol{J}$ of a central body looks similar to the magnetic field $\boldsymbol{B}$ generated by a magnetic moment $\boldsymbol{m}$ in electrodynamics. Frame dragging, $\dot{\boldsymbol{\Omega}}$, is represented on a test gyroscope $\boldsymbol{S}$.

dipole). The precession of the spin axis of a test gyroscope by the angular momentum **J** of the central body is: $\dot{\Omega}_{\text{Spin}} = \frac{G(3(\mathbf{J}\cdot\hat{r})\hat{r}-\mathbf{J})}{c^2r^3}$, where $\dot{\Omega}_{\text{Spin}}$ is the rate of change of the spin vector of the gyroscope, $\hat{r}$ is the position unit-vector of the test gyroscope, and r is its radial distance from the central body.

In general relativity, the gravitomagnetic field, **H**, due to the angular momentum **J** of a central body is, in the weak field and slow motion approximation:

$$\mathbf{H} = \nabla \times \mathbf{h} \cong 2G\left[\frac{\mathbf{J} - 3(\mathbf{J}\cdot\hat{x})\,\hat{x}}{c^3r^3}\right],$$

where r is the radial distance from the central body, $\hat{x}$ is the position unit-vector and **h** is the so-called gravitomagnetic vector potential (equal to the nondiagonal, space and time, part of the metric), see Figure 1.

Since frame dragging is due to the additional spacetime curvature generated by the rotation of a mass, the Pontryagin pseudoinvariant (built using the Riemann curvature tensor) has been proposed for a precise characterization of this phenomenon: see the next section, [29, 30] and Section 6.11 of [10]). For discussions of the meaning of frame dragging and gravitomagnetism, see [29–38] and Section 6.11 of [10].

3 Tests of string theory and the LAGEOS and LARES space experiments

Chern–Simons gravity is one of the most often discussed modified gravity models as extensions of general relativity. It naturally emerges from string theory as a required anomaly cancelling term to conserve unitarity, and also from loop quantum gravity, as a counter term for the anomaly [39, 40].

The Chern–Simons model is involved in the interpretation of a variety of cosmological and astrophysical phenomena such as dark energy, the dynamics of binary pulsars, gravitational wave emission by binary back holes, and the properties of the jets observed in active galactic nuclei and other properties of the accretion into massive black holes; see [41–46] and the references therein.

While those observational phenomena still do not allow the obtaining of reasonable constraints on Chern–Simons gravity, the forthcoming high precision measurements of frame dragging by LARES (LAser RElativity Satellite) can, as we will discuss below.

Chern–Simons gravity is defined by the correction to the Einstein–Hilbert action in the form (e.g. [41, 43])

$$S_{CS} = \int d^4x\sqrt{-g}\left[\frac{l}{12}\theta R\overline{R} - \frac{1}{2}(\partial\theta)^2 - V(\theta)\right], \tag{3.1}$$

where the units $c = G = 1$ are used and θ is a scalar field, R is the Ricci scalar, and l is a coupling constant. Here $R\overline{R}$ is the Pontryagin density [49]

$$R\overline{R} = R^{b\ \ cd}_{\ a}\overline{R}^{a}_{\ bcd}, \tag{3.2}$$

where the dual of the Riemann tensor is given as

$$\overline{R}^{a}_{bcd} = \frac{1}{2}\epsilon_{efcd}R^{a\ ef}_{\ b}, \tag{3.3}$$

ϵ_{efcd} is the totally antisymmetric Levi–Civita tensor, the roman indices takes values 0, 1, 2, 3, and repeated indices mean the Einstein summation rule.

The dynamical equation for the scalar field θ is

$$\Box\theta = \frac{dV}{d\theta} - \frac{1}{12}lR\overline{R}, \tag{3.4}$$

where $\Box$ is the d'Alembertian operator and l is a new length scale, a parameter of the theory. The modified gravitational field equations are

$$G^{ab} - \frac{16}{3}lC^{ab} = 8\pi T^{ab}, \tag{3.5}$$

where G^{ab} is the Einstein tensor and C^{ab} is the Cotton–York tensor

$$C^{ab} = v_l\left(\epsilon^{lacd}\nabla_c R^b_d + \epsilon^{lbcd}\nabla_c R^a_d\right) + v_{lk}\left(\overline{R}^{kalb} + \overline{R}^{kbla}\right). \tag{3.6}$$

and

$$v_l = \partial_l\theta = \nabla_l\theta\,, \tag{3.7}$$

$$v_{lk} = \nabla_l v_k = \nabla_l\nabla_k\theta\,. \tag{3.8}$$

are respectively Chern–Simons velocity and acceleration. T^{ab} includes the contributions of the matter stress–energy tensor T^{ab}_{mat} and stress–energy tensor T^{ab}_{θ} of the scalar field

$$T^{ab}_{\theta} = (\nabla_a\theta)\,(\nabla_b\theta) - \frac{1}{2}g_{ab}\,(\nabla_a\theta)\,(\nabla^a\theta) - g_{ab}V\,(\theta)\,. \tag{3.9}$$

For dark energy models, e.g. with dark energy density expressed via Ricci scalar R [47]

$$\rho = -\frac{\alpha}{16\pi}R\,, \tag{3.10}$$

where α is a constant, the Friedmann equation reads

$$\left(\frac{\ddot{a}}{a}\right)^2 + \frac{k}{a^2} = \alpha\left(\ddot{H} + 2H^2 + \frac{k}{a^2}\right) + \frac{4\pi}{3}\ddot{\theta}^2\,, \tag{3.11}$$

and $\ddot{H} + 2H^2 = \ddot{a}/a + (\ddot{a}/a)^2$, where a is the scale factor of the space geometry and H the Hubble "constant". It yields

$$\alpha\frac{\ddot{a}}{a} + \left(\frac{\ddot{a}}{a}\right)^2(\alpha - 1) + \frac{k}{a^2}(\alpha - 1) + \frac{4\pi}{3}\ddot{\theta}^2 = 0\,. \tag{3.12}$$

The data of the LAGEOS satellites have enabled to obtain a lower limit 0.01 km^{-1} for the Chern–Simons parameter [41]

$$m_{CS} = -3/(8\pi l\dot{\theta})\,,$$

as the ratio of the drag rates for Chern–Simons theory and for general relativity yields

$$\frac{\dot{\Omega}_{\text{CS}}}{\dot{\Omega}_{\text{GR}}} = 15\frac{a_L^2}{R_\oplus^2}j_2(m_{\text{CS}}R_\oplus)y_1(m_{\text{CS}}a_L)\,, \tag{3.13}$$

where $R_\oplus$ is the radius of Earth, a_L the semimajor axis of the LAGEOS satellite, and $j_\ell(x)$ and $y_\ell(x)$ are, respectively, the first and the second kind spherical Bessel functions [41].

LARES is expected to produce higher accuracy data and one can obtain the expected constraint on the Chern–Simons parameter

$$\dot{\Omega}_{\text{LARES}} = \dot{\Omega}_{\text{GR}}(1 + n)\,, \tag{3.14}$$

in the form of the dependence of $n = \frac{\dot{\Omega}_{\text{CS}}}{\dot{\Omega}_{\text{GR}}}$ (in percent) versus the lower limit of the Chern–Simons parameter m_{CS} (in km^{-1}); see Figure 2 [48].

Thus, the expected data of LARES can have a direct impact on the interpretation of the dark energy data within Chern–Simons gravity models, as well as on various astrophysical phenomena.

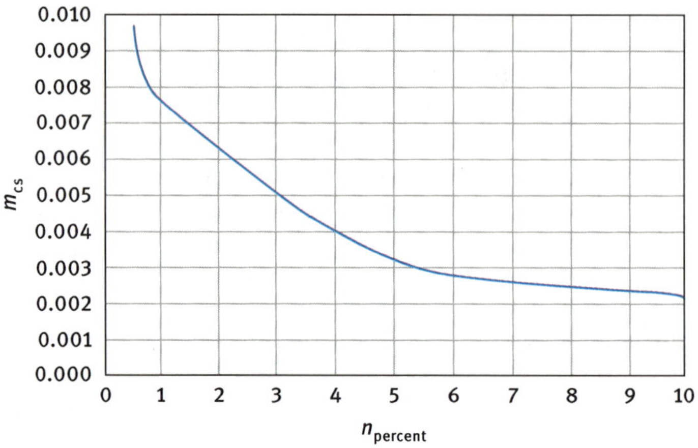

Fig. 2. The dependence of the Chern–Simons constant m_{CS} on the frame-dragging measurement parameter n as expected by the LARES mission [48].

4 LAGEOS and Gravity Probe B: two independent space experiments measuring frame dragging

The effect of frame dragging was hypothesized in 1896, i.e. before the theory of general relativity was conceived. Researchers, influenced by the ideas of Ernst Mach, attempted to measure frame-dragging effects generated by the rotation of Earth on torsion balances [51] and ground gyroscopes [52]. The first quantitative derivation was obtained by de Sitter in 1916, on the basis of Einstein's general relativity, which calculated the precession of the perihelion of Mercury due to the angular momentum of the Sun. In 1918, Lense and Thirring [25] derived the formula describing, in the case of weak field and slow motion, the frame-dragging orbital effects on a test particle orbiting a rotating body, which is now known as the Lense–Thirring effect (see Section 2).

Around 1959, a space experiment (Gravity Probe B) was proposed to test the frame dragging of orbiting gyroscope [23, 24]. Later on, another space experiment based on laser-ranged satellite was proposed [61] and succesfully launched in 2012 [50]. General relativity predicts that, at the altitude of the Gravity Probe B spacecraft, the frame-dragging secular precession of the four gyroscopes, owing to the rotation of the Earth, is approximately 39 milliarcsec per year (i.e. 0.000011 deg per year) around an axis contained in the polar orbital plane of Gravity Probe B. On the 20th of April 2004, after more than 40 years of preparation, the experiment was launched into a polar orbit at an altitude of about 642 km. Gravity Probe B [53] (see http://einstein.stanford.edu/) consisted of a spacecraft carrying four gyroscopes and a telescope pointing towards the guide star IM Pegasi (HR8703). It was designed to measure the drifts of the gyroscopes predicted by general relativity (frame dragging and geodetic precession) with respect to the far away "fixed" stars. On the 14th of April 2007, after approximately 18 months of data analysis the first results of Gravity Probe B were presented: The Gravity Probe B experiment was suffering from an unexpectedly large drift of the rotation axes of the gyroscopes produced by unexpected classical torques. The team of the Gravity Probe B [54] (see also [55]) ascribed these large drifts to electrostatic patches on the surface of the spheres of the gyroscopes and supports, and estimated the non-modeled systematic errors to be of the order of 100 milliarcsec per year, corresponding to an uncertainty of more than 250% of the frame dragging induced by the Earth's rotation. In 2011, after approximately 4 years of data analysis, the Gravity Probe B team claimed that by including in the estimation the effect of the charge patches, the uncertainty in the measurement of frame dragging had been reduced to about 19% [53].

An alternative approach to the measurement of the dragging of inertial frames is the use of laser-ranged satellites. Since frame dragging is extremely small for Solar System bodies and Earth's satellites, it is necessary to measure the position of a spacecraft with high accuracy. This is possible thanks to the technique of laser-ranging, which is the most accurate technique for measuring distances to the Moon and to artificial satellites such as LAGEOS (LAser GEOdynamics Satellite) and LAGEOS 2 [56]. Short laser pulses emitted by ground stations and reflected back from retro-reflectors placed on the Moon or on artificial satellites allow, through the measurement of their time of flight, an extremely precise determination of the distance of the retro-reflectors from the ground station. The positioning of the LAGEOS type satellites can reach an accuracy of few millimeters [57]. The combined use of ranging measurements from a number of stations around the globe allows accurately reconstructing the orbit of these satellites and in particular the nodal longitude of LAGEOS and LAGEOS 2 with an accuracy of a fraction of a milliarcsec per year [58–60]. Table 1 shows some information about the LAGEOS satellites.

Since using laser-ranging makes it possible to determine the orbits of the LAGEOS satellites with an accuracy of a few centimeters [58–60] and their nodal rate with an accuracy of a fraction of milliarcsec per year, provided that all their orbital perturbations are modeled well enough [58, 61, 62], it is possible to measure the Lense–Thirring effect and frame dragging. The LAGEOS satellites are indeed solid metal spheres with low surface-to-mass ratios, so that atmospheric particles and photons can only slightly perturb their orbits and in particular can only very slightly change the orientation of their orbital planes [58, 62–64]. Nevertheless, the main secular displacement of their

Table 1. The LAGEOS satellites [56, 57].

	Launch (year)	Semimajor axis (km)	Inclination (deg)	Institution	Frame dragging: Milliarcsec per year	Frame dragging: Meters per year
LAGEOS	1976	≅ 12 270	≅ 52.65	NASA	≅ 31	≅ 1.9
LAGEOS 2	1992	≅ 12 160	≅ 109.9	ASI and NASA	≅ 31.5	≅ 1.9

orbital planes is by far that from the Earth's deviations from spherical symmetry and in particular to the Earth's even zonal spherical harmonics, which are the main source of error in the measurement of frame dragging [65]. Recall that the gravitational field of the Earth and its gravitational potential can be expanded in spherical harmonics and the even zonal harmonics are those of even degree and zero order. The even zonals are those deviations from spherical symmetry of the gravitational potential of Earth that are axially symmetric and that are also symmetric with respect to its equatorial plane. They produce large secular drifts of the nodes of the LAGEOS satellites. The coefficients that measure the size of the even zonal harmonics are denoted by J_{2n}, where $2n$ is their degree. In particular, the flattening of the gravitational potential of the Earth, corresponding to the second degree zonal harmonic, J_2, describes the quadrupole moment of the Earth and is by far the largest source of error in the measurement of frame dragging because it produces the highest secular perturbation of

Fig. 3. The GFZ EIGEN-6C Geoid Model from 2011. EIGEN-6C (European Improved Gravity model of the Earth by New techniques – model 6C) is a high-resolution global gravity field model. It is the first combination model that includes GOCE data. In this figure the geoid is plotted with respect to the Earth's reference ellipsoid to remove the dominant Earth's quadrupole component. Its role is fundamental in geodesy and Earth sciences and ranges from practical purposes, like orbit determination, to scientific applications, like the investigation of the density structure of the Earth's interior. The recent EIGEN released in 2012 is called EIGEN-6C2 and has been created from a combination of a multitude of data (courtesy of GFZ-Potsdam).

the nodes of the LAGEOS [61, 66] and LARES satellites. However, thanks to the observations of the geodetic satellites, the Earth's gravitational field is extremely well known. For example, the flattening of the gravitational potential of Earth is currently measured [67] with an uncertainty of about one part in 10^7, which is, however, still not sufficient for testing frame dragging. To eliminate the orbital uncertainties due to the errors in the models of the Earth's gravitational field, the use of both the LAGEOS and LAGEOS 2 satellites has been proposed [66]. Nevertheless, it has not been easy to evaluate the accuracy of some previous measurements [68] of the Lense–Thirring effect with the LAGEOS satellites, given the limitations in the knowledge of the real uncertainty of the Earth's gravitational models available in 1998.

In March 2002, the problem of the uncertainties in the Earth's gravitational field was overcome with the launch of two twin GRACE (Gravity Recovery and Climate Experiment) satellites [69, 70] of NASA/USA and DLR/Germany, 200–250 km apart, into a near-polar orbit at an altitude of about 400 km. The spacecraft range to each other by radar and collect Global Positioning System (GPS) data for positioning and timing

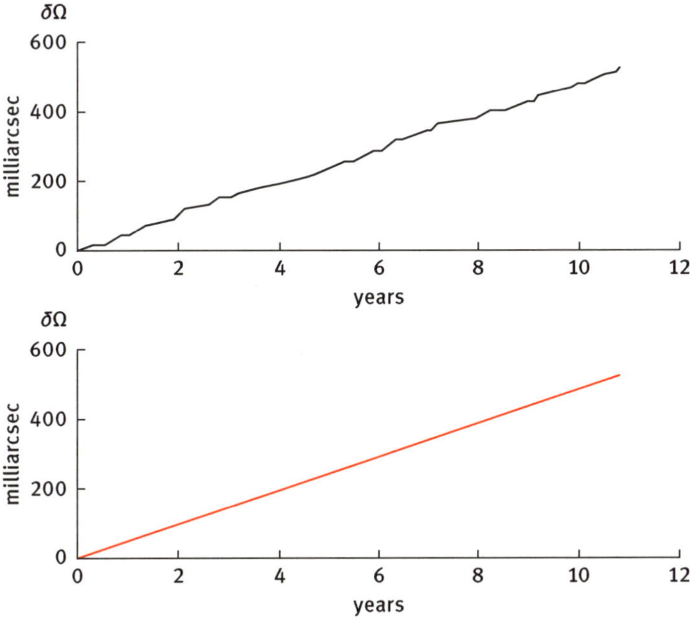

Fig. 4. Measurement of frame dragging obtained in 2004 analyzing the LAGEOS and LAGEOS 2 orbits with GEODYN, using the 2004 GRACE determination of the Earth's gravitational field [71, 77]. The top figure (in black) shows the cumulative residual of the nodal longitudes, $\delta\Omega$, of the LAGEOS satellites combined to eliminate the uncertainty in the quadrupole moment of Earth. The best fit line (not reported here for sake of clarity) through this observed residual has a slope of 47.9 milliarcsec per year. The bottom figure (in red) shows the general relativistic theoretical predictions of the Lense–Thirring effect for the combination of the nodal longitudes of the LAGEOS satellites, i.e. a line with a slope of 48.2 milliarcsec per year (figure taken from [71]).

information. The GRACE satellites have greatly improved our knowledge of the Earth's gravitational field.

The method that we proposed [62, 66] to measure the Lense–Thirring effect was to use the two observables provided by the two nodes of the LAGEOS satellites, for the two unknowns: the Lense–Thirring effect and the uncertainty in the Earth quadrupole moment J_2. Then, by combining the Earth's gravitational determinations obtained by GRACE with the observations of the two LAGEOS satellites, it would be possible to measure the Lense–Thirring effect. In fact, in 2004, nearly eleven years of laser ranging data were analyzed and this analysis allowed the measurement of the Lense–Thirring effect with an accuracy [15, 71–73] of about 10%. The largest uncertainty in the gravitational field of Earth, owing to its quadrupole moment J_2, was in fact eliminated by the use of the two LAGEOS satellites, see Figure 4. However, that measurement was still affected by the uncertainties of the Earth's even zonal harmonics of degree

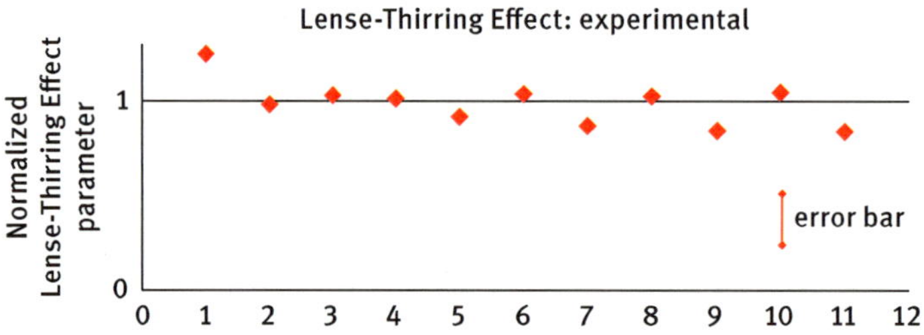

Fig. 5. Measurement of frame dragging with the LAGEOS and LAGEOS 2 satellites obtained independently in 2008 by the CSR team of the University of Texas at Austin using UTOPIA and the GRACE models: EIGEN-GRACE02S, GGM02S, EIGEN-CG03C, GIF22a, JEM04G, EIGEN-GL04C, JEM01-RL03B, GGM03S, ITG-GRACE03S and EIGEN-GL05C. The average value of frame dragging measured by Ries et al. using these models is 0.99, where 1 is the value predicted by general relativity. The total error budget of the CSR-UT team is about 12% of the frame dragging effect. Adapted from [74, 75].

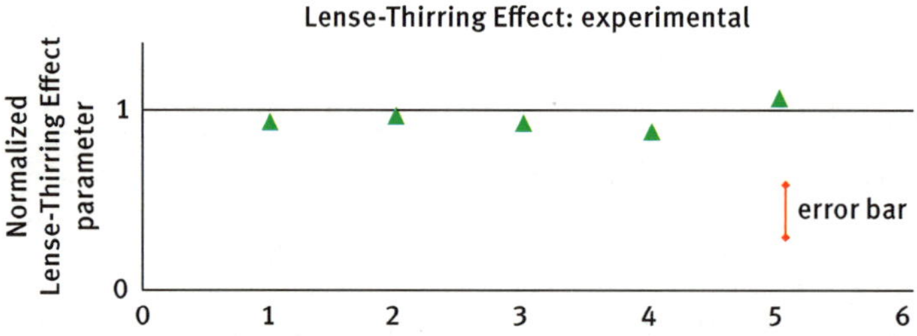

Fig. 6. Measurement of frame dragging with the LAGEOS and LAGEOS 2 satellites obtained independently in 2012 by the German GFZ team of Potsdam–Munich using EPOS-OC and the GRACE models: EIGEN-6C, EIGEN-6C (without considering the annual and semiannual variations in the Earth's gravitational field), EIGEN-6S, EIGEN-51C, and EIGEN-GRACE03S. The average value of frame dragging measured by Koenig et al. using these models is 0.95, where 1 is the value predicted by general relativity. Adapted from [76].

greater than two, and especially of the zonal harmonic of degree four, i.e. J_4. After 2004, more accurate determinations of the Earth's gravitational field by GRACE were published (see e.g. Figure 3). The analyses of the LAGEOS orbital data were repeated independently using these new gravitational field determinations, for a longer period and using three different orbital programs: GEODYN, developed by NASA Goddard, UTOPIA by the University of Texas at Austin [74, 75], see Figure 5, and EPOS-OC by the GFZ, German Research Centre for Geosciences of Potsdam [76], see Figure 6. The recent measurements of frame dragging [72–76] carried out by a team of Universities (in Lecce, Rome, Maryland, and Texas) and research centers (NASA Goddard and GFZ Potsdam) have confirmed the previous determination of the Lense–Thirring effect, obtained in 2004, with an uncertainty of about 10%. The predictions of general relativity were consistent with these measurements of frame dragging carried out with the two LAGEOS satellites.

5 The LARES mission

With the launch of the LAser Relativity Satellite – LARES – of the Italian Space Agency (ASI), see Figure 12, a new orbital observable is available that will improve the accuracy of the measurement of frame dragging. LARES was successfully launched on the February 13th, 2012 by the maiden flight of the VEGA launcher of the European Space Agency (ESA), which was developed by ELV (Avio–ASI) [78, 79].

The availability of this new, additional, observable along with those provided by LAGEOS and LAGEOS 2, will enable eliminating the error source due to the Earth's zonal harmonic of degree four, J_4, which introduced an uncertainty as large as 10% in the measurement of frame dragging using only the LAGEOS and LAGEOS 2 satellites [80]. Thus, the uncertainty in the measurement of frame dragging using LARES, plus LAGEOS and LAGEOS 2, and the determination of the Earth gravitational field by the GRACE space mission, has been shown to be approximately 1% [73, 80, 81]; see Sections 5.4 and 5.5. Figure 7 shows an artistic representation of the LARES space experiment.

The GRACE space mission [69, 70] provides very accurate determinations of the Earth's gravitational field, which are required to be able to test the frame-dragging effect due to the Earth's rotation with an uncertainty of a few percent. The combination of the three satellites will also allow other tests of fundamental physics [73, 80, 81], see Section 3. The Lense–Thirring drag of the orbital planes of LARES, theoretically predicted by general relativity, is approximately 118 milliarcsec per year, corresponding, at the altitude of LARES, to about 4.5 m/year. LARES is well observed by the global network of laser stations of the ILRS [83] and the measured orbital elements of LARES are shown in Table 2.

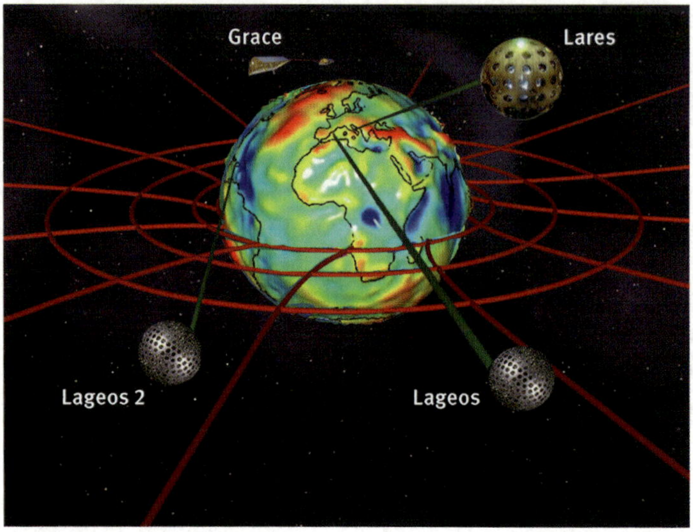

Fig. 7. Concept of the LARES space experiment, also shown in the figure are the LAGEOS, LAGEOS 2 and GRACE satellites that play a key role in the experiment. The spacetime distortion due to frame-dragging induced by the Earth's rotation is represented by the distortion of the radial curves. The Earth's colors represent the EIGEN-GRACE02S anomalies of the Earth's gravitational field obtained with GRACE.

Table 2. LARES orbit data.

Semimajor axis (km)	Orbital eccentricity	Orbital inclination (deg)
7820	≅ 0.0008	69.5

In the following five subsections we describe, respectively: the LARES satellite (5.1), LARES and geodesic motion (5.2), preliminary orbital results (5.3), error analyses (5.4), and Monte Carlo simulations (5.5) of the LARES experiment.

5.1 The LARES satellite

LARES is spherical and covered with 92 retroreflectors (CCR). It has a radius of 182 mm and a total mass of 386.8 kg. It has the highest average density of any single known object orbiting in the Solar System [84].

With the qualification launch of VEGA its altitude could not exceed 1500 km. This limitation resulted in a very thorough design of the satellite, aimed at minimizing the effects of nongravitational perturbations, and in particular those induced by the presence of the denser atmosphere at that altitude and by thermal thrust. To that end, it was decided to push the surface to mass ratio to a value about 2.6 times lower than

Fig. 8. LARES semifinished sphere mounted on the lathe machine.

the already extremely small such ratio of the LAGEOS satellites. This design choice, in turn, resulted in several technological complications and in the choice of a material never before used for the structure of a satellite: a tungsten alloy [85]. Thermal thrust is another subtle perturbation that was further reduced by adopting a single-piece spherical design, in other words the satellite has been machined out of a single piece of tungsten alloy. Figure 8 shows the semifinished sphere of the LARES satellite of radius 186.00 ± 0.01 mm. Another important aspect has been the verification of the CCR behavior under a simulated space environment [87]. In fact, the high temperature reached by the satellite during the operative life, might induce deformations on the dihedral angle of the CCRs which in turn might cause a modification on the Far Field Diffraction Pattern (FFDP) and consequently the possibility that the laser return might miss the laser ground station.

Pure tungsten is very difficult, if not impossible, to machine. Therefore, an alloy made with the liquid phase sintering process was used. In spite of the granular nature

Table 3. Comparison of mechanical characteristics of LARES tungsten alloy, steel used on the brackets of the separation system and the material used for the shells of LAGEOS.

	LARES Tungsten THA-18	*Separation system* 15-5-PH 1025	*LAGEOS* Aluminum 6061-T6
Nominal density (kg/m^3)	18 000	7800	2700
Composition	95%W, 3%Ni, 2%Cu		Min 95.8%Al, 0.8%Mg, 0.4%Si
Elastic modulus (GPa)	345	196	68.9
Tensile strength (MPa)	758	1069	310
Yeld strength (MPa)	586	1000	276
Poisson ratio	0.31	0.27	0.33

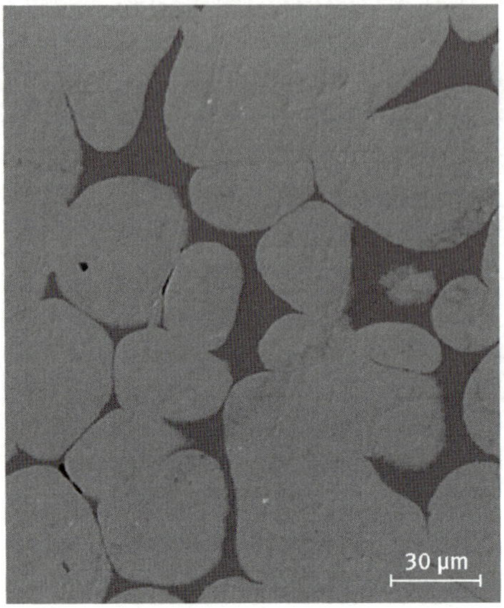

Fig. 9. Micrograph of LARES material. The material structure consists of tungsten particles surrounded by a Ni–Cu phase (picture taken from [85]).

of the material (95% of tungsten particles embedded in a 5% nickel–copper matrix), see Figure 9, its mechanical characteristics were very good (Table 3).

The tolerances of the manufactured parts were equal to or better than those obtained on the aluminum parts of the LAGEOS satellites. We recall here that the design of LAGEOS and of the almost identical LAGEOS 2 consisted of two hemispherical shells made of an aluminum alloy with an internal cylindrical core of a copper alloy (Figure 10). Another challenge was the separation system because protruding parts

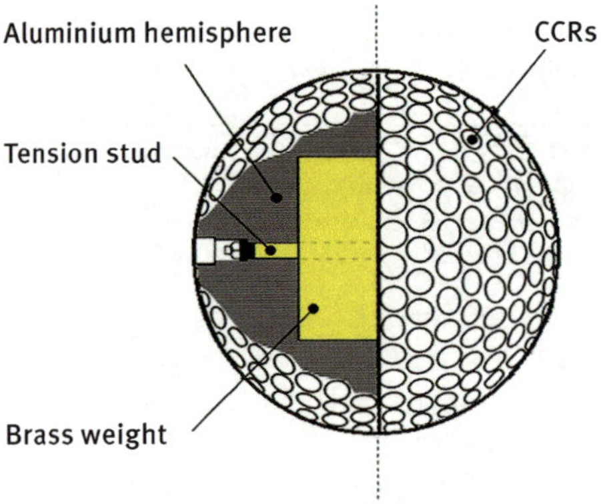

Fig. 10. Cross section of the LAGEOS satellites, showing the internal structure and in particular the tension stud to assemble and constrains the three pieces together.

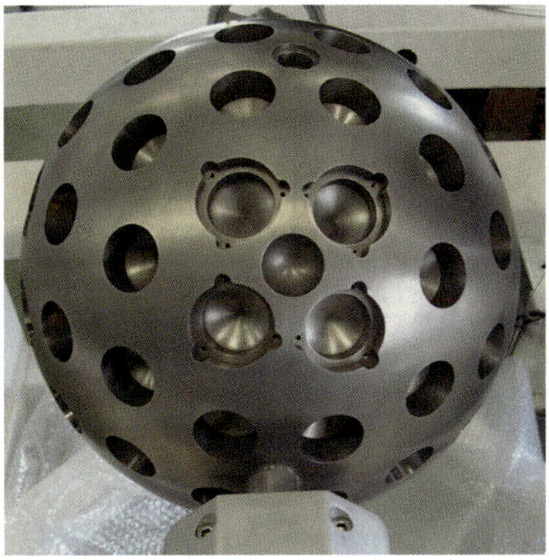

Fig. 11. LARES demonstration model on the integration stand.

from the spherical surface were not acceptable. Therefore, four hemispherical equatorial cavities were manufactured to couple with as many brackets with hemispherical pins [86]. Figure 11 shows the LARES demonstration model. At the center, one of the four hemispherical cavities is clearly shown. It is surrounded by four Cube Corner Reflectors (CCR) cavities. To prevent detachment of the satellite from the separation system during launch, each pin pushed the hemispherical cavity with a very high force. The contact area is small [86], therefore, the pressure at the contact area was very high and was too close to the admissible limit. During the design phase, reducing the contact pressure was considered, either by increasing the diameter of the hemispheres or by pushing the manufacturing tolerances to the lowest technological limit possible. This second approach, though very tough, was preferred, because an increase in the diameter of the hemispherical cavities would have affected the optical design of LARES, i.e. the CCR distribution over the surface of the satellite.

In summary, LARES is entirely made from a tungsten alloy of density equal to $18\,000\,\mathrm{kg/m^3}$, with a total mass of 386.8 kg, and with an average density (including the CCRs made of glass and related mounting systems) equal to approximately $15\,000\,\mathrm{kg/m^3}$. This resulted in a ratio of the cross-sectional area to mass which is about 2.6 times smaller than that of the two LAGEOS satellites [79], which had so far the lowest ratio of any artificial satellite. This ratio is proportional to the effect of the nongravitational perturbations that have to be minimized to accurately measure frame-dragging. In fact, both the extremely low value of this ratio for LARES, equal only to $0.00027\,\mathrm{m^2/kg}$, and its particular structure, a one-piece solid ball with high thermal conductivity, ensure that, despite the lower altitude of LARES compared to that of LAGEOS, the effect of the unmodeled nongravitational orbital perturbations is extremely small. In practice these nongravitational perturbations are smaller than those

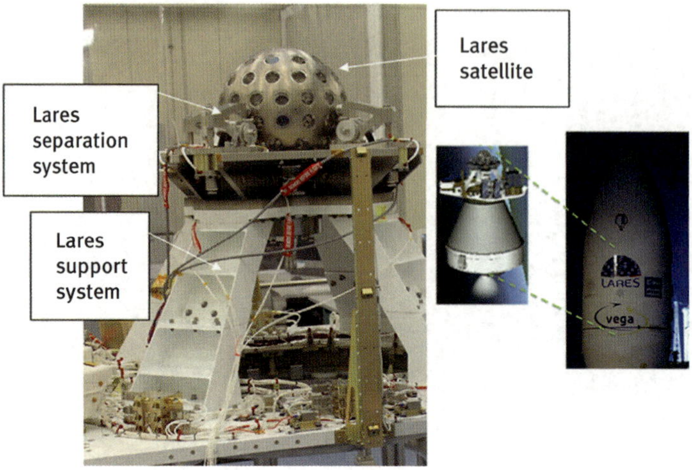

Fig. 12. (left) LARES satellite mounted on the separation and support systems. (center) LARES system on the upper stage, (right) VEGA payload compartment (fairing).

of any other satellite, as already confirmed by a few months of laser observations [78]. Figure 12 shows the LARES system, the cone adapter and the payload compartment (fairing) of VEGA.

5.2 The LARES satellite and geodesic motion

In general relativity spacetime curvature, describing the gravitational interaction, is generated by the distribution and the currents of mass–energy via the Einstein field equations [6, 10, 88]: a test particle then moves along a spacetime geodesic. A timelike geodesic path in spacetime's Lorentzian geometry is one that locally maximizes proper time, in analogy with the length-minimizing property of Euclidean straight lines.

Depending on the physical context, a star, a planet, or a satellite can approach the behavior of a small test particle. For example, the gravitational pull of the Earth on its Moon or on artificial satellites is explained by general relativity through the spacetime curvature generated by the mass of the Earth. These Earth satellites can be considered as test particles whose orbits describe spacetime geodesics with deviations from ideal geodesic paths due to their finite size and to the nongravitational forces acting on them. Geodesic motion is also used to calculate the advance of the perihelion of a planet's orbit and the dynamics of a binary pulsar system.

Geodesic motion is at the basis of general relativity and of other theories in which the gravitational interaction is described by the spacetime curvature dynamically generated by mass–energy. Therefore, the realization of the best possible approximation for the free motion of a test particle, a spacetime geodesic, is an important result for experiments dedicated to studying the spacetime geometry in the vicinity of a body, yielding high-precision tests of general relativity and constraints on alternative gravitational theories.

A fundamental issue is that regarding the approximation to a geodesic that is provided by the motion of an actually extended body. In general relativity [89, 90] the problem of an extended body is subtle, not only due to the nonlinearity of the equations of motion, but also to the need to deal with the internal structure of the compact body, constructed of a continuous medium, where kinetic variables and thermodynamic potentials are involved. Further, there may be intrinsically nonlocal effects arising from the internal structure of the extended body, such as tidal influences. Moreover, there are problems concerning the approximations that need to be made to describe a given extended body as a test particle moving along a geodesic. These problems are related to the fact that many of the common Newtonian gravitational concepts such as the "center of mass," "total mass" or "size" of an extended material body do not have well-defined counterparts in general relativity [91].

The Ehlers–Geroch theorem [92] (generalizing the result in [93]) attributes a geodesic to the trajectory of an extended body with a small enough gravitational self-field, if for a Lorentzian metric the Einstein tensor satisfies the so-called dominant energy condition [88], this tensor being nonzero in some neighborhood of the geodesics and vanishing at its boundaries. This theorem, asserting that small massive bodies move on near-geodesics, thus provides a rigorous bridge from general relativity to space experiments with "small" satellites, which suggests a high level of suppression of nongravitational and self-gravitational effects from the satellite's own small gravitational field. This enables us to consider the satellite's motion to be nearly geodesic, and hence provides a genuine testing ground for general relativity's effects.

Given the extreme weakness of the gravitational interaction compared to the other interactions of nature, the space environment is the ideal laboratory to test gravitational and fundamental physics. However, in order to test gravitational physics, a satellite must behave as nearly as possible as a test particle and must be as little as possible affected by nongravitational perturbations such as radiation pressure and atmospheric drag, which shift the orbit of the satellite away from an ideal spacetime geodesic.

There are basically two ways to minimize the effects of these perturbations: a drag-free satellite or a small satellite with a high density and an extremely small surface-to-mass ratio. The Gravity Probe B drag-free satellite achieved an average residual acceleration of $40 \times 10^{-12}\ \mathrm{m/s^2}$ [53]. For a passive satellite (with no drag-free system), the key feature that determines the level of attenuation of the nongravitational perturbations is the ratio of its cross-sectional area to its mass.

In addition, it is necessary to determine the position of the satellite with extreme precision: on bodies of extended dimensions, one must follow the trajectory of a precise point of the orbiting object, for example, the geometric center or the center of mass. On LARES, the retroreflectors allow measuring the instantaneous distance of the satellite with an accuracy of a few millimeters, by measuring the round-trip time of a laser pulse. All these considerations have influenced the design of most laser-ranged satellites and in particular that of LARES, which, at present, appears to be the best realization of an orbiting test particle, as described in the next section.

5.3 Preliminary orbital analyses

The first laser return from the LARES satellite was obtained on the February 17th 2012. Since then, we have performed preliminary data analyses and studied the LARES behavior. The orbital analysis and data reduction were performed using UTOPIA of UT/CSR (Center for Space Research at the University of Texas at Austin), GEODYN II of NASA Goddard [95], and EPOS-OC of GFZ (Helmholtz Centre Potsdam GFZ German Research Centre for Geosciences) [94]. These orbital programs included the state-of-the-art orbital dynamical models. In particular, they included models of the Earth's gravitational field based on GRACE data [69, 70], the best updated models for oceans and solid Earth tides, as well as solar radiation pressure, the Earth's albedo, and atmospheric drag [63, 96–98] with all the post-Newtonian corrections of general relativity. However, in the analysis reported in this section, no model of thermal thrust was used [97, 98]. The laser data of LARES were processed in arcs of 15 days.

For the 105 days analyzed, GEODYN, UTOPIA and EPOS-OC independently determined that the residual along-track accelerations for LARES were only about 0.4×10^{-12} m/s^2, whereas for the two LAGEOS satellites, the acceleration residuals were in the range of 1 to 2×10^{-12} m/s^2. This is particularly impressive since LARES is in an orbit lower by far than that of LAGEOS. The residual acceleration along the orbit of a satellite is a measure of the level of suppression of its nongravitational perturbations: atmospheric drag, solar, and terrestrial radiation pressure and the effects of thermal thrust. The atmospheric drag acts mainly along the velocity vector of the satellite, while the solar radiation pressure, the pressure of terrestrial radiation (visible and infrared radiation from the Earth), and the effects of thermal thrust will have both along-track and nonalong-track contributions. The thermal thrust, or Yarkovsky effect, on a spinning satellite is a thermal acceleration resulting from the anisotropic flux of radiation from the satellite owing to the anisotropic temperature distribution on its surface by solar heating. A variation of this effect, due to the infrared radiation of the Earth, is the Earth–Yarkovsky effect or Yarkovsky–Rubincam effect [97, 98].

The effects of the residual nonmodeled accelerations along the orbit of the laser-ranged satellites LARES, LAGEOS, and STARLETTE (a CNES (Centre national d'études spatiales) laser-ranged satellite launched in 1975) are illustrated in Figure 13 which shows the variation of their distance from their "ideal" orbits caused by the nonmodeled accelerations along their orbits [78]. The vertical axis can be thought of as representing an "ideal" reference world-line of LARES, LAGEOS, and STARLETTE, "ideal" in the sense that all its corresponding orbital perturbations are well known and well

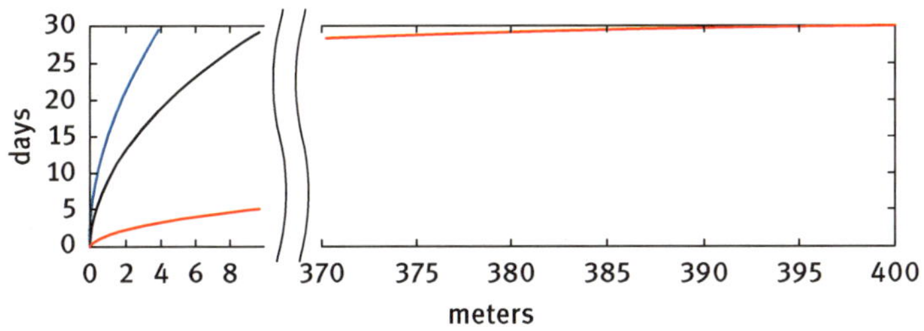

Fig. 13. Deviations in meters, from a spacetime geodesic followed by an ideal "test particle," of the orbit of a satellite perturbed by an average along-track acceleration with magnitude respectively equal to the unmodeled along-track accelerations observed on the LARES (blue curve), LAGEOS (green curve) and STARLETTE (red curve) satellites. The vertical axis may be thought of to represent the spacetime geodesic in a reference frame co-moving with the test particle. The blue, green, and red curves were obtained using an average along-track acceleration with magnitude, respectively, equal to: 0.4×10^{-12} m/s^2, 1×10^{-12} m/s^2, and 40×10^{-12} m/s^2, i.e. the unmodeled residual along-track acceleration observed on each of the three satellites (adapted from [78]).

modeled. In fact, Figure 13 might be thought of as representing the deviations from geodesic motion for LARES, LAGEOS, and STARLETTE due to the nonmodeled or mismodeled accelerations along their orbit, after all the known nongravitational perturbations have been removed to the extent permitted by the current models. Since all the post-Newtonian corrections of general relativity have been included in our orbital analysis, these figures show the level of agreement between the orbits of LAGEOS, LARES, and STARLETTE and the geodesic motion predicted by general relativity.

It has to be be pointed out that assuming an out-of-plane residual acceleration, *constant in direction*, of the same order of magnitude as the nonmodeled along-track acceleration observed in the orbit of LARES, this acceleration will produce a very small secular variation of the longitude of the LARES node, that is, of its orbital angular momentum. For example, by assuming an out-of-plane acceleration with an amplitude of 0.4×10^{-12} m/s^2, *constant in direction*, its effects on the node of LARES would be many orders of magnitude smaller than the tiny secular drift of the node of LARES due to frame dragging [73] of about 118 milliarcsec/year. Therefore, LARES, along with the LAGEOS satellites, and with the determination of the Earth's gravitational field obtained by the GRACE mission, will be used to accurately measure the frame-dragging effect predicted by general relativity, allowing an improvement of about one order of magnitude over the accuracy of previous measurements of frame dragging through the use of only the two LAGEOS satellites [71–73].

In conclusion, LARES offers the best available test particle in the Solar System to test general relativity and gravitational physics, for example, for the accurate measurement of frame dragging, and after modeling its known nongravitational perturbations, its orbit shows the best agreement, among those of all other satellites, with the geodesic motion predicted by general relativity.

5.4 Error analyses

The major uncertainties in the measurement of frame dragging using a test satellite of the Earth are due to the Earth's lowest degree even zonal harmonics. However the largest $n - 1$ uncertainties, due to the lowest degree even zonal harmonics, can be eliminated by using n test satellites [62, 66], thus allowing a measurement of frame dragging not affected by these $n - 1$ uncertainties. In particular, using three observables, namely the three nodes of LARES, LAGEOS, and LAGEOS 2, one can eliminate the two largest uncertainties due to the two lowest degree even zonal harmonics, J_2 and J_4, and measure frame dragging with an uncertainty of only a few percent (see below). The final choice of the inclination of the orbit of LARES (set at 69.5°) was mainly for the minimization of the uncertainties from the even zonal harmonics of degree strictly larger than 4, that is, from δJ_{2n} with $2n > 4$, but also due to security issues connected with the overflight of residential areas and any fallout of the VEGA stages. The final chosen inclination was anyway close to the optimal one (which should have been 70°) for the measurement of the frame dragging. In [73, 80] the errors induced by each even zonal harmonic up to degree 70 are analyzed. As to other gravitational perturbations due to temporal variations of the Earth's gravitational field, and especially to terrestrial tides, we stress that the major tidal effects on the node of LARES are due to the zonal tides with $l = 2$ and $m = 0$ induced by the Moon's node, and to the K_1 tide with $l = 2$ and $m = 1$ (tesseral tide). However, the error caused by the medium and long period zonal tides with $l = 2$ and $m = 0$ will be eliminated, along with the static errors due to J_2 and J_4, using the combination of the three nodes. Also, the uncertainties in the time-dependent secular variations $\dot{J}_2$, $\dot{J}_4$ will be cancelled using this combination of three observables. Furthermore, the tesseral tide K_1 will be fitted over a period that is multiple of the LARES nodal period (see [58] and Chapter 5 of [60]) and therefore this tide will introduce only a small uncertainty in our combination.

Concerning the effects of the nongravitational perturbations, such as solar radiation pressure, albedo, thermal thrust, and atmospheric drag we refer to [58, 62, 72, 77, 80]. Here we only stress that the secular nodal shift of LARES due to atmospheric drag is negligible. This is due to the almost circular orbit of LARES, which has an orbital eccentricity $e \cong 0.0008$, and to the special structure of LARES with an extremely small ratio of its cross-sectional area to its mass. In fact, even assuming that the exosphere were co-rotating with the Earth at any satellite altitude, in the case of zero orbital eccentricity, $e = 0$, the total nodal shift of the satellite would be zero, as calculated in [62]. In the case of a very small orbital eccentricity, the total nodal shift would be proportional to the orbital eccentricity and thus for LARES it would be a very small effect [62] owing once again to the very small ratio of its cross-sectional area to its mass. Detailed examinations of all the error sources, both gravitational and nongravitational, in the measurement of frame dragging using the LAGEOS and LARES satellites have been published in a number of papers including [48, 58–60, 62, 66, 72, 73, 77, 80, 81, 99]. These studies have shown that the uncertainties in the knowledge of the even zonal harmonics of degree strictly higher than four are the main source of error in the LARES experiment. An additional error analysis was recently carried out using a number of Monte Carlo simulations [81] (see 5.5)

To visualize the error in the LARES experiment due to each even zonal harmonic, we report here, in Figure 14, the results published in [73, 80], but not including the errors due to the harmonics of degree 2 and 4 that do not affect the measurement of frame dragging because they are eliminated thanks to the combined use of the three nodes of LARES, LAGEOS, and LAGEOS 2. The gravitational field model used is EIGEN-GRACE02S (2004), today outdated by more recent and more accurate models.

Figure 14 clearly shows that the error due to each even zonal harmonic with degree greater than 4 is much less than 1% and, in particular, that the error is practically zero for even zonal harmonics of degree higher than or equal to 26 (panel left). The results of Figure 14 are based on the calibrated uncertainties (i.e. including systematic errors) of the model EIGEN-GRACE02S (GFZ, Potsdam, 2004) used in [71]). By using EIGEN-GRACE02S and GGM02S (see [80]), and including the even zonal harmonics up to degree 70, the total error in the measurement of the Lense–Thirring effect due to the uncertainties in the even zonals is found to be equal to 1.4% and 2.1%, respectively. Figure 14 also shows the upper bound of the percentage error of each even zonal harmonic obtained by taking the difference between the values of the same harmonic in the two different models EIGEN-GRACE02S and GGM02S (right panel). The total error in the LARES measurement of frame dragging using EIGEN-GRACE02S is 1.4% while using as uncertainty the difference between the same coefficient in the two different models, it is 3.4%. The evaluation of the upper bound of the error in each harmonic coefficient by using the difference between the uncertainties of that coefficient in two different models is routinely done in space geodesy to estimate the reliability of the published errors of a model, however, with the basic warning that the two models considered must be comparable, i.e. that they must have comparable accuracy. Then, the upper bound of the error in the measurement of frame dragging with EIGEN-GRACE02S, i.e. 3.4%, is

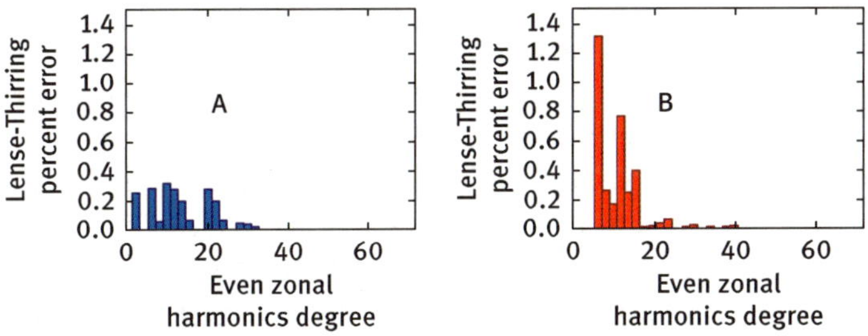

Fig. 14. Percentage error in the measurement of frame dragging by the three satellites LARES, LAGEOS, and LAGEOS 2 corresponding to each even zonal harmonic using the published uncertainties of the EIGEN-GRACE02S model (left panel) and using, as uncertainties, the difference between the coefficients of the models EIGEN-GRACE02S and GGM02S (right panel). From [50]

about two or three times larger than the error obtained using its published uncertainties, i.e. 1.4%. However, one has to consider that EIGEN-GRACE02S was just a preliminary 2004 GRACE determination of the Earth's gravitational field based on less than 365 days of observations after the 2002 launch of GRACE. Today's (2014) gravitational models of the Earth are much more accurate than EIGEN-GRACE02S, especially thanks to the much longer period of observations of GRACE. Therefore, there will be a substantial improvement in the accuracy of the frame-dragging measurements obtained using the new GRACE gravitational models available now and in the next few years, i.e. during the period of the first LARES data analyses to measure frame dragging. In addition, other space missions, such as GOCE [100, 101], have further improved the knowledge of the Earth's gravitational field over the 2004 models.

The error in the determination of frame dragging owing to the uncertainties in the Earth's gravitational field is greatly mitigated by the use of three satellites. On the other hand, a minimization of the nongravitational orbital perturbations was possible by optimizing the design of the three satellites; the LARES and LAGEOS satellites are indeed spherical and very dense, designed with a very small surface-to-mass ratio. This ratio is a measure of the minimization of the nongravitational perturbations [62]. Thanks to this demanding LARES design, optimal performance has been obtained as has been confirmed by the preliminary results shown above, i.e. the orbital effects of the nonmodeled perturbations are smaller for LARES, in spite of its lower orbit, than for the LAGEOS satellites.

5.5 Monte Carlo simulations

In order to obtain an extremely solid error estimate of the LARES test of frame dragging and to eventually confirm its previous error analyses, we used a completely independent method of analysis, i.e. a number of Monte Carlo simulations of the LARES experiment. In [81] the detailed results of these simulations are reported, and they are also summarized below.

One hundred simulations were designed and performed to take into account the uncertainties in the physical parameters of the LARES experiment and to quantitatively evaluate the effect of these uncertainties on the final measurement of frame dragging. We first selected the main physical parameters whose uncertainties have a critical impact on the accuracy of the measurement of frame dragging using LARES, LAGEOS, and LAGEOS 2, namely, the main parameters describing the Earth's gravitational field and the parameters related to the radiation pressure orbital perturbations. Then, using the orbital estimator EPOS-OC, we simulated one hundred times the orbits of LARES, LAGEOS, and LAGEOS 2 by generating by chance the values of GM, i.e. the gravitational constant times the mass of Earth, of its five largest even zonal harmonics, J_2, J_4, J_6, J_8, and J_{10}, of the secular rate of change of the two largest even zonal harmonics $\dot{J}_2$ and $\dot{J}_4$, and of the radiation pressure coefficients of the LARES,

LAGEOS, and LAGEOS 2 satellites. Frame dragging was always kept equal to its value predicted by the theory of general relativity. Finally, we performed an analysis of their simulated laser-ranging observations.

The results of the 100 simulations of the LARES space experiment confirmed the previous extensive analyses of the experiment. In fact, the standard deviation of the 100 simulated measurements of frame dragging, obtained corresponding to each of the 100 Monte Carlo simulations, was equal to 1.4% of the frame-dragging effect predicted by general relativity. The mean value of the 100 simulated measurements of frame dragging was equal to 100.24% of its relativistic value. Thus, the Monte Carlo simulations have confirmed an error budget of approximately 1% in the future measurement of frame dragging using LARES, LAGEOS, and LAGEOS 2, and the Earth's gravitational field determined by GRACE.

6 Conclusions

In this chapter, we first described frame dragging, an intriguing phenomenon predicted by general relativity that has relevant astrophysical effects in the vicinities of rotating black holes. We then reviewed the previous experimental tests of frame dragging obtained using the LAGEOS satellites and with the dedicated Gravity Probe B space mission, that respectively reported an accuracy of about 10% and 19% in their tests. To confirm the robustness of the results obtained with the LAGEOS satellites, we reported similar results for the measurement of frame dragging obtained *independently* by three different groups: the Universities of Salento, Maryland, and Rome Sapienza; the University of Texas at Austin; and GFZ Potsdam. The three groups used in their analyses the three completely independent orbital programs GEODYN, UTOPIA, and EPOS-OC, respectively. In this chapter, we also reported the use of the LAGEOS measurement of frame dragging to set limits on string theories of Chern–Simons type.

The second part of the chapter was devoted to the LARES space experiment, which together with LAGEOS, LAGEOS 2 and the determinations of the Earth's gravitational field by GRACE, will improve the accuracy of the test of frame dragging by an order of magnitude and improve the limits on string theories of Chern–Simons type. We first briefly reported a description of the main characteristics of the LARES satellite and of its special design to minimize the orbital effects of nongravitational perturbations. The first few months of LARES observations and orbital analyses confirmed that, thanks to its special design, the orbit of LARES is closer to the theoretical motion of a test particle predicted by general relativity than that of any other satellite. Finally, we reported the results of extensive error analyses and of one hundred Monte Carlo simulations confirming an error budget of approximately 1% in the test of frame dragging with the LARES space experiment.

Acknowledgement: The authors gratefully acknowledge the International Laser Ranging Service for providing high-quality laser ranging tracking of the LARES satellites. I. Ciufolini and A. Paolozzi gratefully acknowledge the support of the Italian Space Agency, grants I/043/08/0, I/016/07/0, I/043/08/1 and I/034/12/0. J. C. Ries the support of NASA Contract NNG06DA07C, and E. C. Pavlis and R. A. Matzner the support of NASA Grant NNX09AU86G.

References

[1] S. Perlmutter, Supernovae, dark energy, and the accelerating universe. *Phys. Today* **56**, 53–60 (2003).
[2] R. R. Caldwell, "Dark Energy". *Physics World*, 37–42 (2004).
[3] A. Riess et al., Observational evidence from supernovae for an accelerating universe and a cosmological constant. *Astron. J.* **116**, 1009–1038 (1998).
[4] S. Perlmutter et al., Measurements of Ω and Λ from 42 high-redshift supernovae. *Astrophys. J.* **517**, 565–586 (1999).
[5] Planck Collaboration, Planck 2013 results, to appear in *Astronomy and Astrophysics*, (2013).
[6] C. W. Misner, K. S. Thorne and J. A. Wheeler, *Gravitation*, Freeman, San Francisco (1973).
[7] S. Turyshev, Experimental tests of general relativity: Recent Progress and Future Directions. *Physics-Uspekhi* **52**, 1–27 (2009).
[8] C. M. Will, *Theory and Experiment in Gravitational Physics 2nd edn*, Cambridge Univ. Press, Cambridge, UK (1993).
[9] T. Damour and J. H. Taylor, Strong-field tests of relativistic gravity and binary pulsars. *Phys. Rev. D* **45**, 1840–1868 (1992).
[10] I. Ciufolini and J. A. Wheeler, *Gravitation and Inertia*, Princeton Univ. Press, USA (1995).
[11] S. Kopeikin, M. Efroimsky and G. Kaplan, *Relativistic Celestial Mechanics of the Solar System*, Wiley VCH, Berlin (2011).
[12] M. Kamionkowski, in *Visions of Discovery*, Cambridge Univ. Press, 247–293 (2007).
[13] R. Penrose, Gravitational collapse and space-time singularities. *Phys. Rev. Lett.* **14**, 57–59 (1965).
[14] A. De Felice and S. Tsujikawa, f(R) Theories. *Living Rev. Relativity* **13**, 3 (2010).
[15] I. Ciufolini, Dragging of inertial frames. *Nature* **449**, 41–47 (2007).
[16] S. Weinberg, *Gravitation and Cosmology: Principles and Applications of the General Theory of Relativity*, Wiley, New York (1972).
[17] A. Einstein, Letter to Ernst Mach. Zurich, 25 June 1913, in Ref. [6], p. 544.
[18] L. D. Landau and E. M. Lifshitz, *The Classical Theory of Fields 3rd rev.*, English edn., Pergamon, London (1971).
[19] Ya. B. Zeldovich and I. D. Novikov, *Relativistic Astrophysics, Vol. I Stars and Relativity*, Univ. Chicago Press, Chicago (1971).
[20] I. Ciufolini and F. Ricci, Time delay due to spin and gravitational lensing. *Class. Quantum Grav.* **19**, 3863–3874 (2002).
[21] I. Ciufolini and F. Ricci, Time delay due to spin inside a rotating shell. *Class. Quantum Grav.* **19**, 3875–3881 (2002).
[22] I. Ciufolini, F. Ricci, S. Kopekin and B. Mashhoon, On the gravitomagnetic time delay. *Phys. Lett. A* **308**, 101–109 (2003).

[23] G. E. Pugh, Proposal for a satellite test of the Coriolis prediction of general relativity. *Weapons Systems Evaluation Group Research Memorandum N.* **11**, The Pentagon, Washington (1959).
[24] L. I. Schiff, Motion of a gyroscope according to Einstein's theory of gravitation. *Proc. Nat. Acad. Sci.* **46**, 871–82 (1960); Possible new test of general relativity theory. *Phys. Rev. Lett.* **4**, 215–7 (1960).
[25] J. Lense and H. Thirring, Über den Einfluss der Eigenrotation der Zentralkörper auf die Bewegung der Planeten und Monde nach der Einsteinschen Gravitationstheorie. *Phys. Z.* **19** 156–163 (1918). See also English translation by B Mashhoon, F W Hehl, D S Theiss *Gen. Relativ. Gravit.* **16**, 711–750 (1984).
[26] J. M. Bardeen and J. A. Petterson, The Lense–Thirring effect and accretion disks around Kerr black holes. *Astrophysical J.* **195**, L65–67 (1975).
[27] K. S. Thorne, R. H. Price and D. A. Macdonald, *The Membrane Paradigm*, Yale Univ. Press, New Haven (1986).
[28] R. P. Kerr, Gravitational field of a spinning mass as an example of algebraically special metrics. *Phys. Rev. Lett.* **11**, 237–238 (1963).
[29] I Ciufolini, Gravitomagnetism and status of the LAGEOS III experiment. *Class. Quantum Grav.* **11**, A73–A81 (1994).
[30] I. Ciufolini, Frame-dragging, gravitomagnetism, and lunar laser ranging. *New Astronomy* **15**, 332–337 (2010).
[31] N. Ashby and B. Shahid-Saless, Geodetic precession or dragging of inertial frames? *Phys. Rev. D* **42**, 1118–1122 (1990).
[32] R.F. O'Connell, A note on frame dragging. *Class. Quant. Grav.* **22**, 3815–3816 (2005).
[33] B. M. Barker and R.F. O'Connel, The gravitational interaction: Spin, rotation, and quantum effects: A review. *Gen. Rel. Grav.* **11**, 149–175 (1979).
[34] A. R. Khan and R.F. O'Connell, Gravitational analogue of magnetic force. *Nature* **261**, 480–481 (1976).
[35] T. W. Murphy Jr, K. Nordtvedt and S. G. Turyshev, Gravitomagnetic influence on gyroscopes and on the lunar orbit. *Phys. Rev. Lett.* **98**, 071102-1-4 (2007).
[36] S. M. Kopeikin, Comment on "Gravitomagnetic influence on gyroscopes and on the lunar orbit". *Phys. Rev. Lett.* **98**, 229001 (2007).
[37] T. W. Murphy Jr., K. Nordtvedt and S. G. Turyshev, Murphy, Nordtvedt, and Turyshev Reply. *Phys. Rev. Lett.* **98**, 229002 (2007).
[38] S. M. Kopeikin, The gravitomagnetic influence on Earth-orbiting spacecrafts and on the lunar orbit, in *John Archibald Wheeler and general relativity* I. Ciufolini and R. Matzner (eds.), 337–343, Springer Verlag (2010).
[39] R. Jackiw and S.-Y. Pi, Chern–Simons modification of general relativity. *Phys. Rev. D* **68**, 104012 (2003).
[40] S. Alexander and N. Yunes, Chern–Simons modified general relativity. *Phys. Rep.* **480**, 1–55 (2009).
[41] T. L. Smith, A. Erickcek, R. Caldwell and M. Kamionkowski, Effects of Chern–Simons gravity on bodies orbiting the Earth. *Phys. Rev. D* **77**, 024015 (2008).
[42] B. A. Campbell, M. J. Duncan, N. Kaloper and K. A. Olive, Gravitational dynamics with lorentz Chern–Simons terms. *Nuclear Physics B* **351**, 778–792 (1991).
[43] K. Yagi, N. Yunes and T. Tanaka, Slowly rotating black holes in dynamical Chern–Simons gravity: Deformation quadratic in the spin. *Phys. Rev.D* **86**, 044037 (2012).
[44] S. Alexander, A. Marciano and D. Spergel, Chern–Simons Inflation and Baryogenesis. arXiv:1107.0318.

[45] T. Harko, Z. Kovács, F. S. N. Lobo, Thin accretion disk signatures in dynamical Chern–Simons modified gravity. *Class. Quant. Grav.* **27**:105010 (2010).

[46] K. Yagi, N. Yunes and T. Tanaka, Gravitational waves from quasicircular black-hole binaries in dynamical Chern–Simons gravity. *Phys. Rev. Lett.* **109**, 251105 (2012).

[47] J. G. Silva, A. F. Santos, Ricci dark energy in Chern–Simons modified gravity. *Eur. Phys. J. C.* **73**, 2500 (2013).

[48] V. G. Gurzadyan, I. Ciufolini, S. Sargsyan, G. Yegorian, S. Mirzoyan and A. Paolozzi, Satellite probing general relativity and its extensions and Kolmogorov Analysis. *EPL* **102**, 60002-p1-p4 (2013).

[49] A. Z. Petrov, New methods in general relativity, Nauka, Moscow. English edn: Einstein Spaces. Pergamon Press (1969).

[50] I. Ciufolini, A. Paolozzi, R. Koenig, E. C. Pavlis, J. Ries, R. Matzner, V. Gurzadyan, R. Penrose, G. Sidoni and C. Paris, Fundamental physics and general relativity with the LARES and LAGEOS satellites. *Nuclear Physics B – Proceedings Supplements* **243–244**, 180–193 (2013).

[51] B. Friedländer and I. Friedländer, *Absolute und Relative Bewegung?* Simion-Verlag, Berlin (1896).

[52] A. Föppl, Über einen Kreiselversuch zur Messung der Umdrehungsgeschwindigkeit der Erde. *Sitzb. Bayer. Akad. Wiss.* **34**, 5–28 (1904), Phys. Z. **5**, 416; see also A. Föppl, Über Absolute und Relative Bewegung. *Sitzb. Bayer. Akad. Wiss.* **34**, 383–395 (1904).

[53] C. W. Everitt et al., Gravity probe B: Final results of a space experiment to test general relativity. *Phys. Rev. Lett.* **106**, 22110 (2011).

[54] D. K. Gill and S. Buchman, Evidence for Patch Effect Forces On the Gravity Probe B Gyroscopes. Stanford Univ., Stanford (April 2007); poster at http://einstein.stanford.edu/content/aps_posters/EvidenceForPatchEffectForces.eps.

[55] B. M. Barker and R. F. O'Connel, The gyroscope test of general relativity. *Nature* **312**, 314 (1984).

[56] S. C. Cohen and P. J. Dunn (eds.), LAGEOS scientific results. *J. Geophys. Res.* **90** (B11), 9215 (1985).

[57] R. Noomen, S. Klosko, C. Noll and M. Pearlman (eds.), *13th International Workshop on Laser Ranging, Proceedings From the Science Session and Full Proceedings CD-ROM*, NASA CP 2003-212248, NASA Goddard, Greenbelt, MD (2003).

[58] B. Tapley, J. C. Ries, R. J. Eanes and M. M. Watkins, *NASA-ASI Study on LAGEOS III*, CSR-UT publication n. CSR-89-3, Austin, Texas (1989) and I. Ciufolini et al., *ASI-NASA Study on LAGEOS III*, CNR, Rome, Italy (1989). See also: I. Ciufolini et al., *INFN study on LARES/WEBER-SAT* (2004).

[59] J. C. Ries, *Simulation of an Experiment to Measure the Lense–Thirring Precession using a Second LAGEOS Satellite*, Ph.D. thesis, Univ. of Texas, Austin (1989).

[60] G. E. Peterson, *Estimation of the Lense–Thirring Precession Using Laser-Ranged Satellites*, Ph.D. thesis, Univ. of Texas, Austin (1997).

[61] I. Ciufolini, Measurement of the Lense–Thirring drag on high-altitude laser-ranged artificial satellites. *Phys. Rev. Lett.* **56**, 278–281 (1986).

[62] I. Ciufolini, A comprehensive introduction to the Lageos gravitomagnetic experiment: from the importance of the gravitomagnetic field in physics to preliminary error analysis and error budget. *Int. J. Mod. Phys. A* **4**, 3083–3145 (1989).

[63] D. P. Rubincam, On the secular decrease in the semimajor axis of Lageos's orbit. *Celest. Mech.* **26**, 361–382 (1982).

[64] D. M. Lucchesi, Reassessment of the error modelling of non-gravitational perturbations on LAGEOS 2 and their impact in the Lense–Thirring determination. Part I. *Planet. Space Sci.* **49**, 447–463 (2001).

[65] W. M. Kaula, *Theory of Satellite Geodesy*, Blaisdell, Waltham (1966).

[66] I. Ciufolini, On a new method to measure the gravitomagnetic field using two orbiting satellites. *Nuovo Cimento A* **109**, 1709–1720 (1996).

[67] G. Petit and B. Luzum (eds.), *IERS Conventions*, Verlag des Bundesamts für Kartographie und Geodaesie, Frankfurt am Main (2010).

[68] I. Ciufolini, E. C. Pavlis, F. Chieppa, E. Fernandes-Vieira and J. Perez-Mercader, Test of general relativity and measurement of the Lense–Thirring effect with two Earth satellites. *Science* **279**, 2100–2103 (1998).

[69] Ch. Reigber, F. Flechtner, R. Koenig, U. Meyer, K. Neumayer, R. Schmidt, P. Schwintzer and S. Zhu, GRACE orbit and gravity field recovery at GFZ Potsdam – First experiences and perspectives. *Eos. Trans. AGU*, **83**(47), Fall Meet. Suppl., Abstract G12B-03 (2002).

[70] B. D. Tapley, The GRACE Mission: Status and Performance Assessment. *Eos. Trans. AGU*, **83**(47), Fall Meet. Suppl., Abstract G12B-01 (2002).

[71] I. Ciufolini and E. C. Pavlis, A confirmation of the general relativistic prediction of the Lense–Thirring effect. *Nature* **431**, 958–960 (2004).

[72] I. Ciufolini, E. C. Pavlis, J. Ries, R. Koenig, G. Sindoni, A. Paolozzi and H. Newmayer, Gravitomagnetism and its measurement with laser ranging to the LAGEOS satellites and GRACE earth gravity models. In *John Archibald Wheleer and General Relativity* I. Ciufolini and R. Matzner (eds.), 371–434, Springer Verlag (2010).

[73] I. Ciufolini, A. Paolozzi, E. C. Pavlis, J. Ries, R. Koenig, R. Matzner, G. Sindoni and H. Neumayer, Testing gravitational physics with satellite laser ranging. *The European Physical Journal Plus* **126**, 72 (2011).

[74] J. C. Ries, R. J. Eanes and M. M. Watkins, Confirming the frame-fragging effect with satellite laser ranging. *16th International Workshop on Laser Ranging*, Poznan, Poland, 13–17 October 2008.

[75] J. C. Ries, Relativity in Satellite Laser Ranging. American Astronomical Society, *IAU Symposium 261. Relativity in Fundamental Astronomy: Dynamics, Reference Frames, and Data Analysis*, Virginia Beach, VA, USA, 27 April–1 May 2009.

[76] R. Koenig, B. Moreno Monge and G. Michalak, Some aspects and perspectives of measuring Lense–Thirring with GNSS and geodetic satellites. *Second International LARES Science Workshop*, Accademia dei Lincei, Rome, September 2012.

[77] I. Ciufolini, E. C. Pavlis and R. Peron, Determination of frame-dragging using Earth gravity models from CHAMP and GRACE. *New Astronomy* **11**, 527–550 (2006).

[78] I. Ciufolini, A. Paolozzi, E. C. Pavlis, J. Ries, V. Gurzadyan, R. Koenig, R. Matzner, R. Penrose and G. Sindoni, Testing general relativity and gravitational physics using the LARES satellite. *The European Physical Journal Plus* **127**, 1–7 (2012).

[79] A. Paolozzi and I. Ciufolini, LARES successfully launched in orbit: Satellite and mission description. *Acta Astronautica* **91**, 313–321 (2013).

[80] I. Ciufolini, A. Paolozzi, E. C. Pavlis, J. Ries, R. Koenig, R. Matzner and G. Sindoni, The LARES Space Experiment: LARES Orbit, Error Analysis and Satellite Structure, in *John Archibald Wheeler and general relativity* I. Ciufolini and R. Matzner (eds.), 371–434, Springer Verlag (2010).

[81] I. Ciufolini, B. Moreno Monge, A. Paolozzi, R. Koenig, G. Sindoni and G. Michalak, Monte Carlo Simulations of the LARES space experiment to test general relativity and fundamental physics. *Class. Quantum Grav.* **30**, 235009 (2013), doi:10.1088/0264-9381/30/23/235009. See also [82].

[82] B. Moreno Monge, R. Koenig, G. Michalak, I. Ciufolini, A. Paolozzi and G. Sindoni, Preliminary study for the measurement of the Lense- Thirring effect with the GALILEO satellites. *Acta Futura* **7**, 87–96 (2013).

[83] M. R. Pearlman, J. J. Degnan and J. M. Bosworth, The international laser ranging service. *Adv. Space Res.* **30**, 135–143, doi:10.1016/S0273-1177(02)00277-6 (2002).

[84] A. Paolozzi, I. Ciufolini, C. Vendittozzi, Engineering and scientific aspects of LARES satellite. *Acta Astronautica* **69**, 127–134 (2011).

[85] A. Paolozzi, I. Ciufolini, F. Felli, A. Brotzu, D. Pilone, Issues on LARES Satellite material. *Proceedings of 60th International Astronautical Congress IAC 2009*, Paper: IAC-09.C2.4.5, Daejeon, Republic of Korea, 12–16 October 2009.

[86] A. Paolozzi, I. Ciufolini, G. Caputo, L. Caputo, F. Passeggio, F. Onorati, C. Paris, A. Chiodo, LARES satellite and separation system. *Proceedings of 63rd International Astronautical Congress IAC 2012*, Paper: IAC-12- C2.9.5.x15247, Naples, Italy, 1–5 October 2012.

[87] A. Paolozzi, I. Ciufolini, C. Paris, G. Sindoni, D. Spano, Qualification tests on the optical retroreflectors of LARES satellite. *Proceedings of 63rd International Astronautical Congress IAC 2012*, Paper: IAC-12-C2.1.17.p1, Naples, Italy, 1–5 October 2012.

[88] S. W. Hawking and G. F. R. Ellis, *The Large Scale Structure of Space-Time*, Cambridge University Press (1975).

[89] J. B. Hartle, *Gravity: An Introduction to Einstein's General Relativity*, Addison Wesley, San Francisco (2003).

[90] W. Rindler, *Relativity: Special, General, and Cosmological*, Oxford University Press, Oxford (2001).

[91] J. Ehlers, Survey of general relativity theory, in *Relativity, Astrophysics and Cosmology*, W. Israel (ed.), 1–125, Reidel Publishing (1973).

[92] J. Ehlers and R. Geroch, Equation of motion of small bodies in relativity. *Ann. Phys.* **309**, 232–236 (2004).

[93] R. Geroch and P. S. Jang, Motion of a body in general relativity. *J. Math. Phys.* **16**, 65–67 (1975).

[94] S. Zhu, Ch. Reigber and R. Koenig, Integrated adjustment of CHAMP, GRACE, and GPS data. *J. Geodesy* **78**, 103–108 (2004).

[95] D. E. Pavlis et al., *GEODYN operations manuals*, Contractor Report, Raytheon, ITSS, Landover MD (1998).

[96] C. F. Martin and D. P. Rubincam, Effects of Earth albedo on the LAGEOS I satellite. *J. Geophys. Res. B* **101**, 3215–3226 (1996).

[97] D. P. Rubincam, Yarkovsky Thermal Drag on LAGEOS. *J. Geophys. Res. B* **93**, 13805–13810 (1988).

[98] D. P. Rubincam, Drag on the LAGEOS satellite. *J. Geophys. Res.* **95** (B11), 4881–4886 (1990).

[99] J. Ries, I. Ciufolini, E. Pavlis, A. Paolozzi, R. Koenig, R. Matzner, G. Sindoni and H. Neumayer, The Earth's frame-dragging via laser-ranged satellites: a response to "Some considerations on the present-day results for the detection of frame-dragging after the final outcome of GP-B" by L. Iorio. *EPL* **96**, 1–5 (2011).

[100] M. R. Drinkwater, R. Floberghagen, R. Haagmans, D. Muzi and A. Popescu, GOCE: ESA's first Earth Explorer Core mission, in *Earth Gravity Field From Space – From Sensors to Earth Science*, Space Sci. Ser. ISSI, Vol. 18, G. Beutler et al. (eds.), 419–432, Kluwer Acad., Dordrecht, Netherlands (2003).

[101] R. Pail et al., Combined satellite gravity field model GOCO01S derived from GOCE and GRACE. *Geophys. Res. Lett.* **37**, L20314 (2010), doi:10.1029/2010GL044906.

Toshio Fukushima

Elliptic functions and elliptic integrals for celestial mechanics and dynamical astronomy

1 Introduction

The elliptic functions and the elliptic integrals are one of the most complicated special functions [68]. Their textbooks are Akhiexer [1], Bowman [3], Cayley [25], Hancock [57, 58], Lawden [60], Thompson [66] and their references are Abramowitz and Stegun [2], Byrd and Friedman [12], Oldham et al. [61], Olver et al. [62], Wolfram [69]. The reference [62] is freely accessible from its Website: //http:/dlmf.nist.gov/. For practical purposes, we limit ourselves with the case where all the input arguments and the function/integral values are real-valued hereafter.

Among the various forms of elliptic functions and elliptic integrals, the Jacobian elliptic functions and Legendre's form of elliptic integrals are most popular. In celestial mechanics and dynamical astronomy, the Jacobian elliptic functions frequently appear in the analytical expressions of the variables of some dynamical systems. For example, the rapidly convergent series required in the orbital dynamics of planets and satellites are constructed by using the Jacobian elliptic functions [4–6, 26, 59, 65, 67]. This is because the basic orbit is an ellipse, and therefore, the elliptic functions naturally appear in its coordinate description, especially when the independent variable is set as the arc length of an ellipse, which is represented by Legendre's form of incomplete elliptic integral of the second kind.

Also, the Jacobian elliptic functions are essentially needed in describing the torque-free rotation of the rigid body [31, 39–41, 43, 55]. In fact, a Gaussian formulation for non canonical elements of rotational dynamics requires their evaluation frequently in the forward and backward transformation between the elements and the standard variables such as the combination of the moving coordinate triad and the body-fixed components of the angular velocity vector [40]. Similarly, a canonical element formulation calls the functions repeatedly [41]. Indeed, some complicated problems on rotational dynamics are exactly solved in terms of the Jacobian elliptic functions [29].

On the other hand, the elliptic integrals naturally appear in the computation of gravitational field of a celestial body with a sort of symmetry. This is because the Green kernel of the Newtonian gravitational field is the reciprocal of the mutual dis-

Toshio Fukushima: National Astronomical Observatory of Japan, 2-21-1 Osawa, Mitaka, Tokyo 181-8588, Japan

tance which is expressed as the square root of a certain function of coordinates, and therefore, the resulting integral reduces to the elliptic integrals when the inside of the square root is a simple function, say a polynomial of the integral variable of degree 4 at most. Good examples are that of a homogeneous ring or disk [45], and of homogeneous triaxial ellipsoid [12, Introduction].

Another application of the incomplete elliptic integrals is the analytical solution expression of angle variables of the torque-free rotation [39–41, 43]. Indeed, Encke's method for rotational dynamics assumes an easy access to computing routines of the reference solution as a function of time, which is usually taken as the torque-free solution osculating at the initial epoch [31]. These situations recall us that the solution of Kepler's equations of various kinds plays the key role in the perturbed two-body problems in orbital dynamics [30, 32–36]. These applications are not only important in astronomy but also in physics and chemistry, especially in the symplectic integration of the rotational dynamics of molecules.

However, it is also true that the elliptic functions and the elliptic integrals are difficult to approach, understand, and utilize. Therefore, in order to enhance their accessibility, we first present their numerous notations and remark some pitfalls to understand the written materials in Section 2. Next, we explain their properties, key formulas, and internal relationships for the elliptic functions in Section 3 and for the elliptic integrals in Section 4, respectively. Finally, in Section 5, we present the essence of the new procedures of their numerical computation recently developed by us [42, 44–54].

2 Notations

The elliptic functions and the elliptic integrals are classic subjects of mathematics and have a long history of researches. Since there are quite a few different notations in the literature [69], we first summarize them and then give some remarks.

2.1 Glossary

- a: the second argument of Jacobi's form of incomplete elliptic integral of the third kind.
- $\operatorname{am}(u|m)$: Jacobi's amplitude function.
- b: the main variable in the duplication of Jacobian elliptic functions.
- $c = \operatorname{cn} u = \operatorname{cn}(u|m)$: the cosine amplitude function.
- `cel`: Bulirsch's general form of complete elliptic integral of all kinds.
- `cel1`: Bulirsch's form of complete elliptic integral of the first kind.
- `cel2`: Bulirsch's general form of complete elliptic integral of the second kind.
- `cel3`: Bulirsch's form of complete elliptic integral of the third kind.

- $d = \mathrm{dn}\, u = \mathrm{dn}(u|m)$: the delta amplitude function.
- `el`: Bulirsch's general form of incomplete elliptic integral of all kinds.
- `el1`: Bulirsch's form of incomplete elliptic integral of the first kind.
- `el2`: Bulirsch's general form of incomplete elliptic integral of the second kind.
- `el3`: Bulirsch's form of incomplete elliptic integral of the third kind.
- g_2, g_3: the invariants of Weierstrass elliptic function.
- h: a parameter in the addition theorem of incomplete elliptic integral of the third kind.
- k: the modulus.
- k_c: the complementary modulus.
- m: the parameter.
- m_c: the complementary parameter.
- n: the characteristic.
- n_c: the complementary characteristic.
- $\wp$: Weierstrass elliptic function.
- $q(m)$: Jacobi's nome.
- $s = \mathrm{sn}\, u = \mathrm{sn}(u|m)$: the sine amplitude function.
- $t(u, v)$: a common factor in the addition theorems of incomplete elliptic integrals.
- u: the argument.
- x: the second main variable of the half-argument transformation.
- y: the first main variable of the half-argument transformation.
- $B(m)$: the first associate complete elliptic integral of the second kind.
- $B(\varphi|m)$: the first associate incomplete elliptic integral of the second kind.
- $B_u(u|m)$: Jacobi's form of first associate incomplete elliptic integral of the second kind.
- $D(m)$: the second associate complete elliptic integral of the second kind.
- $D(\varphi|m)$: the second associate incomplete elliptic integral of the second kind.
- $D_u(u|m)$: Jacobi's form of second associate incomplete elliptic integral of the second kind.
- $E(m)$: Legendre's form of complete elliptic integral of the second kind.
- $E(\varphi|m)$: Legendre's form of incomplete elliptic integral of the second kind.
- $E_u(u|m)$: Jacobi's form of incomplete elliptic integral of the second kind.
- $F(\varphi|m)$: Legendre's form of incomplete elliptic integral of the first kind.
- $J(n|m)$: the associate complete elliptic integral of the third kind.
- $J(\varphi, n|m)$: the associate incomplete elliptic integral of the third kind.
- $J_u(u, n|m)$: Jacobi's form of associate incomplete elliptic integral of the third kind.
- $K(m)$: Legendre's form of complete elliptic integral of the first kind.
- R_D: Carlson's symmetric form of elliptic integral of the second kind.
- R_F: Carlson's symmetric form of elliptic integral of the first kind.
- R_J: Carlson's symmetric form of elliptic integral of the third kind.
- $S(m)$: the special complete elliptic integral of the second kind.
- $T(t, h)$: the universal arc tangent function.

- $\alpha \equiv \sin^{-1} k$: the modular angle.
- $\alpha \equiv \sqrt{n}$: the alternative characteristic.
- $\varepsilon(u|m)$: Jacobi's Epsilon function.
- φ: the amplitude.
- ν: the negative characteristic.
- $\Delta(\theta|m)$: Jacobi's delta function.
- $\Pi(n|m)$: Legendre's form of complete elliptic integral of the third kind.
- $\Pi(\varphi, n|m)$: Legendre's form of incomplete elliptic integral of the third kind.
- $\Pi_u(u, n|m)$: Jacobi's form of incomplete elliptic integral of the third kind.
- $\Pi_1(a|m)$: Jacobi's original complete elliptic integral of the third kind.
- $\Pi_1(u, a|m)$: Jacobi's original incomplete elliptic integral of the third kind.

2.2 First input argument: φ, u, and x

A source of confusion is the difference in the first argument of the elliptic functions and the elliptic integrals: (1) the amplitude φ in Legendre's notation, (2) the argument $u \equiv F(\varphi|m)$ in Jacobi's notation, and (3) the tangent amplitude $x \equiv \tan\varphi$ in Bulirsch's notation. In the followings, we discriminate the first two carefully by explicitly attaching the subscript u to Jacobi's notation such as

$$E_u(u|m) = E(\varphi|m) . \tag{2.1}$$

2.3 Second input argument: k, m, and α

Another confusion comes from the different notations in describing the second input argument. Traditionally, it is written as k and called the modulus [12]. For example, the classic expression of the complete elliptic integral of the first kind is

$$K = \int_0^{\pi/2} \frac{d\theta}{\sqrt{1 - k^2 \sin^2\theta}} . \tag{2.2}$$

Of course, if all numbers are regarded to be complex valued, this is acceptable. However, if only real numbers are dealt with as in the practical applications, k is not appropriate since the integral is real-valued even when k is a pure imaginary. Therefore, the parameter $m \equiv k^2$ is preferred recently [69]. The third option is the modular angle α [2] defined such that

$$\alpha \equiv \sin^{-1} k , \tag{2.3}$$

assuming that k is real and $0 \le k < 1$. In order to discriminate whether k, m, or α is used, the different separators are selected in front of them such as

$$F(\varphi, k) = F(\varphi|m) = F(\varphi \setminus \alpha) . \tag{2.4}$$

Namely, (1) if the separator is the comma ",", the following input argument means k, (2) if the separator is the vertical bar "|", the following input argument means m, and (3) if the separator is the backslash "\", the following input argument means α.

Still, there remains a source of confusion in expressing single-argument quantities such as the complete integrals K, E, B, and D. In this case, there is no apparent difference among $K(k)$, $K(m)$, and $K(\alpha)$, especially if the argument is expressed numerically or by the symbols other than k, m, and α.

2.4 Sign of n

A notorious confusion is in the sign of n appearing in the incomplete elliptic integral of the third kind:

$$\Pi = \int_0^{\varphi} \frac{d\theta}{(1 \pm n \sin^2\theta)\sqrt{1 - k^2 \sin^2\theta}} . \tag{2.5}$$

Traditionally, the plus sign "+" is adopted [2, 11, 62]. This is probably because the rotation angle of a triaxial rigid body in the short axis mode requires the integral of the third kind with the plus sign in the place of n [43]. Some authors use different symbol ν for the plus sign:

$$\Pi = \int_0^{\varphi} \frac{d\theta}{(1 + \nu \sin^2\theta)\sqrt{1 - k^2 \sin^2\theta}} . \tag{2.6}$$

In the followings, we adopt n with the negative sign "–" [69].

2.5 Third input argument: n and α

Quite confusingly, the same symbol α as the modular angle also appears as an alternative of the characteristic n as

$$\alpha^2 \equiv \nu = \pm n \geq 0 . \tag{2.7}$$

Again, there are both choices of the sign of n, which makes the problem more complicated.

2.6 Ordering of arguments

Another type of confusion is brought by different conventions in the ordering of the input arguments, especially in expressing the incomplete elliptic integral of the third kind. The usual convention is the order of (1) the argument (or its alternative) first,

(2) the characteristic (or its alternative) second, and (3) the parameter (or its alternative) last. An example is

$$\Pi(\varphi, -\alpha^2, k) = \Pi_u(u, n|m) . \tag{2.8}$$

On the other hand, Mathematica [69] puts the characteristic first as

```
EllipticPi[n, phi, m]
```

2.7 Omission of parameters

In the literature, some input arguments like k, m, or n are frequently omitted such as (1) K in place of $K(k)$ or $K(m)$, and (2) $\operatorname{sn} u$ in place of $\operatorname{sn}(u, k)$ or $\operatorname{sn}(u|m)$. This is just a convention to simplify the expression of lengthy formulas and equations. Nevertheless, one should be careful that hidden parameters exist behind the scene. This becomes important especially when conducting the partial differentiation of these abbreviated quantities.

3 Elliptic functions

3.1 General elliptic function

Historically, an elliptic function was introduced as the inverse of a standard elliptic integral [12, p.18]:

$$I(x) = \int_a^x \frac{dz}{\sqrt{f(z)}} , \tag{3.1}$$

where $f(z)$ is a cubic or quartic polynomial of z with distinct roots. If the argument of the integral x is regarded as a function of the integral value denoted by

$$u \equiv I(x) , \tag{3.2}$$

then x is called an elliptic function of argument u and is expressed as

$$x = p(u) . \tag{3.3}$$

We explicitly write $f(z)$ as

$$f(z) = a_0 z^4 + 4a_1 z^3 + 6a_2 z^2 + 4a_3 z + a_4 , \tag{3.4}$$

under the condition

$$a_0 \neq 0 \quad \text{and/or} \quad a_1 \neq 0 . \tag{3.5}$$

Thus, by definition, the first-order derivative of p with respect to u is expressed as

$$p' \equiv \frac{dp}{du} = 1 \bigg/ \frac{du}{dp} = \sqrt{f(p)} = \sqrt{a_0 p^4 + 4a_1 p^3 + 6a_2 p^2 + 4a_3 p + a_4} \,. \tag{3.6}$$

Namely, if a certain dynamical variable p is governed by a first-order ordinary differential equation (ODE) and if the right-hand side of ODE is expressed as the square root of a polynomial of p of degree 4 at most, then the variable is analytically expressed in terms of a sort of elliptic function. This is the first-type application of elliptic functions.

Square the above first-order derivative expression, differentiate both sides of it with respect to u, and divide them by the common factor, $2p'$, which is assumed to be nonzero. Then, the expression of the second-order derivative of p is obtained as

$$p'' \equiv \frac{d^2 p}{du^2} = 2a_0 p^3 + 6a_1 p^2 + 6a_2 p + 2a_3 \,. \tag{3.7}$$

This means that, if the right-hand side of the equation of motion of a certain dynamical variable, p, is expressed as its polynomial of degree 3 at most, then the variable is analytically given by a kind of elliptic function. This is the second type application of elliptic functions.

3.2 Weierstrass elliptic function

Weierstrass's $\wp$ function [62, Chapter 23] is a special case of $p(u)$ such that

$$u = \int_{\wp}^{\infty} \frac{dz}{\sqrt{4z^3 - g_2 z - g_3}} \,. \tag{3.8}$$

This is the case when

$$a_0 = a_2 = 0\,, \quad a_1 = 1\,, \quad a_3 = -\left(\frac{g_2}{4}\right)\,, \quad a_4 = -g_3\,, \tag{3.9}$$

where the newly introduced coefficients, g_2 and g_3, are called the invariants. Thus, its derivatives are expressed [62, Section 23.3(ii)] as

$$\wp' = \sqrt{4\wp^3 - g_2\wp - g_3}\,, \tag{3.10}$$

$$\wp'' = 6\wp^2 - \frac{g_2}{2}\,. \tag{3.11}$$

It is rare to use $\wp$ in the practical applications. This is true also in celestial mechanics and dynamical astronomy.

3.3 Jacobian elliptic functions

Move to the Jacobian elliptic functions [62, Section 22.1]. The three principal Jacobian elliptic functions,

$$s(u) \equiv \mathrm{sn}(u|m)\,, \tag{3.12}$$

$$c(u) \equiv \mathrm{cn}(u|m)\,, \tag{3.13}$$

$$d(u) \equiv \mathrm{dn}(u|m)\,, \tag{3.14}$$

are special cases of $p(u)$ when $f(z)$ is a quadratic polynomial of z^2 [62, Section 22.15(ii)] as

$$u = \int_0^s \frac{dz}{\sqrt{(1-z^2)(1-mz^2)}}\,, \tag{3.15}$$

$$u = \int_c^1 \frac{dz}{\sqrt{(1-z^2)(m_c+mz^2)}}\,, \tag{3.16}$$

$$u = \int_d^1 \frac{dz}{\sqrt{(1-z^2)(z^2-m_c)}}\,, \tag{3.17}$$

where m is a parameter simply called the parameter, and

$$m_c \equiv 1 - m\,, \tag{3.18}$$

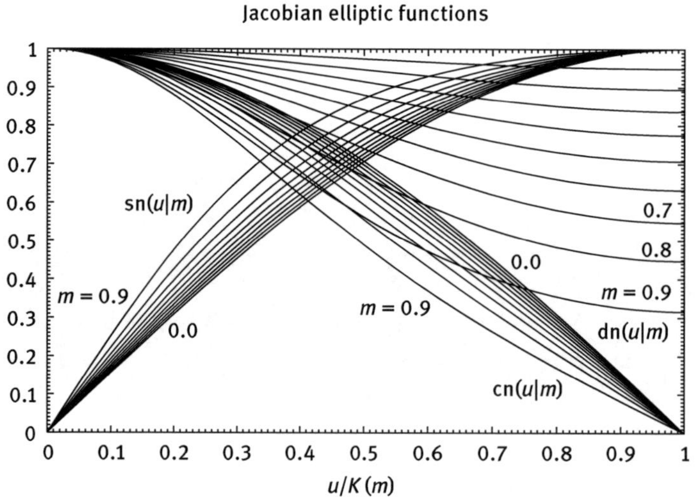

Fig. 1. Sketch of Jacobian elliptic functions: local. Illustrated are the graphs of three principal Jacobian elliptic functions, $\mathrm{sn}(u|m)$, $\mathrm{cn}(u|m)$, and $\mathrm{dn}(u|m)$ for u in its standard domain $0 \le u \le K(m)$ for various values of m as $m = 0.0, 0.1, \ldots, 0.9$.

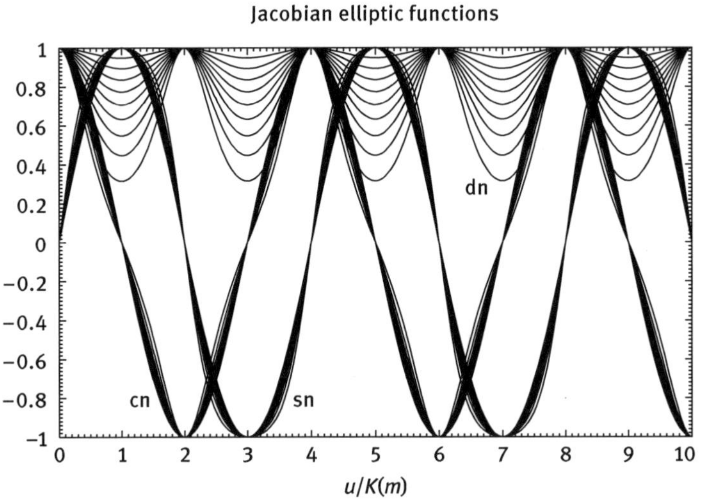

Fig. 2. Sketch of Jacobian elliptic functions: global. Same as Figure 1 but for a longer period.

is the complementary parameter. See Carlson [23] for the hidden symmetry in these elliptic functions. Traditionally, in place of m, another parameter called the modulus and defined as

$$k \equiv \sqrt{m}\,, \tag{3.19}$$

is frequently used.

Refer to Figures 1 and 2 for the behavior of the functions. It is noteworthy that $\mathrm{sn}(u|m)$ and $\mathrm{cn}(u|m)$ resemble $\sin[(\pi u)/(2K(m))]$ and $\cos[(\pi u)/(2K(m))]$ unless $m \approx 1$.

Classically, their numerical values are computed by the descending Landen transformation [8]. A faster algorithm using their conditional duplication is recently established [51].

3.4 Jacobi's amplitude function

Consider the partial derivative of the principal Jacobian elliptic functions with respect to the argument u while the parameter m is fixed. Using the general formula, Equation (3.6), we immediately obtain their expressions such as

$$s' \equiv \left(\frac{\partial s}{\partial u}\right)_m = \sqrt{\left(1-s^2\right)\left(1-ms^2\right)}\,. \tag{3.20}$$

However, these are not so useful. Instead, Jacobi noticed the similarity of the implicit definition of $s(u)$, Equation (3.15), with that of the sine function,

$$u = \int_0^{\sin u} \frac{dz}{\sqrt{1-z^2}} \,. \tag{3.21}$$

Then, he wrote the arc sine of $s(u)$ as

$$\varphi(u) \equiv \sin^{-1}[s(u)] \,, \tag{3.22}$$

and called it the amplitude. In the modern notation, it is written as

$$\varphi(u) = \mathrm{am}(u|m) \,. \tag{3.23}$$

At any rate, using the amplitude, Jacobi redefined $s(u)$ and introduced two other principal functions as its trigonometric function values as

$$s(u) \equiv \sin \varphi(u) \,, \tag{3.24}$$

$$c(u) \equiv \cos \varphi(u) \,, \tag{3.25}$$

$$d(u) \equiv \Delta(\varphi(u)|m) = \sqrt{1 - m \sin^2 \varphi(u)} \,, \tag{3.26}$$

where $\Delta(\varphi|m)$ is Jacobi's delta function. These definitions immediately lead to the following identity relations:

$$[s(u)]^2 + [c(u)]^2 = 1 \,, \tag{3.27}$$

$$m\,[s(u)]^2 + [d(u)]^2 = 1 \,. \tag{3.28}$$

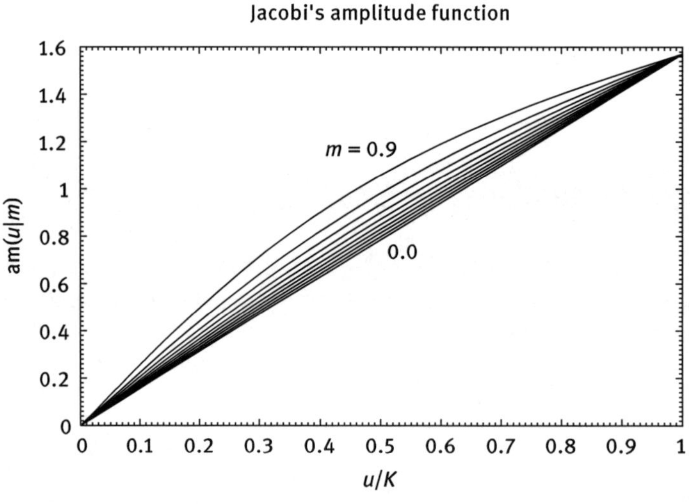

Fig. 3. Sketch of Jacobi's amplitude function: local. Same as Figure 1 but for $\mathrm{am}(u|m)$.

Jacobi called $\mathrm{sn}(u|m)$, $\mathrm{cn}(u|m)$, and $\mathrm{dn}(u|m)$ the sine amplitude function, the cosine amplitude function, and the delta amplitude function, respectively. Figure 3 illustrates Jacobi's amplitude function in the standard domain of the argument, $0 \le u \le K(m)$, for various values of m.

3.5 Differential equations of Jacobian elliptic functions

Return to the derivative expressions of the Jacobian elliptic functions. From the not-so-useful expression of s', Equation (3.20), the derivative expression of Jacobi's amplitude function is obtained as

$$\varphi' = d\,. \tag{3.29}$$

This naturally leads to the derivative expression of the sine amplitude function and the cosine amplitude function [62, Table 22.13.1] as

$$s' = cd\,, \tag{3.30}$$

$$c' = -sd\,. \tag{3.31}$$

Also, from the identity relation of $s(u)$ and $d(u)$, the derivative expression of $d(u)$ is obtained as

$$d' = -msc\,. \tag{3.32}$$

Indeed, s, c, and d can be defined as the fundamental solution of a system of first order ODEs of three components:

$$\frac{dx}{du} = yz\,, \tag{3.33}$$

$$\frac{dy}{du} = -xz\,, \tag{3.34}$$

$$\frac{dz}{du} = -mxy\,, \tag{3.35}$$

where m is a constant being independent on u. This is the third-type application of elliptic functions.

It is noteworthy that the introduction of Jacobi's amplitude function φ can be regarded as a sort of regularization of the rotational dynamics described by the above system of ODE equations [37, 38].

3.6 Addition theorem of Jacobian elliptic functions

Among many properties of the elliptic functions, the most important one is the availability of the addition theorems with respect to their argument. Indeed, the existence of the addition theorems is a key feature to characterize the elliptic function among many special functions.

In the case of the Jacobian elliptic functions, the theorems are written [12, Formula 123.01] as

$$\operatorname{sn}(u+v) = \frac{\operatorname{sn} u \ \operatorname{cn} v \ \operatorname{dn} v + \operatorname{sn} v \ \operatorname{cn} u \ \operatorname{dn} u}{1 - m \ \operatorname{sn}^2 u \ \operatorname{sn}^2 v}, \tag{3.36}$$

$$\operatorname{cn}(u+v) = \frac{\operatorname{cn} u \ \operatorname{cn} v - \operatorname{sn} u \ \operatorname{sn} v \ \operatorname{dn} u \ \operatorname{dn} v}{1 - m \ \operatorname{sn}^2 u \ \operatorname{sn}^2 v}, \tag{3.37}$$

$$\operatorname{dn}(u+v) = \frac{\operatorname{dn} u \ \operatorname{dn} v - m \ \operatorname{sn} u \ \operatorname{sn} v \ \operatorname{cn} u \ \operatorname{cn} v}{1 - m \ \operatorname{sn}^2 u \ \operatorname{sn}^2 v}, \tag{3.38}$$

where m as the second argument is omitted for it is common to all the functions. As a by-product, the double argument formulas are obtained as

$$\operatorname{sn} 2u = \frac{2 \operatorname{sn} u \ \operatorname{cn} u \ \operatorname{dn} u}{1 - m \ \operatorname{sn}^4 u}, \tag{3.39}$$

$$\operatorname{cn} 2u = \frac{\operatorname{cn}^2 u - \operatorname{sn}^2 u \ \operatorname{dn}^2 u}{1 - m \ \operatorname{sn}^4 u}, \tag{3.40}$$

$$\operatorname{dn} 2u = \frac{\operatorname{dn}^2 u - m \ \operatorname{sn}^2 u \ \operatorname{cn}^2 u}{1 - m \ \operatorname{sn}^4 u}, \tag{3.41}$$

which play an important rule in the numerical computation of the Jacobian elliptic functions.

The double argument formulas are inverted as

$$\operatorname{sn} \frac{u}{2} = \sqrt{\frac{1 - \operatorname{cn} u}{1 + \operatorname{dn} u}}, \tag{3.42}$$

$$\operatorname{cn} \frac{u}{2} = \sqrt{\frac{\operatorname{cn} u + \operatorname{dn} u}{1 + \operatorname{dn} u}}, \tag{3.43}$$

$$\operatorname{dn} \frac{u}{2} = \sqrt{\frac{\operatorname{cn} u + \operatorname{dn} u}{1 + \operatorname{cn} u}}, \tag{3.44}$$

which are called the half-argument formulas. Since the first of them suffers from the cancellation problem when $|u|$ is small, it is rewritten as

$$\operatorname{sn}^2 \frac{u}{2} = \frac{\operatorname{sn}^2 u}{(1 + \operatorname{cn} u)(1 + \operatorname{dn} u)}, \tag{3.45}$$

which becomes important in the numerical computation of all the incomplete elliptic integrals.

3.7 Jacobi's form of incomplete elliptic integrals

Jacobi regarded incomplete elliptic integrals as functions of Legendre's form of incomplete elliptic integral of the first kind,

$$u \equiv F(\varphi|m) \,. \tag{3.46}$$

In his notation, some incomplete elliptic integrals are expressed as the integrals of rational functions of Jacobian elliptic functions [12, Formulas 110.02 through 110.04]:

$$E_u(u|m) \equiv E(\mathrm{am}(u|m)|m) = \int_0^u \mathrm{dn}^2(v|m)dv \,, \tag{3.47}$$

$$\Pi_u(u,n|m) \equiv \Pi(\mathrm{am}(u|m),n|m) = \int_0^u \frac{dv}{1 - n\,\mathrm{sn}^2(v|m)} \,, \tag{3.48}$$

$$B_u(u|m) \equiv B(\mathrm{am}(u|m)|m) = \int_0^u \mathrm{cn}^2(v|m)dv \,, \tag{3.49}$$

$$D_u(u|m) \equiv D(\mathrm{am}(u|m)|m) = \int_0^u \mathrm{sn}^2(v|m)dv \,, \tag{3.50}$$

$$J_u(u,n|m) \equiv J(\mathrm{am}(u|m),n|m) = \int_0^u \frac{\mathrm{sn}^2(v|m)\,dv}{1 - n\,\mathrm{sn}^2(v|m)} \,, \tag{3.51}$$

where n is called the characteristic.

Although the standard notation is without the suffix u, we dare attach it in order to discriminate them from Legendre's form of incomplete elliptic integrals, $E(\varphi|m)$, $\Pi(\varphi,n|m)$, $B(\varphi|m)$, $D(\varphi|m)$, and $J(\varphi,n|m)$, which will be discussed in Section 4 later, since the meaning of the first argument is different. Among them, $E_u(u|m)$ is known as Jacobi's Epsilon function and sometimes denoted by $\varepsilon(u)$ [62, Section 22.16(ii)]. See Figure 4 plotting its curves in the standard domain of the argument, $0 \le u \le K(m)$, for various values of m. Also, $J_u(u,n|m)$ is tightly related with Jacobi's original incomplete elliptic integral of the third kind $\Pi_1(u)$ [12, p.233]. See Figure 5 for its sketch as a function of u for various values of m when $n = 0.5$. The numerical value of Jacobi's form of incomplete elliptic integrals are efficiently computed by the procedure in our work [42].

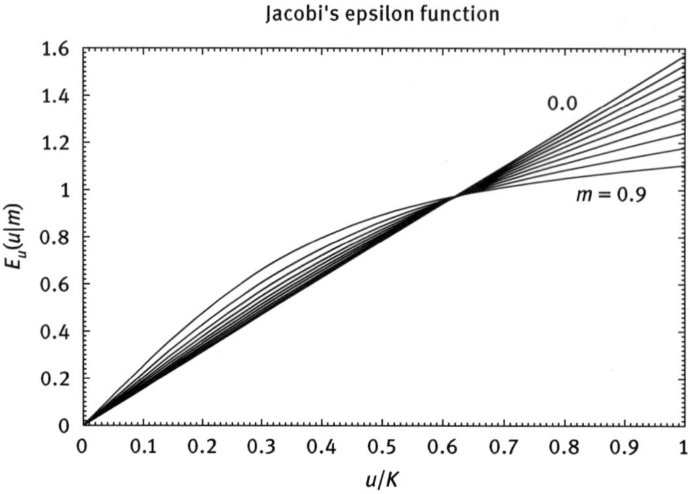

Fig. 4. Sketch of Jacobi's Epsilon function: local. Same as Figure 1 but for $E_u(u|m)$.

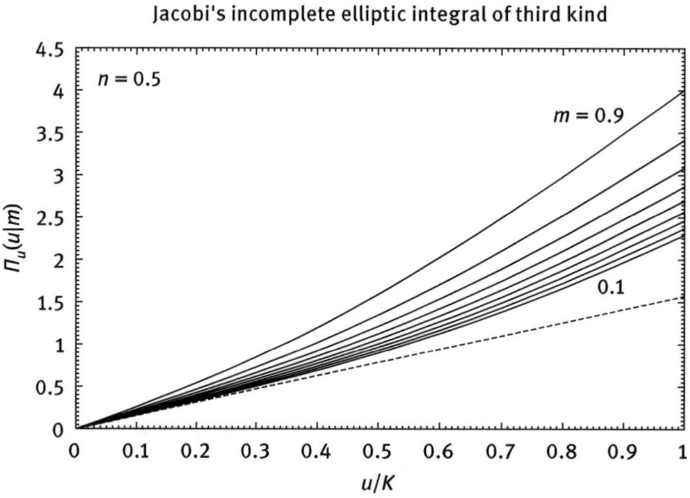

Fig. 5. Sketch of Jacobi's form of incomplete elliptic integral of the third kind: local. Same as Figure 1 but for $\Pi_u(u, n|m)$ when $n = 0.5$. Added is the straight line u plotted by a broken line for reference.

When a set of coordinate variables are expressed in terms of the Jacobian elliptic function, $p(u)$, their conjugate momentum variables are sometimes expressed by their integral with respect to u. In that case, these Jacobi's form of incomplete elliptic integrals appear as the fourth-type application of elliptic functions.

3.8 Addition theorem of incomplete elliptic integrals

The addition theorems of Jacobi's form of incomplete elliptic integrals are also available as

$$E_u(u+v) = E_u(u) + E_u(v) - m \ \mathrm{sn}\, u \ \mathrm{sn}\, v \ \mathrm{sn}(u+v)\,, \tag{3.52}$$

$$\Pi_u(u+v) = \Pi_u(u) + \Pi_u(v) + n\, T(t(u,v),h)\,, \tag{3.53}$$

$$B_u(u+v) = B_u(u) + B_u(v) - \ \mathrm{sn}\, u \ \mathrm{sn}\, v \ \mathrm{sn}(u+v)\,, \tag{3.54}$$

$$D_u(u+v) = D_u(u) + D_u(v) + \ \mathrm{sn}\, u \ \mathrm{sn}\, v \ \mathrm{sn}(u+v)\,, \tag{3.55}$$

$$J_u(u+v) = J_u(u) + J_u(v) + T(t(u,v),h)\,, \tag{3.56}$$

where m and/or n are omitted, for they being common to all the integrals,

$$t(u,v) \equiv \frac{\mathrm{sn}\, u \ \mathrm{sn}\, v \ \mathrm{sn}(u+v)}{1 - n\left[\mathrm{sn}^2(u+v) - \mathrm{sn}\, u \ \mathrm{sn}\, v \ \mathrm{cn}(u+v) \ \mathrm{dn}(u+v)\right]}\,, \tag{3.57}$$

$$h \equiv n n_c (n-m)\,, \tag{3.58}$$

and $T(t,h)$ is the universal arc tangent function defined [49, Equation (25)] as

$$T(t,h) \equiv t \sum_{j=0}^{\infty} \frac{(-ht^2)^j}{2j+1} = \begin{cases} \tan^{-1}\left(t\sqrt{h}\right)/\sqrt{h} & (h>0) \\ t & (h=0) \\ \tanh^{-1}\left(t\sqrt{-h}\right)/\sqrt{-h} & (h<0) \end{cases}\,, \tag{3.59}$$

the fast computation of which is described in [49, § 3.7].

As a natural consequence, the double argument formulas are derived from these as

$$E_u(2u) = 2E_u(u) - m \ \mathrm{sn}^2 u \ \mathrm{sn}\, 2u\,, \tag{3.60}$$

$$\Pi_u(2u) = 2\Pi_u(u) + n\, T(t(u,u),h)\,, \tag{3.61}$$

$$B_u(2u) = 2B_u(u) - \ \mathrm{sn}^2 u \ \mathrm{sn}\, 2u\,, \tag{3.62}$$

$$D_u(2u) = 2D_u(u) + \ \mathrm{sn}^2 u \ \mathrm{sn}\, 2u\,, \tag{3.63}$$

$$J_u(2u) = 2J_u(u) + T(t(u,u),h)\,, \tag{3.64}$$

where $t(u,u)$ is simplified as

$$t(u,u) = \frac{\mathrm{sn}^2 u \ \mathrm{sn}\, 2u}{1 - n\left[\mathrm{sn}^2 2u - \mathrm{sn}^2 u \ \mathrm{cn}\, 2u \ \mathrm{dn}\, 2u\right]}\,, \tag{3.65}$$

These formulas are useful in the numerical computation of these integrals. Among the above formulas, the most fundamental are those of J_u since (1) those of B_u are equivalent with those of E_u if $m = 1$, (2) those of J_u become the same as those of Π_u when $n = 1$, (3) those of D_u are obtained from those of J_u by letting $n = 0$. and (4) those of B_u and D_u are practically the same since $B_u(u) + D_u(u) = u$.

3.9 Jacobi's original form of incomplete elliptic integral of the third kind

In place of $\Pi_u(u, n|m)$, Jacobi preferred a different form as the incomplete elliptic integral of the third kind [58, p.420]:

$$\Pi_1(u, a|m) \equiv m \ \operatorname{sn}(a|m) \ \operatorname{cn}(a|m) \ \operatorname{dn}(a|m) \int_0^u \frac{\operatorname{sn}^2(v|m)\, dv}{1 - m \ \operatorname{sn}^2(a|m) \ \operatorname{sn}^2(v|m)}, \quad (3.66)$$

where a is a variable replacing n such that

$$n = m \ \operatorname{sn}^2(a|m) . \quad (3.67)$$

Although it is denoted by Π_1, the integral is essentially the same as $J_u(u, n|m)$ since

$$\Pi_1(u, a|m) = m \ \operatorname{sn}(a|m) \ \operatorname{cn}(a|m) \ \operatorname{dn}(a|m) \ J_u(u, m \operatorname{sn}^2(a|m)|m) . \quad (3.68)$$

The complete elliptic integral is similarly written as

$$\Pi_1(a|m) \equiv \Pi_1(K(m), a|m) = m \ \operatorname{sn}(a|m) \ \operatorname{cn}(a|m) \ \operatorname{dn}(a|m) \ J_u(m \operatorname{sn}^2(a|m)|m) . \quad (3.69)$$

The usefulness of this form is in the existence of the not-so-popular addition theorem with respect to a [58, p.428]:

$$\begin{aligned}\Pi_1(u, a + b|m) &= \Pi_1(u, a|m) + \Pi_1(u, b|m) - m \ u \ \operatorname{sn}(a|m) \ \operatorname{sn}(b|m) \ \operatorname{sn}(a + b|m) \\ &\quad - \tfrac{1}{2} \log \left(\frac{1 - m \ \operatorname{sn}(a|m) \ \operatorname{sn}(b|m) \ \operatorname{sn}(u|m) \ \operatorname{sn}(a + b - u|m)}{1 + m \ \operatorname{sn}(a|m) \ \operatorname{sn}(b|m) \ \operatorname{sn}(u|m) \ \operatorname{sn}(a + b + u|m)} \right) .\end{aligned} \quad (3.70)$$

The complete version becomes

$$\Pi_1(a+b|m) = \Pi_1(a|m)+\Pi_1(b|m)-m \ K(m) \ \operatorname{sn}(a|m) \ \operatorname{sn}(b|m) \ \operatorname{sn}(a+b|m). \quad (3.71)$$

As a result, the double argument formula with respect to a is derived as

$$\Pi_1(2a|m) = 2\Pi_1(a|m) - m \ K(m) \ \operatorname{sn}^2(a|m) \ \operatorname{sn}(2a|m) . \quad (3.72)$$

This becomes useful in computing $J(n|m)$ efficiently [53].

4 Elliptic integrals

4.1 General elliptic integral

An elliptic integral was naturally introduced in discussing an integral of a general function $g(z)$ as

$$G(x) = \int_a^x g(z)\, dz . \quad (4.1)$$

If $g(z)$ is a rational function, $G(x)$ becomes a sum of a rational function and a product of the logarithm and another rational function. The next simplest case is to add a single square root of a polynomial as its irrational part:

$$G(x) = \int_a^x g\left(z; \sqrt{f(z)}\right) dz \,. \tag{4.2}$$

If $f(z)$ is a polynomial of degree 2 at most, the resulting integral contains the inverse trigonometric function and the logarithm.

The difficulty increases for higher degrees such as when encountered in investigating an arc length of a general quadratic curve including an ellipse or a hyperbola. If $f(z)$ is a polynomial of degree 4 at most as described in Equation (3.4), $G(x)$ is called a general elliptic integral. Since g is a rational function of z when f is regarded as a constant, $G(x)$ can be split into a sum of a rational function $R(x)$ and a special kind of elliptic integral as

$$G(x) = R(x) + \int_a^x \frac{P(z)}{Q(z)\sqrt{f(z)}} dz \,, \tag{4.3}$$

where $P(z)$ and $Q(z)$ are polynomials.

4.2 Legendre's form of incomplete elliptic integrals

It is very complicated to simplify the general elliptic integral, $G(x)$. See the compact reduction tables in terms of the symmetric integrals [17–21]. Legendre finally succeeded to resolve its elliptic part into a linear combination of three fundamental integrals [62, Section 19.2]:

$$F(\varphi|m) \equiv \int_0^\varphi \frac{d\theta}{\Delta(\theta|m)} \,, \tag{4.4}$$

$$E(\varphi|m) \equiv \int_0^\varphi \Delta(\theta|m)\, d\theta \,, \tag{4.5}$$

$$\Pi(\varphi, n|m) \equiv \int_0^\varphi \frac{d\theta}{(1 - n\sin^2\theta)\,\Delta(\theta|m)} \,, \tag{4.6}$$

where n is a parameter called the characteristic. They are called Legendre's form of incomplete elliptic integral of the first, the second, and the third kind, respectively. These are termed "incomplete" since the upper end point of the integral is general. Figures 6 and 7 show the curves of $F(\varphi|m)$ and $E(\varphi|m)$ as functions of φ for various values of m.

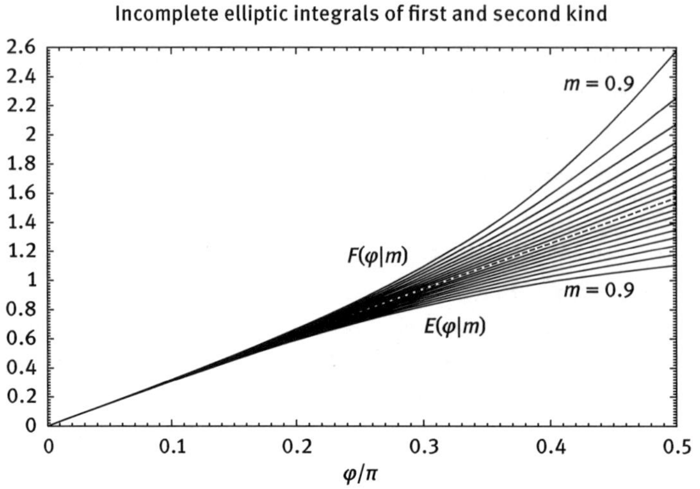

Fig. 6. Sketch of Legendre's form of incomplete elliptic integrals: local. Plotted are the graphs of $F(\varphi|m)$ and $E(\varphi|m)$ as functions of φ in its standard domain, $0 \le \varphi \le \pi/2$ for various values of m as $m = 0.1, 0.2, \ldots, 0.9$. Notice inequalities $E(\varphi|0.9) < E(\varphi|0.8) < \cdots < F(\varphi|0.8) < F(\varphi|0.9)$. The results when $m = 0$ become the same straight line, $F(\varphi|0) = E(\varphi|0) = \varphi$, which is shown by a broken line.

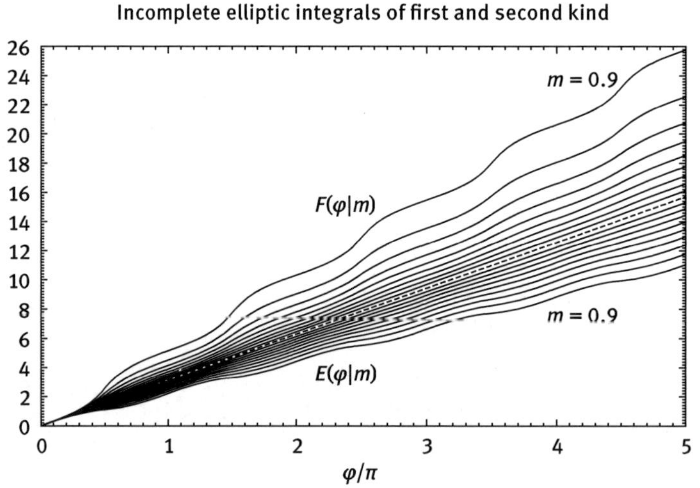

Fig. 7. Sketch of Legendre's form of incomplete elliptic integrals: global. Same as Figure 6 but for a longer period.

4.3 Associate incomplete elliptic integrals

In many applications, Legendre's form of incomplete elliptic integrals, $F(\varphi|m)$, $E(\varphi|m)$, and/or $\Pi(\varphi, n|m)$ are used in combination. For example, the length of hyperbola requires both of $F(\varphi|m)$ and $E(\varphi|m)$ [12, Introduction]. Also, the rotation angle of the torque-free rotation of a rigid body is described by a linear combination of $F(\varphi|m)$ and $\Pi(\varphi, n|m)$ [39]. Further, both of $F(\varphi|m)$ and $E(\varphi|m)$ are needed in computing their partial derivatives with respect to m [62]. These partial derivatives are necessary in computing the force field of potentials expressed using $F(\varphi|m)$. Similar needs arise in evaluating the partial derivatives of Jacobian elliptic functions with respect to u and/or m [12].

From a practical viewpoint, however, rather important is not $F(\varphi|m)$, $E(\varphi|m)$, and $\Pi(\varphi, n|m)$ but a trio of associate incomplete elliptic integrals [8, 48, 49]:

$$B(\varphi|m) \equiv \int_0^{\varphi} \frac{\cos^2\theta \, d\theta}{\Delta(\theta|m)}, \tag{4.7}$$

$$D(\varphi|m) \equiv \int_0^{\varphi} \frac{\sin^2\theta \, d\theta}{\Delta(\theta|m)}, \tag{4.8}$$

$$J(\varphi, n|m) \equiv \int_0^{\varphi} \frac{\sin^2\theta \, d\theta}{(1 - n\sin^2\theta)\,\Delta(\theta|m)}. \tag{4.9}$$

The practical usefulness of $B(\varphi|m)$, $D(\varphi|m)$, and $J(\varphi, n|m)$ are well described [8, 47–49, 52, 53]. See Figures 8 and 9 for the plots of $J(\varphi, n|m)$ as function of φ for various values of m when n is fixed as $n = 0.5$.

The numerical values of these incomplete elliptic integrals are effectively computed by the following procedures: (1) $F(\varphi|m)$ by `elf` [45], (2) $B(\varphi|m)$ and $D(\varphi|m)$ simultaneously by `elbd` [48], and (3) $B(\varphi|m)$, $D(\varphi|m)$, and $J(\varphi, n|m)$ simultaneously by `elbdj` [49]. The last procedure, `elbdj`, is the most general one since $F(\varphi|m)$, $E(\varphi|m)$, and $\Pi(\varphi, n|m)$ are computed from $B(\varphi|m)$, $D(\varphi|m)$, and $J(\varphi, n|m)$ without suffering from the precision loss as

$$F(\varphi, n|m) = B(\varphi|m) + D(\varphi|m)\,, \tag{4.10}$$

$$E(\varphi, n|m) = B(\varphi|m) + m_c D(\varphi|m)\,, \tag{4.11}$$

$$\Pi(\varphi, n|m) = B(\varphi|m) + D(\varphi|m) + nJ(\varphi, n|m)\,, \tag{4.12}$$

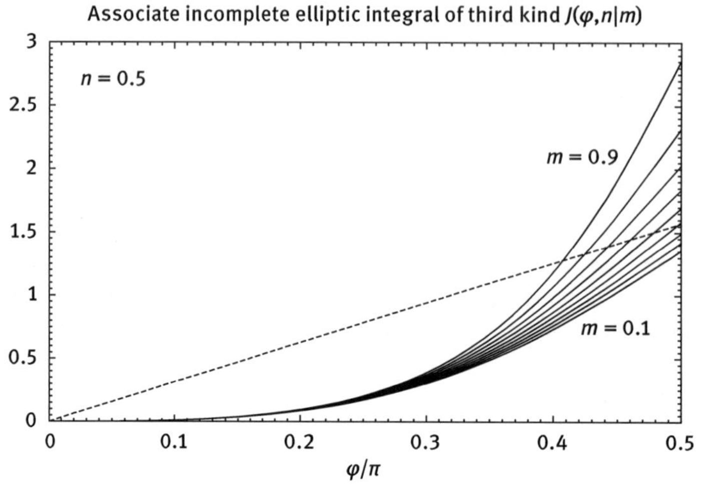

Fig. 8. Sketch of associate incomplete elliptic integral of the third kind: local. Same as Figure 6 but for $J(\varphi, n|m)$ when $n = 0.5$. For reference, a straight line φ is shown by a broken line again.

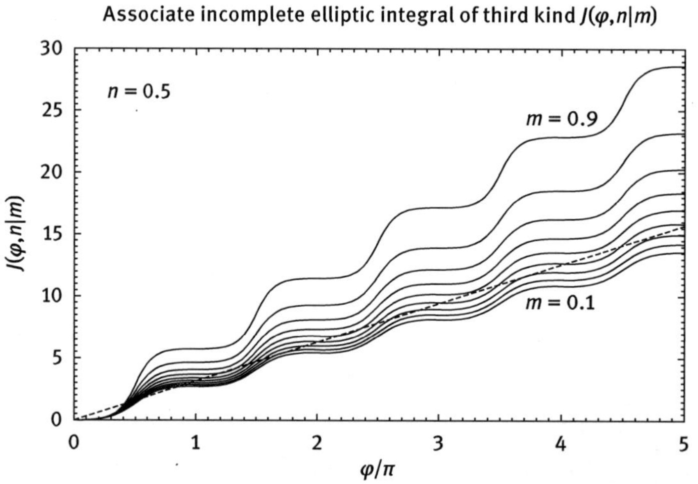

Fig. 9. Sketch of associate incomplete elliptic integral of the third kind: global. Same as Figure 8 but for a longer period.

respectively. The reverse procedures face with the problem of small divisors, m and n as

$$B(\varphi|m) = \frac{E(\varphi|m) - m_c F(\varphi|m)}{m}, \tag{4.13}$$

$$D(\varphi|m) = \frac{F(\varphi|m) - E(\varphi|m)}{m}, \tag{4.14}$$

$$J(\varphi, n|m) = \frac{\Pi(\varphi, n|m) - F(\varphi|m)}{n}. \tag{4.15}$$

This is problematic in the actual applications since

$$m \approx +1 \times 10^{-7}, \quad n \approx -7 \times 10^{-4}, \tag{4.16}$$

in the case of the Earth rotation [40, 41].

4.4 Complete elliptic integrals

The incomplete elliptic integrals are called "complete" when $\varphi = \pi/2$:

$$K(m) \equiv F(\pi/2|m) = \int_0^{\pi/2} \frac{d\theta}{\Delta(\theta|m)}, \tag{4.17}$$

$$E(m) \equiv E(\pi/2|m) = \int_0^{\pi/2} \Delta(\theta|m)\, d\theta, \tag{4.18}$$

$$\Pi(n|m) \equiv \Pi(\pi/2, n|m) = \int_0^{\pi/2} \frac{d\theta}{(1 - n\sin^2\theta)\,\Delta(\theta|m)}, \tag{4.19}$$

$$B(m) \equiv B(\pi/2|m) = \int_0^{\pi/2} \frac{\cos^2\theta\, d\theta}{\Delta(\theta|m)}, \tag{4.20}$$

$$D(m) \equiv D(\varphi|m) = \int_0^{\pi/2} \frac{\sin^2\theta\, d\theta}{\Delta(\theta|m)}, \tag{4.21}$$

$$J(n|m) \equiv J(\pi/2, n|m) = \int_0^{\pi/2} \frac{\sin^2\theta\, d\theta}{(1 - n\sin^2\theta)\,\Delta(\theta|m)}. \tag{4.22}$$

The numerical values of these complete integrals are effectively computed by the following procedures: (1) $K(m)$ and $E(m)$ by `celk` and `cele`, respectively [45], (2) $B(m)$ and $D(m)$ simultaneously by `celbd` [48], and (3) $B(m)$, $D(m)$, and $J(n|m)$ simultaneously by `celbdj` [51]. The last procedure `celbdj` is the most general one since $K(m)$, $E(m)$, and $\Pi(n|m)$ are computed from $B(m)$, $D(m)$, and $J(n|m)$ without suffering from the loss of information as

$$K(m) = B(m) + D(m), \tag{4.23}$$

$$E(m) = B(m) + m_c D(m), \tag{4.24}$$

$$\Pi(n|m) = B(m) + D(m) + nJ(n|m). \tag{4.25}$$

During the course of investigation of the gravitational acceleration field due to a uniform ring, we faced with the round-off errors even if using $B(m)$ and $D(m)$. To resolve

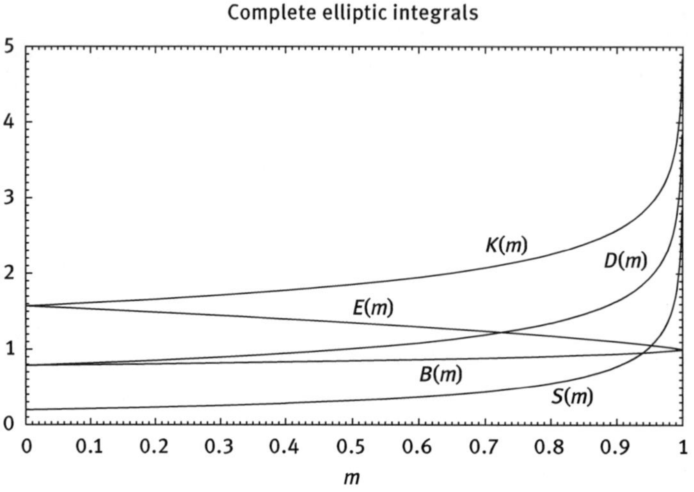

Fig. 10. Sketch of five complete elliptic integrals. Illustrated are the graphs of five complete elliptic integrals, $K(m)$, $E(m)$, $B(m)$, $D(m)$, and $S(m)$ for m in its standard domain $0 \leq m < 1$.

it, a special complete elliptic integral is introduced [46] as

$$S(m) \equiv \frac{-1}{m} \int_0^{\pi/2} \frac{\cos 2\theta \, d\theta}{\Delta(\theta|m)} = \frac{D(m) - B(m)}{m} = \frac{(2-m)K(m) - 2E(m)}{m^2} . \tag{4.26}$$

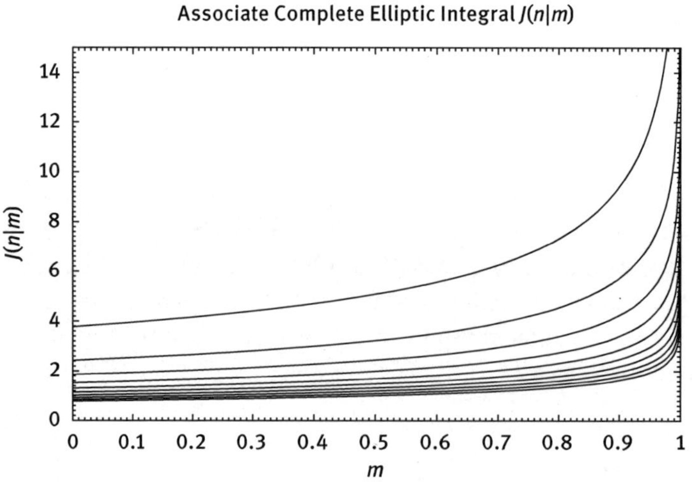

Fig. 11. Sketch of $J(n|m)$. Same as Figure 10 but of $J(n|m)$ for various values of n as $n = 0.0, 0.1, \ldots, 0.9$. Notice that $J(n|m)$ is monotonically increasing with respect to n when m is fixed.

The pair of $B(m)$ and $S(m)$ is more fundamental than that of $B(m)$ and $D(m)$ since $D(m)$ is computed from them without information loss as

$$D(m) = B(m) + mS(m) . \tag{4.27}$$

Figure 10 illustrates the behavior of these five complete elliptic integrals as functions of m in the standard domain, $0 \le m < 1$. Also, Figure 11 shows the curves of $J(n|m)$ as functions of m for various values of n in the standard domain, $0 \le n < 1$.

4.5 Generalized elliptic integrals

From a practical viewpoint of numerical computation, Bulirsch generalized Legendre's form of elliptic integrals by changing the main variable from φ to $x \equiv \tan\varphi$ and adding a few parameters [8, 11]. Bulirsch's form of incomplete elliptic integrals of the first, second, and third kind are defined as

$$\mathtt{el1}\,(x, k_c) \equiv \int_0^x \frac{d\xi}{\sqrt{(1+\xi^2)\,(1+k_c^2\xi^2)}} = \int_0^\varphi \frac{d\theta}{\sqrt{\cos^2\theta + k_c^2 \sin^2\theta}} , \tag{4.28}$$

$$\mathtt{el2}\,(x, k_c, a, b) \equiv \int_0^x \frac{a + b\xi^2}{\sqrt{(1+\xi^2)\,(1+k_c^2\xi^2)}} d\xi = \int_0^\varphi \frac{a\cos^2\theta + b\sin^2\theta}{\sqrt{\cos^2\theta + k_c^2 \sin^2\theta}} d\theta , \tag{4.29}$$

$$\begin{aligned}\mathtt{el3}\,(x, k_c, p) &\equiv \int_0^x \frac{d\xi}{(1+p\xi^2)\sqrt{(1+\xi^2)\,(1+k_c^2\xi^2)}} \\ &= \int_0^\varphi \frac{d\theta}{(\cos^2\theta + p\sin^2\theta)\sqrt{\cos^2\theta + k_c^2 \sin^2\theta}} ,\end{aligned} \tag{4.30}$$

where

$$k_c \equiv \sqrt{m_c} = \sqrt{1-m} , \tag{4.31}$$

is the complementary modulus. Their computing procedures are described in [8, 9, 11]. Originally, he intended to generalize these integrals further [10, Equation (4.1.14)]:

$$\begin{aligned}\mathtt{el}\,(x, k_c, p, a, b) &\equiv \int_0^x \frac{a + b\xi^2}{(1+p\xi^2)\sqrt{(1+\xi^2)\,(1+k_c^2\xi^2)}} d\xi \\ &= \int_0^\varphi \frac{a\cos^2\theta + b\sin^2\theta}{(\cos^2\theta + p\sin^2\theta)\sqrt{\cos^2\theta + k_c^2 \sin^2\theta}} ,\end{aligned} \tag{4.32}$$

since it covers all of the incomplete elliptic integrals of three kinds as

$$\texttt{el1}\left(x,k_c\right) = \texttt{el}\left(x,k_c,p,1,p\right), \quad (p\text{: arbitrary}) \tag{4.33}$$

$$\texttt{el2}\left(x,k_c,a,b\right) = \texttt{el}\left(x,k_c,1,a,b\right)\ , \tag{4.34}$$

$$\texttt{el3}\left(x,k_c,p\right) = \texttt{el}\left(x,k_c,p,1,1\right)\ . \tag{4.35}$$

Bulirsch did not provide its computing algorithm. It was later presented by us [56]. At any rate, Bulirsch's form of incomplete elliptic integrals are related with Legendre's and the associate incomplete elliptic integrals as

$$F(\varphi|m) = \texttt{el1}\left(\tan\varphi,k_c\right)\ , \tag{4.36}$$

$$E(\varphi|m) = \texttt{el2}\left(\tan\varphi,k_c,1,m_c\right)\ , \tag{4.37}$$

$$B(\varphi|m) = \texttt{el2}\left(\tan\varphi,k_c,1,0\right)\ , \tag{4.38}$$

$$D(\varphi|m) = \texttt{el2}\left(\tan\varphi,k_c,0,1\right)\ , \tag{4.39}$$

$$\Pi(\varphi,n|m) = \texttt{el3}\left(\tan\varphi,k_c,n_c\right)\ , \tag{4.40}$$

$$J(\varphi,n|m) = \texttt{el}\left(\tan\varphi,k_c,n_c,0,1\right)\ , \tag{4.41}$$

where

$$n_c \equiv 1-n\ , \tag{4.42}$$

is the complementary characteristic. Unfortunately, Bulirsch's algorithms to compute el2, el3, and el numerically suffer from the loss of precision when $|\varphi|$ is small [48, 49]. Therefore, we do not recommend their usage in the practical computations.

Bulirsch's general form of complete integrals of the first, the second, and the third kind are defined as

$$\texttt{cel1}\left(k_c\right) \equiv \int_0^\infty \frac{d\xi}{\sqrt{\left(1+\xi^2\right)\left(1+k_c^2\xi^2\right)}}\ , \tag{4.43}$$

$$\texttt{cel2}\left(k_c,a,b\right) \equiv \int_0^\infty \frac{a+b\xi^2}{\sqrt{\left(1+\xi^2\right)\left(1+k_c^2\xi^2\right)}} d\xi\ , \tag{4.44}$$

$$\texttt{cel3}\left(k_c,p\right) \equiv \int_0^\infty \frac{d\xi}{\left(1+p\xi^2\right)\sqrt{\left(1+\xi^2\right)\left(1+k_c^2\xi^2\right)}} d\xi\ . \tag{4.45}$$

Bulirsch succeeded to generalize these as

$$\texttt{cel}\left(k_c,p,a,b\right) \equiv \int_0^\infty \frac{a+b\xi^2}{\left(1+p\xi^2\right)\sqrt{\left(1+\xi^2\right)\left(1+k_c^2\xi^2\right)}} d\xi\ , \tag{4.46}$$

These Bulirsch's form of complete elliptic integrals are related with Legendre's and the associate complete elliptic integrals as

$$K(m) = \mathtt{cel1}\left(k_c\right) , \tag{4.47}$$
$$E(m) = \mathtt{cel2}\left(k_c, 1, m_c\right) , \tag{4.48}$$
$$B(m) = \mathtt{cel2}\left(k_c, 1, 0\right) , \tag{4.49}$$
$$D(m) = \mathtt{cel2}\left(k_c, 0, 1\right) , \tag{4.50}$$
$$\Pi(n|m) = \mathtt{cel3}\left(k_c, n_c\right) , \tag{4.51}$$
$$J(n|m) = \mathtt{cel}\left(k_c, n_c, 0, 1\right) . \tag{4.52}$$

4.6 Symmetric elliptic integrals

Carlson reconstructed the theory of elliptic integrals by introducing their symmetric forms [62, Section 19.16 (i)]:

$$R_F(x, y, z) \equiv \frac{1}{2} \int_0^\infty \frac{dt}{\sqrt{(t+x)(t+y)(t+z)}} . \tag{4.53}$$

$$R_D(x, y, z) \equiv \frac{3}{2} \int_0^\infty \frac{dt}{\sqrt{(t+x)(t+y)(t+z)^3}} = R_J(x, y, z, z) , \tag{4.54}$$

$$R_J(x, y, z, p) \equiv \frac{3}{2} \int_0^\infty \frac{dt}{(t+p)\sqrt{(t+x)(t+y)(t+z)}} . \tag{4.55}$$

As its subprocedures, R_J requires R_F and a nonelliptic symmetric integral defined as

$$R_C(x, y) \equiv \frac{1}{2} \int_0^\infty \frac{dt}{(t+y)\sqrt{t+x}} = R_F(x, y, y) . \tag{4.56}$$

Legendre's form of incomplete elliptic integrals and the associate incomplete elliptic integrals are expressed in terms of these symmetric integrals as

$$F(\varphi|m) = sR_F\left(c^2, d^2, 1\right) , \tag{4.57}$$
$$B(\varphi|m) = \frac{1}{3} m_c s^3 R_D\left(c^2, 1, d^2\right) + \frac{sc}{d} , \tag{4.58}$$
$$D(\varphi|m) = \frac{1}{3} s^3 R_D\left(c^2, d^2, 1\right) , \tag{4.59}$$
$$J(\varphi, n|m) = \frac{1}{3} s^3 R_J\left(c^2, d^2, 1, 1 - ns^2\right) , \tag{4.60}$$

where

$$s \equiv \sin\varphi\,, \tag{4.61}$$

$$c \equiv \cos\varphi\,, \tag{4.62}$$

$$d \equiv \sqrt{1 - ms^2}\,. \tag{4.63}$$

are nothing but the three principal Jacobian elliptic functions if φ is regarded as Jacobi's amplitude function. Similarly, the complete elliptic integrals are expressed in terms of the symmetric integrals as

$$K(m) = R_F\left(0, m_c, 1\right)\,, \tag{4.64}$$

$$B(m) = \frac{1}{3} m_c R_D\left(0, 1, m_c\right)\,, \tag{4.65}$$

$$D(m) = \frac{1}{3} R_D\left(0, m_c, 1\right)\,, \tag{4.66}$$

$$J(n|m) = \frac{1}{3} R_J\left(0, m_c, 1, n_c\right)\,. \tag{4.67}$$

5 Numerical computation of elliptic functions and elliptic integrals

5.1 Overview

Although the detailed theoretical and some numerical aspects of the computation of elliptic functions and elliptic integrals had been well discussed from the early days of Euler, Landen, and Gauss, it is only after Bulirsch's pioneer works [8–11] when their precise numerical computation becomes available [63]. Nevertheless, some of his procedures computing elliptic integrals turned out to face with a loss of precision in practical applications.

Then, Carlson made a breakthrough in the theory of elliptic integrals by introducing their symmetric forms [13–16, 22]. His duplication method [24] is free from the precision loss, and therefore, has been regarded as the standard numerical method [64]. Nevertheless, Carlson's procedures are fairly slow as will be seen later.

Therefore, we have conducted a series of studies [42, 44, 45, 47–54] to develop the algorithms computing various elliptic functions, elliptic integrals, and their inverses. The obtained procedures `scd2`, `elf`, `elbd`, `elbdj`, `celk`, `celbd`, and `celbdj` are (1) greatly faster than Carlson's procedures `rf`, `rd`, and `rj`, (2) significantly faster than Bulirsch's procedures `sncndn`, `el1`, `cel1`, `cel2`, and `cel`, (3) without suffering from the precision loss experienced by Bulirsch's algorithms `el2`, `el3`, and our implementation of `el` [56], and (4) as precise as the best existing procedures.

5.2 Transformation method

All the methods of Bulirsch, Carlson, and ours belong to the same category of computing method: the transformation method. Its spirit is similar to the canonical transformation method in the analytical dynamics. Let us explain its essence.

Consider the computation of a given function

$$p = f(q) \, . \tag{5.1}$$

Regard the pair of the function value and the argument, (p, q), as a sort of "conjugate" pair of variables. Notice that the dimension of p and q are not the same in general. For example, in the case of the simultaneous computation of the principal Jacobian elliptic functions, the dimension of p is three as $p = (s, c, d)$ while that of q is two as $q = (u, m)$.

Assume that there exists a one-to-one transformation of (p, q) as

$$p^* = P^*(p, q) \, , \quad q^* = Q^*(p, q) \, , \tag{5.2}$$

$$p = P(p^*, q^*) \, , \quad q = Q(p^*, q^*) \, , \tag{5.3}$$

such that the transformed pair satisfies the same functional relation as

$$p^* = f(q^*) \, . \tag{5.4}$$

If the direct evaluation of $f(q)$ is more difficult than that of $f(q^*)$, a method of computation of $f(q)$ is constructed as a three-step-method based on the transformation as

(1) Forward transformation of the argument: Equation (5.2)
(2) Direct evaluation of the function value for the transformed argument: Equation (5.4)
(3) Backward transformation of the obtained function value: Equation (5.3)

In general, the difficulty of the direct computation is not sufficiently reduced by a single application of the transformation. In that case, a series of transformations are applied in both the forward and the backward directions such as

(1) Forward stage: repeat the execution of Equation (5.2) until Equation (5.4) is easily computed.
(2) Evaluation stage: compute Equation (5.4).
(3) Backward stage: repeat the execution of Equation (5.3) as many as in the forward stage.

From the computational viewpoint, the transformation method is effective only when the computational labor of the direct computation of the untransformed argument is significantly larger than the sum of the computational labor of these three stages.

5.3 Example of transformation method

Let us show a simple example of the transformation method of the function computation. Consider the computation of the exponential function, $p = e^q$. An example of suitable transformations in this case is the half-argument transformation:

$$p^* = \sqrt{p}, \quad q^* = q/2, \tag{5.5}$$

$$p = (p^*)^2, \quad q = 2q^*. \tag{5.6}$$

If $|q|$ is sufficiently small, e^q is precisely and quickly computed by its Maclaurin series

$$e^q = 1 + q + \frac{q^2}{2} + \frac{q^3}{3!} + \cdots \tag{5.7}$$

Thus, the resulting algorithm of computation becomes

(1) Halve the argument repeatedly until its magnitude becomes sufficiently small.
(2) Evaluate the function value of the reduced argument by the truncated Maclaurin series.
(3) Repeat squaring the obtained function value as many times as the number of half-argument transformations.

Notice that only the forward transformation of q and the backward transformation of p are needed. The computational labor of both of them are small: one multiplication each. Therefore, except the aspect of the computational error accumulation, the transformation method can be fairly efficient.

5.4 Simultaneous computation of Jacobian elliptic functions

Begin with the simultaneous computation of the three principal Jacobian elliptic functions,

$$s \equiv \mathrm{sn}(u|m), \quad c \equiv \mathrm{cn}(u|m), \quad d \equiv \mathrm{dn}(u|m). \tag{5.8}$$

Once s and c are obtained, Jacobi's amplitude function can be evaluated from them as

$$\mathrm{am}(u|m) = \mathrm{atan2}(s, c), \tag{5.9}$$

where $\mathrm{atan2}(y, x)$ is the two-argument arc tangent function computing $\tan^{-1}(y/x)$ while taking the sign of arguments into account. Therefore, its discussion will be omitted hereafter.

A simple forward transformation is the half-argument transformation [42, 44]. It is a transformation halving the argument u as

$$u^* = \frac{u}{2}, \tag{5.10}$$

while keeping the parameter m the same. When $|u|$ becomes sufficiently small, the function values s, c, and d are evaluated by the Maclaurin series expansion of the Jacobian elliptic functions as

$$s = u - \frac{1+m}{6}u^3 + \frac{1+14m+m^2}{120}u^5 - \cdots , \tag{5.11}$$

$$c = 1 - \frac{1}{2}u^2 + \frac{1+4m}{24}u^4 - \cdots , \tag{5.12}$$

$$d = 1 - \frac{m}{2}u^2 + \frac{4m+m^2}{24}u^4 - \cdots . \tag{5.13}$$

The expansion coefficients for up to the term u^{16} are explicitly given in Table 2 of Fukushima [42]. Those of higher order ones are easily obtained by recursion [54].

The corresponding backward transformation is the double argument transformation of s, c, and d, Equations (3.39) through (3.41), as

$$s = \frac{2s^* c^* d^*}{1 - m\,(s^*)^4} , \tag{5.14}$$

$$c = \frac{(c^*)^2 - (s^*)^2\,(d^*)^2}{1 - m\,(s^*)^4} , \tag{5.15}$$

$$d = \frac{(d^*)^2 - m\,(s^*)^2\,(c^*)^2}{1 - m\,(s^*)^4} . \tag{5.16}$$

Since the algorithm needs only the four arithmetic operations, it runs fairly fast. This is the reason why the present algorithm using a linear transformation is faster than Bulirsch's algorithm based on the Landen transformation, which must be fast in principle since it is quadratic but requires the costly operations such as the square root. The resulting procedure is named `scd` [44].

5.5 Better computation of Jacobian elliptic functions

Later, the simultaneous computation scheme described in the previous subsection turns out to suffer from the accumulation of round-off errors in the course of successive application of the backward transformation. Then, another method is developed [52]. As the new main variable, we adopted the one's complement of c defined as

$$b \equiv 1 - c . \tag{5.17}$$

Its Maclaurin series is practically the same as that of c as

$$b = \frac{1}{2}u^2 - \frac{1+4m}{24}u^4 + \cdots . \tag{5.18}$$

while its double argument transformation is expressed as

$$b = \frac{2y^*\left(1 - my^*\right)}{1 - m\left(y^*\right)^2}\,, \tag{5.19}$$

where y is an auxiliary variable defined and computed as

$$y \equiv s^2 = b(2 - b)\,. \tag{5.20}$$

When both of y and m become sufficiently large, the computation of b may face with the information loss in the process to compute $1 - my^*$. In that case, the main variable is switched from b to c. Its double argument transformation is rewritten as

$$c = \frac{\left(mx^* + 2m_c\right)x^* - m_c}{m_c + mx^*\left(2 - x^*\right)}\,, \tag{5.21}$$

where x is another auxiliary variable defined as

$$x \equiv c^2\,. \tag{5.22}$$

Once the backward transformation is finished, the corresponding value of s, c, and d are computed from b and y as

$$s = \sqrt{y}\,, \tag{5.23}$$

$$c = 1 - b\,, \tag{5.24}$$

$$d = \sqrt{1 - my}\,. \tag{5.25}$$

When the main variable is switched, s and d are computed from c and x as

$$s = \sqrt{1 - x}\,, \tag{5.26}$$

$$d = \sqrt{m_c + mx}\,. \tag{5.27}$$

The revised procedure is named `scd2` [51]. Although it calls two square roots in the final step, the number of main variable is reduced from three to one when compared with `scd` such that the total amount of computational labor significantly reduced [51].

5.6 Computation of Jacobi's form of incomplete elliptic integrals

The double argument transformation is also effective in computing Jacobi's form of incomplete elliptic integrals [42]. Abbreviate them simply as

$$E \equiv E_u(u|m), \quad \Pi \equiv \Pi_u(u, n|m), \quad B \equiv B_u(u|m), \quad D \equiv D_u(u|m), \quad J \equiv J_u(u, n|m). \tag{5.28}$$

As remarked earlier, E and Π are computed from B, D, and J. Also, there is a relation between B and D as

$$B + D = u\,. \tag{5.29}$$

Therefore, the more fundamental are J and either B or D. Between B and D, we prefer D since it becomes the smaller when u is small, and therefore the less erroneous in computing the Maclaurin series. Notice that the computation of J includes that of D because

$$J_u(u, 0|m) = D_u(u|m) , \tag{5.30}$$

by definition. In this sense, the computation of J is the most fundamental. At any rate, the forward transformation is the same: $u^* = u/2$. The Maclaurin series of J becomes

$$J = \frac{1}{3}u^3 - \frac{1 + m - 3n}{15}u^5 + \cdots , \tag{5.31}$$

the coefficients of which are explicitly listed in Table 2 of Fukushima [42]. The higher order coefficients can be recursively obtained [54]. Meanwhile, the double argument transformation of J is expressed as

$$J = 2J^* + T(t, h) . \tag{5.32}$$

Once D and J are computed, the other integrals are obtained as

$$E = u - mD , \tag{5.33}$$

$$\Pi = u + nJ , \tag{5.34}$$

$$B = u - D . \tag{5.35}$$

5.7 Computation of Legendre's form of incomplete elliptic integral of the first kind

The computational algorithm of $F(\varphi|m)$ is easily derived from that of Jacobian elliptic functions by reversing the relationship of the argument and the function value [45]. Regard $u \equiv F(\varphi|m)$ as a function of the main variable

$$y \equiv \sin^2 \varphi , \tag{5.36}$$

while m is fixed. Then, the forward transformation becomes the half-argument transformation of y, Equation (3.45), as

$$y^* = \frac{y}{(1 + c)(1 + d)} , \tag{5.37}$$

where c and d are computed from y as

$$c = \sqrt{1 - y} , \tag{5.38}$$

$$d = \sqrt{1 - my} . \tag{5.39}$$

Meanwhile,.the Maclaurin series of u with respect to $s = \sqrt{y}$ becomes

$$u = s + \frac{1+m}{6}s^3 + \frac{3+2m+3m^2}{40}s^5 + \cdots . \tag{5.40}$$

The analytical expression of the expansion coefficients are known and the first 13 coefficients are explicitly given [45, Section 2.3]. On the other hand, the backward transformation is as simple as

$$u = 2u^* . \tag{5.41}$$

Thus, the transformation method to compute $F(\varphi|m)$ is completed.

The above forward transformation has a chance of information loss when y and m are not so small initially. In that case, another main variable is adopted as

$$x \equiv \cos^2\varphi , \tag{5.42}$$

which is assumed to be sufficiently small. This time, the forward transformation becomes

$$x^* = \frac{c+d}{1+d} , \tag{5.43}$$

where c and d are computed from x as

$$c = \sqrt{x} , \tag{5.44}$$

$$d = \sqrt{m_c + mx} . \tag{5.45}$$

Since the half-argument transformation inflates the magnitude of x, soon the computation of x faces with the loss of information. In that case, the main variables is switched to $y = 1 - x$ and the half-argument transformation of y is continued.

5.8 Computation of other incomplete elliptic integrals

Once the computational algorithm of $u = F(\varphi|m)$ is established, it is automatic to obtain those of the other incomplete elliptic integrals [48, 49]. After the forward transformation to obtain u from the given φ and m as explained in the previous subsection, the corresponding values of these integrals are obtained by the same procedure described in Section 5.6 where the abbreviations are now read as

$$E = E(\varphi|m) , \quad \Pi = \Pi(\varphi, n|m) , \quad B = B(\varphi|m) , \quad D = D(\varphi|m) , \quad J = J(\varphi, n|m) . \tag{5.46}$$

5.9 Computation of complete elliptic integrals other than $\Pi(n|m)$ and $J(n|m)$

Consider the computation of Legendre's form of complete elliptic integrals. There exists a difference between (1) the complete integrals of the first kind and the second kind, $K(m)$, $E(m)$, $B(m)$, and $D(m)$, and (2) the complete integrals of the third kind, $\Pi(n|m)$ and $J(n|m)$. The former integrals are single variable functions while the latter ones are not. Let us begin with the first group [44, 47].

The meaningful domain of the parameter m for the complete integrals is as limited as $m < 1$ since the integrals are no longer real-valued otherwise. By means of various kind of transformation formulas, this is further reduced to the so-called standard domain, $0 \le m < 1$. A single variable function can be quickly computed *if* the function is sufficiently smooth [27, 28]. The complete elliptic integrals, $K(m)$, $E(m)$, $B(m)$, and $D(m)$, are sufficiently smooth except near their common logarithmic singularity, $m = 1$. Following the policy of "divide and rule", we first split the standard domain into 11 subdomains as $[0, 0.1), [0.1, 0.2), \ldots, [0.7, 0.8), [0.8, 0.85), [0.85, 0.9)$, and $[0.9, 1.0)$. For the first 10 subintervals, the integrals are effectively approximated by their Taylor series expansion such as

$$K(m) \approx \sum_j K_j (m - m_0)^j \,, \tag{5.47}$$

where K_j is the jth order Taylor series coefficient of $K(m)$ at $m = m_0$, the mid point of the subinterval. The coefficients required in the double precision computation are explicitly listed in the tables of [44, 47]. If necessary, the higher order coefficients can be efficiently obtained by recursion [54].

In the last subinterval, $0.9 \le m < 1$, the integrals are expressed as a sum of two parts: the regular part and the logarithmically singular one [47]. For example, $K(m)$ is split into

$$K(m) = K_X X + K_0 \,, \tag{5.48}$$

where

$$X \equiv -\log\left(\frac{m_c}{16}\right) \,, \tag{5.49}$$

$$K_X \equiv \frac{K(m_c)}{\pi} \,, \tag{5.50}$$

$$K_0 \equiv K_X \left[-\log\left(\frac{16q(m_c)}{m_c}\right)\right] \,, \tag{5.51}$$

while $q(m)$ is Jacobi's nome defined as

$$q(m) \equiv \exp\left(\frac{-\pi K(m_c)}{K(m)}\right) . \tag{5.52}$$

Both of K_X and K_0 are regular around $m = 1$, namely around $m_c = 0$. Thus, they are well approximated by their Taylor series expansion even in the problematic interval, $0.9 \le m < 1$. Similar expansions are obtained for $E(m)$ as

$$E(m) = E_X X + E_0 \,, \tag{5.53}$$

where

$$E_X \equiv \left[1 - \left(\frac{E(m_c)}{K(m_c)}\right)\right] K_X \tag{5.54}$$

$$E_0 \equiv \frac{\pi}{2K(m_c)} + E_X K_0 , \tag{5.55}$$

In the case of $B(m)$ and $D(m)$, not themselves but their modifications are easily split as

$$B^*(m) \equiv \frac{B(m)}{m} = B^*_X X + B^*_0 , \tag{5.56}$$

$$D^*(m) \equiv \frac{D(m)}{m} = D^*_X X + D^*_0 . \tag{5.57}$$

The division by m is harmless near the singularity where $m \approx 1$. The corresponding coefficients are derived from those of $K(m)$ and $E(m)$ as

$$B^*_X = E_X - m_c K_X , \tag{5.58}$$

$$B^*_0 = E_0 - m_c K_0 , \tag{5.59}$$

$$D^*_X = K_X - E_X , \tag{5.60}$$

$$D^*_0 = K_0 - E_0 . \tag{5.61}$$

The tables of the Taylor series coefficients of the basic quantities, $K_X, K_0, E_X, E_0, B^*_X, B^*_0, D^*_X$, and D^*_0, are found in Fukushima [44, 47].

5.10 Computation of complete elliptic integrals of the third kind

On the other hand, the computation of the complete elliptic integral of the third kind is a difficult problem. Since it is a bivariate function, the technique of the series expansion is not easily applicable even if using the symmetric integral forms [50]. Fortunately, there exists a transformation with respect to n: the double argument formula with respect to a in Jacobi's original form of complete elliptic integral of the third kind, $\Pi_1(a|m)$, Equation (3.72). The resulting algorithm [53] is explained below.

First, by using various reduction formulas [53, Appendix A], the domain of n and m are reduced as

$$0 < m < 1 , \quad -k < n < \frac{m}{1 + k_c} . \tag{5.62}$$

Next, the forward transformation takes the same form as the half-argument transformation described in Section 5.7 by adopting a new variable

$$y \equiv \frac{n}{m} , \tag{5.63}$$

and its complement

$$x \equiv 1 - y = \frac{m - n}{m} . \tag{5.64}$$

Third, the Maclaurin series expansion with respect to y becomes

$$J(n|m) = D_0(m) + D_1(m)y + D_2(m)y^2 + \cdots , \tag{5.65}$$

where $D_j(m)$ are computed recursively as

$$D_j(m) = \left(\frac{2j}{2j+1}\right)(1+m)D_{j-1}(m) - \left(\frac{2j-1}{2j+1}\right) mD_{j-2}(m) , \tag{5.66}$$

while their starting values are computed from $D(m)$ and $B(m)$ as

$$D_0(m) = D(m) , \quad D_1(m) = \frac{(1+2m)D(m) - B(m)}{3} . \tag{5.67}$$

Finally, the backward transformation is the double argument transformation of $J(n|m)$, which is obtained by rewriting that of $\Pi_1(a|m)$ in a cancellation-free form as

$$J = \frac{2(c+d)J^* - yK}{cd(1+c)(1+d)} , \tag{5.68}$$

where K is the abbreviation of $K(m)$. It remains as a constant throughout the transformations because both the forward and backward transformations do not change the value of m, and therefore, $K(m)$ too. This is a benefit of the double argument transformation with respect to a although it is a linear transformation.

5.11 CPU time comparison

As already mentioned, there exist no significant difference in the computational errors of the existing procedures to compute the Jacobian elliptic integrals and Legendre's form of complete and incomplete elliptic integrals *except* Bulirsch's `el2`, `el3`, and our implementation of Bulirsch's `el`. Then, the question of concern is the computational speed.

We conducted the CPU time measurements of the existing precise procedures: (1) Bulirsch's `sncndn` and our `scd2` for the simultaneous computation of $\mathrm{sn}(u|m)$, $\mathrm{cn}(u|m)$, and $\mathrm{dn}(u|m)$, (2) Bulirsch's `cel1`, Carlson's R_F, and our `celk` for the computation of $K(m)$, (3) Bulirsch's `cel2`, Carlson's R_D, and our `celbd` for the simultaneous computation of $B(m)$ and $D(m)$, (4) Bulirsch's `cel`, Carlson's R_C, R_D, R_F, and R_J, and our `celbdj` for the simultaneous computation of $B(m)$, $D(m)$, and $J(n|m)$, (5) Bulirsch's `el1`, Carlson's R_F, and our `elf` for the computation of $F(\varphi|m)$, (6) Carlson's R_D and our `elbd` for the simultaneous computation of $B(\varphi|m)$ and $D(\varphi|m)$, and (7) Carlson's R_C, R_D, R_F, and R_J, and our `elbdj` for the simultaneous computation of $B(\varphi|m)$, $D(\varphi|m)$, and $J(\varphi, n|m)$.

All the computation codes were (1) written in Fortran 77/90, (2) compiled by the Intel Visual Fortran Composer XE 2011 update 8 with the level 3 optimization, and (3) executed at a PC with an Intel Core i7-2675QM processor running at 2.20 GHz under Windows 7 OS. Therefore, the one machine clock cycle is 0.455 ns at the PC used for comparison.

Table 1 compares the averaged CPU times of the above 7 computations in the single and double precision environments, respectively. The averages were taken over the standard domain of the input arguments: $0 < \varphi < \pi/2$, $0 < m < 1$, $0 < n < 1$, and $0 < u < K(m)/2$. The sampling points are evenly distributed with the separations (1) $\Delta m = 2^{-14}$ and $\Delta u = 2^{-15}K(m)$ for the simultaneous computation of $\mathrm{sn}(u|m)$, $\mathrm{cn}(u|m)$, and $\mathrm{dn}(u|m)$, (2) $\Delta m = 2^{-28}$ for the computation of $K(m)$ and the simultaneous computation of $B(m)$ and $D(m)$, (3) $\Delta m = 2^{-14}$ and $\Delta\varphi = 2^{-15}\pi$ for the computation of $F(\varphi|m)$ and the simultaneous computation of $B(\varphi|m)$ and $D(\varphi|m)$, and (4) $\Delta m = \Delta n = 2^{-9}$ and $\Delta\varphi = 2^{-11}\pi$ for the simultaneous computation of $B(\varphi|m)$, $D(\varphi|m)$, and $J(\varphi, n|m)$, respectively. As a result, all the total execution times are of the comparable order. The results of comparison show the superiority of our computing procedures except the single precision computation of three Jacobian elliptic functions where Bulirsch's `sncndn` is the faster.

Table 1. Averaged CPU times to compute various elliptic functions and integrals. The unit of CPU time is ns at a PC with an Intel Core i7-2675QM processor running at 2.20 GHz where one machine clock cycle is 0.455 ns.

Functions/Integrals	Method	Procedure	Single	Double
$\mathrm{sn}(u\|m)$, $\mathrm{cn}(u\|m)$, $\mathrm{dn}(u\|m)$	Bulirsch	`sncndn`	99	165
	Fukushima	`scd2`	111	116
$K(m)$	Carlson	R_F	137	330
	Bulirsch	`cel1`	43	62
	Fukushima	`celk`	22	62
$B(m)$, $D(m)$	Carlson	R_D	384	922
	Bulirsch	`cel2`	109	186
	Fukushima	`celbd`	23	62
$B(m)$, $D(m)$, $J(n\|m)$	Carlson	R_C, R_D, R_F, R_J	697	1871
	Bulirsch	`cel`	192	383
	Fukushima	`celbdj`	121	198
$F(\varphi\|m)$	Carlson	R_F	119	313
	Bulirsch	`el1`	105	178
	Fukushima	`elf`	90	173
$B(\varphi\|m)$, $D(\varphi\|m)$	Carlson	R_D	553	712
	Fukushima	`elbd`	182	213
$B(\varphi\|m)$, $D(\varphi\|m)$, $J(\varphi, n\|m)$	Carlson	R_C, R_D, R_F, R_J	759	1426
	Fukushima	`elbdj`	349	464

5.12 Software

The computing procedures described in the previous subsections, namely `scd2`, `celk`, `celbd`, `celbdj`, `elf`, `elbd`, and `elbdj`, are available from the author's personal WEB page at ResearchGate:

https://www.researchgate.net/profile/Toshio_Fukushima/

with their test drivers and subprograms as well as sample output files.

6 Conclusion

Trigonometric functions and hyperbolic functions are daily used in the science and technology. Of course, there are many reasons of their popularity. Nevertheless, one of them is the easy availability of their functional values. Indeed, the functions are incorporated in the standard mathematical library of computers. Even some electronic calculators support them.

On the other hand, the elliptic functions and the elliptic integrals are not so popular despite that some scientific problems become simpler if they are deployed. Now that their fast and precise numerical procedures are available, we hope that these difficult-to-approach but deep-and-beneficial mathematical tools will be more utilized in celestial mechanics and dynamical astronomy.

References

[1] N. I. Akhiezer, *Elements of the Theory of Elliptic Functions*, H. H. McFaden (translated), Amer. Math. Soc., Providence, 1990.

[2] M. Abramowitz and I. A. Stegun (eds.), *Handbook of Mathematical Functions with Formulas, Graphs, and Mathematical Tables*, Chapter 17, National Bureau of Standards, Washington, 1964.

[3] F. Bowman, *Introduction to Elliptic Functions with Applications*, Dover Publications, New York, 1961.

[4] V. A. Brumberg and E. V. Brumberg, *Celestial Dynamics at High Eccentricities*, Gordon & Breach Sci. Publ., UK, 1999.

[5] E. V. Brumberg, V. A. Brumberg, Th. Konrad and M. Soffel, *Analytical Linear Perturbation Theory for Highly Eccentric Satellite Orbits*, Celest. Mech. Dyn. Astron. 61 (1995) 369.

[6] E. Brumberg and T. Fukushima, *Expansions of Elliptic Motion based on Elliptic Function Theory*, Celest. Mech. Dyn. Astron., 60 (1994) 69–89.

[7] V. A. Brumberg and S. A. Klioner, *Numerical Efficiency of the Elliptic Function Expansions of the First-order Intermediary for General Planetary Theory*, in: S. Ferraz-Mello, B. Morando, J. E. Arlot (eds.), *Dynamics, Ephemerides and Astrometry in the Solar System*, Kluwer, Dordrecht, (1995) 101.

[8] R. Bulirsch, *Numerical Computation of Elliptic Integrals and Elliptic Functions*, Numer. Math., 7 (1965a) 78–90.
[9] R. Bulirsch, *Numerical Computation of Elliptic Integrals and Elliptic Functions II*, Numer. Math., 7 (1965b) 353–354.
[10] R. Bulirsch, *An Extension of the Bartky-Transformation to Incomplete Elliptic Integrals of the Third Kind*, Numer. Math., 13 (1969a) 266–284.
[11] R. Bulirsch, *Numerical Computation of Elliptic Integrals and Elliptic Functions III*, Numer. Math., 13 (1969b) 305–315.
[12] P. F. Byrd and M. D. Friedman, *Handbook on Elliptic Integrals for Engineers and Physicists, 2nd edn.*, Springer-Verlag, Berlin, 1971.
[13] B. C. Carlson, *On Computing Elliptic Integrals and Functions*, J. Math. and Phys., 44 (1965) 332–345.
[14] B. C. Carlson, *Elliptic Integrals of the First Kind*, SIAM J. Math. Anal., 8 (1977) 231–242.
[15] B. C. Carlson, *Short Proofs of Three Theorems on Elliptic Integrals*, SIAM J. Math. Anal., 9 (1978) 524–528.
[16] B. C. Carlson, *Computing Elliptic Integrals by Duplication*, Numer. Math., 33 (1979) 1–16.
[17] B. C. Carlson, *A Table of Elliptic Integrals of the Second Kind*, Math. Comp., 49 (1987) 595–606 (Supplement, Math. Comp., 49 (1987) S13–S17).
[18] B. C. Carlson, *A Table of Elliptic Integrals of the Third Kind*, Math. Comp., 51 (1988) 267–280 (Supplement, Math. Comp., 51 (1988) S1–S5).
[19] B. C. Carlson, *A Table of Elliptic Integrals: Cubic Cases*, Math. Comp., 53 (1989) 327–333.
[20] B. C. Carlson, *A Table of Elliptic Integrals: One Quadratic Factor*, Math. Comp., 56 (1991) 267–280.
[21] B. C. Carlson, *A Table of Elliptic Integrals: Two Quadratic Factors*, Math. Comp., 59 (1992) 165–180.
[22] B. C. Carlson, *Numerical computation of Real or Complex Eelliptic Integrals*, Numer. Algorithms, 10 (1995) 13–26.
[23] B. C. Carlson, *Symmetry in c, d, n of Jacobian elliptic functions*, J. Math. Anal. Appl., 299 (2004) 242–253.
[24] B. C. Carlson and E. M. Notis, *Algorithm 577. Algorithms for Incomplete Elliptic Integrals*, ACM Trans. Math. Software, 7 (1981) 398–403.
[25] A. Cayley, *An Elementary Treatise on Elliptic Functions, 2nd edn.*, George Bell and Sons, Cambridge, 1895.
[26] J. Chapront and J.-L. Simon, *Planetary Theories with the Aid of the Expansions of Elliptic Functions*, Celest. Mech. Dyn. Astron., 63 (1996) 171.
[27] W. J. Cody, *Chebyshev Approximations for the Complete Elliptic Integrals K and E*, Math. Comp., 19 (1965a) 105–112.
[28] W. J. Cody, *Chebyshev Polynomial Expansions of Complete Elliptic Integrals K and E*, Math. Comp., 19 (1965b) 249–259.
[29] A. Elipe and V. Lanchares, *Exact Solution of a Triaxial Gyrostat with one Rotor*, Celest. Mech. Dyn. Astron., 101 (2008) 49–68.
[30] T. Fukushima, *A Method Solving Kepler's Equation without Transcendental Function Evaluations*, Celest. Mech. Dyn. Astron., 66 (1996a) 309–319.
[31] T. Fukushima, *Generalization of Encke's Method and its Application to the Orbital and Rotational Motions of Celestial Objects*, Astron. J., 112 (1996b) 1263–1277.
[32] T. Fukushima, *A Fast Procedure Solving Kepler's Equation for Elliptic Case*, Astron. J., 112 (1996c) 2858–2861.
[33] T. Fukushima, *A Method Solving Kepler's Equation for Hyperbolic Case*, Celest. Mech. Dyn. Astron., 68 (1997a) 121–137.

[34] T. Fukushima, *A Procedure Solving the Extended Kepler's Equation for the Hyperbolic Case*, Astron. J., 113 (1997b) 1920–1924.
[35] T. Fukushima, *A Fast Procedure Solving Gauss' Form of Kepler's Equation*, Celest. Mech. Dyn. Astron., 70 (1998) 115–130.
[36] T. Fukushima, *Fast Procedures Solving Universal Kepler's Equation*, Celest. Mech. Dyn. Astron., 75 (1999) 201–226.
[37] T. Fukushima, *New Two-Body Regularization*, Astron. J., 133 (2007a) 1–10.
[38] T. Fukushima, *Numerical Comparison of Two-Body Regularizations*, Astron. J., 133 (2007b) 2815–2824.
[39] T. Fukushima, *Simple, Regular, and Efficient Numerical Integration of Rotational Motion*, Astron. J., 135 (2008a) 2298–2322.
[40] T. Fukushima, *Gaussian Element Formulation of Short–Axis-Mode Rotation of a Rigid Body*, Astron. J., 136 (2008b) 649–653.
[41] T. Fukushima, *Canonical and Universal Elements of Rotational Motion of Triaxial Rigid Body*, Astron. J., 136 (2008c) 1728–1735.
[42] T. Fukushima, *Fast Computation of Jacobian Elliptic Functions and Incomplete Elliptic Integrals for Constant Values of Elliptic Parameter and Elliptic Characteristic*, Celest. Mech. Dyn. Astron., 105 (2009a) 245–260.
[43] T. Fukushima, *Efficient Solution of Initial-Value Problem of Torque-Free Rotation*, Astron. J., 138 (2009b) 210–218.
[44] T. Fukushima, *Fast Computation of Complete Elliptic Integrals and Jacobian Elliptic Functions*, Celest. Mech. Dyn. Astron., 105 (2009c) 305–328.
[45] T. Fukushima, *Fast Computation of Incomplete Elliptic Integral of First Kind by Half Argument Transformation*, Numer. Math., 116 (2010a) 687–719.
[46] T. Fukushima, *Precise Computation of Acceleration due to Uniform Ring or Disk*, Celest. Mech. Dyn. Astron., 108 (2010b) 339–356.
[47] T. Fukushima, *Precise and Fast Computation of General Complete Elliptic Integral of Second Kind*, Math. Comp., 80 (2011a) 1725–1743.
[48] T. Fukushima, *Precise and Fast Computation of a General Incomplete Elliptic Integral of Second Kind by Half and Double Argument Transformations*, J. Comp. Appl. Math., 235 (2011b) 4140–4148.
[49] T. Fukushima, *Precise and Fast Computation of a General Incomplete Elliptic Integral of Third Kind by Half and Double Argument Transformations*, J. Comp. Appl. Math., 236 (2012a) 1961–1975.
[50] T. Fukushima, *Series Expansion of Symmetric Elliptic Integrals*, Math. Comp., 81 (2012b) 957–990.
[51] T. Fukushima, *Precise and Fast Computation of Jacobian Elliptic Functions by Conditional Duplication*, Numer. Math., 123 (2013a) 585–605.
[52] T. Fukushima, *Numerical Inversion of a General Incomplete Elliptic Integral*, J. Comp. Appl. Math., 237 (2013b) 43–61.
[53] T. Fukushima, *Fast Computation of a General Complete Elliptic Integral of Third Kind by Half and Double Argument Transformations*, J. Comp. Appl. Math., 253 (2013c) 142–157.
[54] T. Fukushima, *Recursive Computation of Derivatives of Elliptic Functions and of Incomplete Elliptic Integrals*, Appl. Math. Comp., 221 (2013d) 21–31.
[55] T. Fukushima and H. Ishizaki, *Elements of Spin Motion*, Celest. Mech. Dyn. Astron., 59 (1994a) 149–159.
[56] T. Fukushima and H. Ishizaki, *Numerical Computation of Incomplete Elliptic Integrals of a General Form*, Celest. Mech. Dyn. Astron., 59 (1994b) 237–251.
[57] F. Hancock, *Elliptic Integrals*, Dover Publ., New York, 1958a.

[58] F. Hancock, *Theory of Elliptic Functions*, Dover Publ., New York, 1958b.
[59] S. A. Klioner, A. A. Vakhidov and N. N. Vasiliev, *Numerical Computation of Hansen-like Expansions*, Celest. Mech. Dyn. Astron., 68 (1998) 257–272.
[60] D. F. Lawden, *Elliptic Functions and Applications*, Springer-Verlag, Berlin, 1989.
[61] K. B. Oldham, J. Myland and J. Spanier, *An Atlas of Functions, 2nd edn.*, Hemisphere Publ., Wash. DC, 2009, Chapters 61, 62, and 63.
[62] F. W. J. Olver, D. W. Lozier, R. F. Boisvert and C. W. Clark (eds.), *NIST Handbook of Mathematical Functions*, Cambridge Univ. Press, Cambridge, 2010, Chapters 19 and 22.
[63] W. H. Press, B. P. Flannery, S. A. Teukolsky and W. T. Vetterling, *Numerical Recipes: the Art of Scientific Computing*, Cambridge Univ. Press, Cambridge, 1986.
[64] W. H. Press, S. A. Teukolsky, W. T. Vetterling and B. P. Flannery, *Numerical Recipes: the Art of Scientific Computing, 3rd edn.*, Cambridge Univ. Press, Cambridge, 2007.
[65] A. G. Sokolsky, A. A. Vakhidov and N. N. Vasiliev, *Computation of Elliptic Hansen Coefficients and Their Derivatives*, Celest. Mech. Dyn. Astron., 63 (1996) 357–374.
[66] W. J. Thompson, *Atlas for Computing Mathematical Functions*. John Wiley & Sons, New York, 1997.
[67] A. A. Vakhidov and N. N. Vasiliev, *Development of Analytical Theory of Motion for Satellites with Large Eccentricities*, Astron. J., 112 (1996) 2330–2335.
[68] E. T. Whittaker and G. N. Watson, *A Course of Modern Analysis, 4th edn.*, Cambridge Univ. Press, Cambridge, 1958.
[69] S. Wolfram, *The Mathematica Book, 5th edn.*, Wolfram Research Inc./Cambridge Univ. Press, Cambridge, 2003.

Jean-Louis Simon and Agnes Fienga

Victor Brumberg and the French school of analytical celestial mechanics

1 Introduction

In his state thesis presented in 1970, Jean Chapront introduced a development of the perturbative function based on a formalism developed by V. A. Brumberg in 1967. This was the starting point of a long and fruitful collaboration between Victor Brumberg and the Institut de Mécanique Céleste et de Calcul des Éphémérides (IMCCE) that will last more than forty five years. Victor Brumberg travelled several times in Paris and researchers of the IMCCE (Jean Chapront and Pierre Bretagnon) visited several times Victor in Saint-Petersburg. Thanks to this collaboration, new developments in planetary ephemerides were done by the researchers of the IMCCE using studies of Victor Brumberg not only in the fields of classical celestial mechanics (development of the perturbative function, computation of the Hansen coefficients, construction of general planetary theories, theories of the planetary motion in elliptic variables) but also for his relativistic approaches, especially related to the definition of reference frames such as the KCRS. We will recall some of these works here.

2 Analytical formulism for planetary perturbations

The formalisms for the analytical developments of the planetary perturbations and for the computation of the Hansen coefficients proposed by Victor Brumberg are very efficient tools for the construction of analytical theories of planetary motions. They have been used in particular in [13] for the construction of a literal planetary theory, in [14] for the computation of the secular variations at the first order of the masses for the orbital elements of the four biggest planets and in [24] for the calculation of the perturbations at the second order of the masses for Jupiter and Saturn.

Jean-Louis Simon: Observatoire de Paris, IMCCE-CNRS UMR8028, 77 Av. Denfert-Rochereau, 75014 Paris, France
Agnes Fienga: Observatoire de la Côte d'Azur, Géoazur-CNRS UMR 7329, 250 avenue A. Einstein, 06250 Valbonne, France

2.1 Development of the perturbative function

The method proposed by Victor Brumberg can be describe like this. Let us assume two planets P (the inferior planet) and P' (the superior planet) with the masses m and m' both orbiting the Sun S. In Figure 1, the plane SX, SY is the ecliptic plane, the SX axis being oriented in the direction of the equinoxe. N and N' are the ascending nodes of the orbits of P and P' on the ecliptic, Π and Π', their perihelia.

Let us note that

- a and a' the semimajor axis of the orbits of P and P',
- $\alpha = a/a'$ the ratio of the semimajor axis,
- e and e' the excentricities of the orbits of P and P',
- i and i', the inclinations on the ecliptic of the orbits of P and P',
- γ and γ' such as $\gamma = \sin\frac{i}{2}$, and $\gamma' = \sin\frac{i'}{2}$,
- ω and ω' the arguments of the perihelia,
- Ω and Ω' the longitudes of the nodes,
- l and l' the mean anomalies,
- $\mathbf{r}$ and $\mathbf{r}'$ the vectors $\vec{SP}$ and $\vec{SP'}$ with r and r' as modules,
- Δ, the mutual distance $\Delta = |\mathbf{r} - \mathbf{r}'|$,
- H the angle between $\mathbf{r}$ and $\mathbf{r}'$.

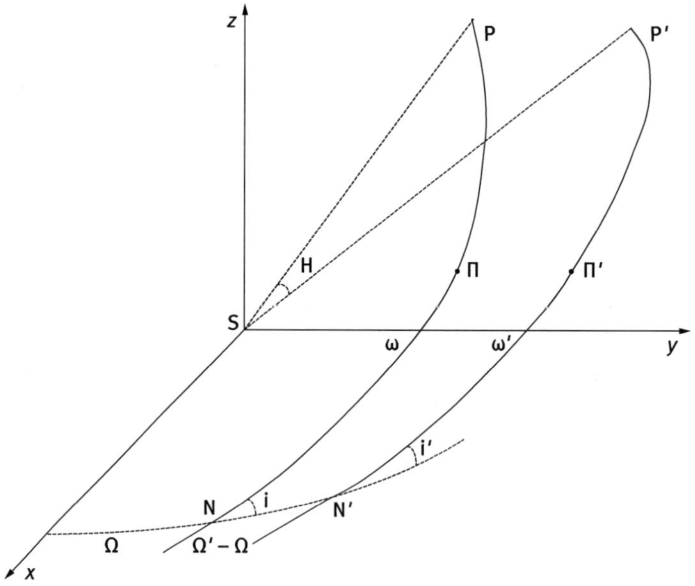

Fig. 1. Planets P and P' orbiting the Sun.

The perturbative function can be written in different ways depending on whether we consider P perturbed by P' or the opposite:

$$R = fm'\left[\frac{1}{\Delta} - \frac{r\cos H}{r'^2}\right] \quad \text{Superior case,}$$

$$R = fm\left[\frac{1}{\Delta} - \frac{r'\cos H}{r^2}\right] \quad \text{Inferior case,} \tag{2.1}$$

where f is the constant of gravitation. For example, for the superior case, one used a development of the perturbative function in series of powers of $\frac{r}{r'}$ such as

$$R = fm'\sum_{k=2}^{\infty}\frac{r^k}{r'^{k+1}}P_k(\cos H)\,, \tag{2.2}$$

where $P_k(\cos H)$ is the Legendre polynomial of order k of $\cos H$. It becomes

$$R = fm'\sum_{qq'ss'j} A_{qq'ss'j}\cos[ql + q'l' + s\omega + s'\omega' + j(\Omega - \Omega')]\,, \tag{2.3}$$

where q, q', s, s' and j are integers and

$$aA_{qq'ss'j} = (2 - \delta_j^0)\sum_{k=\rho}\alpha^{k+1}X_q^{k,s}(e)X_{q'}^{-k-1,s'}(e')\,C_{s,s',j}^k(\gamma,\gamma')\,, \tag{2.4}$$

k varies evenly starting from ρ given by

$$\rho = u + 2v \tag{2.5}$$

with

$$u = |\frac{s - s'}{2}| + |\frac{s + s'}{2}| \tag{2.6}$$

and v differs for the superior and the inferior cases:

$$v = \max\left\{0, E\left(\frac{\max(2,|j|) - u + 1}{2}\right)\right\}\,, \quad \text{Superior case,}$$

$$v = \max\left\{0, E\left(\frac{|j| - u + 1)}{2}\right)\right\}\,, \quad \text{Inferior case.} \tag{2.7}$$

In Equation (2.4), $X_q^{k,s}$ are the Hansen coefficients defined by

$$\left(\frac{r}{a}\right)^k e^{isv} = \sum_{q=-\infty}^{\infty} X_q^{k,s}(e)\,e^{iql}\,, \tag{2.8}$$

v being the true anomaly. Brumberg demonstrates that the coefficients $C_{s,s',j}^k$ can be written as

$$C_{s,s',j}^k = c_{s,s',j}^k\gamma^{|s-j|}\gamma'^{|s'+j|}(1-\gamma^2)^{|\frac{s+j}{2}|}(1-\gamma'^2)^{|\frac{s'-j}{2}|}F_{k,s,j}(\gamma^2)F'_{k,s,j}(\gamma'^2)\,, \tag{2.9}$$

where $c_{s,s',j}$ are constants estimated using the integers k, s, s', and j and the function $\Gamma(n)$. The functions F and F' are hypergeometrical functions of γ^2 and γ'^2 computed also in using the integers k, s, s', and j. Let us define ϕ by

$$\phi = ql + q'l' + s\omega + s'\omega' + j(\Omega - \Omega'), \tag{2.10}$$

and σ by

$$\sigma = |q - s| + |q' - s'| + |s - j| + |s' + j| . \tag{2.11}$$

[13] demonstrates that an optimization of the convergency of the developments of the equation (2.4) can be obtained by factoring $(1 - \alpha^2)^{-\sigma}$. Finaly, for a given ϕ, the Brumberg–Chapront formulism gives $A_{qq'ss'j}$ such as

$$A_{qq'ss'j} = \alpha^{\rho}(1 - \alpha^2)^{-\sigma} e^{|q-s|} e'^{|q'-s'|} \gamma^{|s-j|} \gamma'^{|s'+j|} Q(\alpha^2, e^2, e'^2, \gamma^2, \gamma'^2), \tag{2.12}$$

where Q is given by

$$Q = \sum_{i_1 i_2 i_3 i_4} \frac{P_{i_1 i_2 i_3 i_4}(\alpha^2)}{(1 - \alpha^2)^{i_1+i_2+i_3+i_4}} \gamma^{i_1} \gamma'^{i_2} e^{i_3} e'^{i_4}, \tag{2.13}$$

where $P(\alpha^2)$ is a polynomial of α^2; i_1, i_2, i_3, i_4 are even integers.

2.2 Calculation of the Hansen coefficients

A very useful formulism is given in [6] for the computation of the Hansen coefficients $X_q^{k,s}(e)$ defined in Equation (2.8). The Brumberg equations are different for $q \neq 0$ or $q = 0$ and the Pochhammer symbol

$$(a)_0 = 1 \qquad (a)_m = a(a + 1)\cdots(a + m - 1) \tag{2.14}$$

is used.

- For $q \neq 0$:
  For n positive (or equal to 0) integer, Brumberg defines the polynomials $P_n^{(k,s)}(x)$ with

$$P_n^{(k,s)}(x) = \sum_{r=0}^{r^*} \frac{(-k + s - 1)_r}{(1)_r} \frac{x^{n-r}}{(1)_{n-r}} \tag{2.15}$$

  and

$$\begin{aligned} r^* &= \min\{n, k - s + 1\} \quad &\text{for } k - s + 1 \geq 0 \\ r^* &= n \quad &\text{for } k - s + 1 < 0 \end{aligned} \tag{2.16}$$

  We note $\beta = \frac{e}{1+\sqrt{1-e^2}}$ and $\nu = \frac{q}{1+\beta^2}$. Brumberg writes the coefficients $X_q^{k,s}(e)$ as

$$X_q^{k,s}(e) = \beta^{|s-q|}(1 + \beta^2)^{-k-1} \sum_{n=0}^{\infty} P_{n+\max\{0,q-s\}}^{(k,s)}(\nu)\, P_{n+\max\{0,s-q\}}^{(k,-s)}(-\nu)\beta^{2n} \tag{2.17}$$

- For $q = 0$:
  Brumberg uses the hypergeometric function $F(a, b, c; x)$ defined as

$$F(a,b,c;x) = \sum_{n=0}^{\infty} \frac{(a)_n (b)_n)}{(c)_n} \frac{x^n}{n!} \,. \tag{2.18}$$

  He gives two different equations depending on the values of k:
  - For $k \geq -1$:

$$X_0^{k,s}(e) = (-1)^{|s|} \frac{(k+2)_{|s|}}{(1)_{|s|}} \beta^{|s|} (1+\beta^2)^{-k-1} F(-k-1, -k+|s|-1, 1+|s|; \beta^2) \tag{2.19}$$

  - For $k < -1$:

$$X_0^{k,s}(e) = (-1)^{|s|} \frac{(k+2)_{|s|}}{(1)_{|s|}} \left(\frac{e}{2}\right)^{|s|} (1-e^2)^{k+3/2} \times F\left(\frac{k+|s|+2}{2}, \frac{k+|s|+3}{2}, 1+|s|; e^2\right) \,. \tag{2.20}$$

3 General planetary theory (GPT)

3.1 Introduction

Generally there are two types of analytical theories for the planetary motions: the classic planetary theories (or with secular variations) and the general planetary theories.

The classic planetary theories are used for the construction of the ephemerides by estimating most accurately the motions of the planets on a limited interval of time centered on the present time (usually several thousands of years before and after J2000). Two sets of terms are then needed: the time polynomials called secular terms or mean elements and Poisson series of short-term arguments such as the mean longitudes of planets as defined in Section 4.1.

The GPT aim to describe the planetary motion on the longest interval of time. They do not use secular terms, but they are based only on periodic series of arguments with periods near to the periods of longitudes of the nodes and of perihelia. These arguments are usually called long period arguments.

3.2 General theory by V. Brumberg

The general planetary theory proposed in [5] is based on the ideas of Hill (non-Keplerian indermediate orbit), von Zeipel (separation of slow and fast variables) and Birkhoff (reduction of the dynamical system to a normal form). The expansions are in powers of the planetary masses.

This method was used in the first-order theory of [12]. It starts with replacing the heliocentric rectangular coordinates x_i, y_i, z_i $(i = 1, 2, \ldots, N)$ by the complex variables p_i and the real variables w_i:

$$x_i + \mathring{\imath}\, y_i = a_i(1 - p_i)e^{\mathring{\imath}\bar{\lambda}_i}, \quad z_i = a_i w_i, \text{ where } \mathring{\imath} = \sqrt{-1}. \tag{3.1}$$

These variables represent the small deviations from the circular motion of a planet i, with a semimajor axis a_i and a mean longitude:

$$\bar{\lambda}_i = n_i t + \varepsilon_i. \tag{3.2}$$

The mean motions n_i are fixed, with no relation of commensurability, and related to a_i by Kepler's third law. With these variables, the equations of motion have the form

$$\begin{aligned} \ddot{p}_i + 2\mathring{\imath}\, n_i \dot{p}_i - \frac{3}{2} n_i^2 (p_i + q_i) &= n_i^2 P_i \\ \ddot{w}_i + n_i^2 w_i &= n_i^2 W_i. \end{aligned} \tag{3.3}$$

where q_i is the conjugate of p_i. The right-hand members P_i, W_i are series in power of all the variables p_i, q_i, w_i, the coefficients of which are expressed in terms of Laurent's series with the quantity $\zeta_{ij} = e^{\mathring{\imath}(\bar{\lambda}_i - \bar{\lambda}_j)}$. The final series of the general planetary theory has the form

$$\begin{aligned} p_i &= \sum p_{pqrs}^{(i)} \prod_{j=1}^{N} a_j^{p_j} \bar{a}_j^{q_j} b_j^{r_j} \bar{b}_j^{s_j} \\ w_i &= \sum w_{pqrs}^{(i)} \prod_{j=1}^{N} a_j^{p_j} \bar{a}_j^{q_j} b_j^{r_j} \bar{b}_j^{s_j}. \end{aligned} \tag{3.4}$$

In (3.4), a_j and b_j are complex variables proportional to eccentricities and inclinations of the planetary orbits, respectively. The bar denotes a conjugate quantity. Summation is performed over all nonnegative values of N-indices p, q, r, s. The coefficients of these series are 2π-periodic functions of the differences $\bar{\lambda}_i - \bar{\lambda}_j$, with $j = 1, 2, \ldots, N$ and $j \neq i$. Introducing the slowly variables functions of time α_i and β_i:

$$a_i = \alpha_i e^{\mathring{\imath}\bar{\lambda}_i}, \quad b_i = \beta_i e^{\mathring{\imath}\bar{\lambda}_i}, \tag{3.5}$$

one obtains the secular autonomous system:

$$\begin{aligned} \dot{\alpha} &= \mathring{\imath}\, \mathcal{N}[A\alpha + \Phi(\alpha, \bar{\alpha}, \beta, \bar{\beta})] \\ \dot{\beta} &= \mathring{\imath}\, \mathcal{N}[B\beta + \Psi(\alpha, \bar{\alpha}, \beta, \bar{\beta})]. \end{aligned} \tag{3.6}$$

In (3.6), α and β are the N-vectors of the variables α_i and β_i, $\mathcal{N}$ is the $N \times N$ diagonal matrix of the mean motions, A and B are constant $N \times N$ matrices, and Φ and Ψ are N-vectors of the right-hand members. Solution of (3.6) may be constructed in a pure trigonometric form by Birkhoff normalization [7].

For practical construction of (3.4), Brumberg and Chapront use expansion in the small parameter μ, the ratio of the masses of the planets to the Sun:

$$p_i = p_i^{(0)} + \delta p_i\,, \qquad p_i^{(0)} = \sum_{k=1}^{\infty} \mu^k \underset{k}{p_i^{(0)}}$$

$$\delta p_i = \sum_{k=0}^{\infty} \mu^k \underset{k}{\delta p_i}\,, \qquad w_i = \sum_{k=0}^{\infty} \underset{k}{w_i}\,. \tag{3.7}$$

The intermediary is the particular quasiperiodic solution of equations (3.3):

$$p_i = p_i^{(0)}\,, \qquad w_i = 0\,. \tag{3.8}$$

It depends only on the difference of the mean longitudes and contains $2N$ arbitrary constants: the mean motion n_i and the mean longitudes of the epoch ε_i. Expanding in powers of μ, one gets

$$p_{i1}^{(0)} = \sum_{j=1}^{N} {}^{(i)}\underset{1}{T_j^{(i)}}\,, \tag{3.9}$$

$$p_{i2}^{(0)} = \sum_{j=1}^{N} {}^{(i)}\underset{2}{T_j^{(i)}} + \frac{1}{2}\sum_{k=1}^{N} {}^{(i,j)}\underset{2}{T_{jk}^{(i)}}\,. \tag{3.10}$$

They give the developments of eccentricity and inclination terms δp_i and w_i at the first order of μ. They begin with the two-body solution:

$$\underset{0}{\delta p_i} = \sum_{m=1}^{\infty} \sum_{p+q+r+s=m} \underset{0}{p_{pqrs}}\, a_i^p \bar{a}_i^q b_i^r \bar{b}_i^s\,,$$

$$\underset{0}{\delta w_i} = \sum_{m=1}^{\infty} \sum_{p+q+r+s=m} \underset{0}{w_{pqrs}}\, a_i^p \bar{a}_i^q b_i^r \bar{b}_i^s\,. \tag{3.11}$$

The first order in μ may be represented as

$$\underset{1}{\delta p_i} = \sum_{j=1}^{N} {}^{(i)}\underset{1}{\delta p_{ij}}\,, \qquad \underset{1}{\delta w_i} = \sum_{j=1}^{N} {}^{(i)}\underset{1}{\delta w_{ij}} \tag{3.12}$$

where

$$\underset{1}{\delta p_{ij}} = \sum_{m=1}^{\infty} \sum_{p+q+r+s=m} \underset{1}{p_{pqrs}^{(ij)}} \prod_{I=i,j} a_i^{pI} \bar{a}_i^{qI} b_i^{rI} \bar{b}_i^{sI}\,,$$

$$\underset{1}{\delta w_{ij}} = \sum_{m=1}^{\infty} \sum_{p+q+r+s=m} \underset{1}{w_{pqrs}^{(ij)}} \prod_{I=i,j} a_i^{pI} \bar{a}_i^{qI} b_i^{rI}\,. \tag{3.13}$$

The summation indices p, q, r, s are 2-indices covering the pairs of the planets $I = i, j$. The functions $T_{\underset{1}{j}}^{(i)}, p_{\underset{1}{pqrs}}^{(ij)}, w_{\underset{1}{pqrs}}^{(ij)}$ depend only of one angular element $\lambda_i - \lambda_j$.

Formulae (3.9)–(3.13) give the first-order solution in the form (3.4). The solution of the secular system (3.6) allows us to determine α and β in terms of slowly variable functions of time. Once α and β are obtained for a given epoch of time, we can compute the corresponding values of a and b; then by substitution of those values in (3.4), we find the numerical values of p_i and w_i.

In their paper, Brumberg and Chapront give the practical elaboration of the algorithm for a general first-order theory with some partial results for the pair Jupiter–Saturn.

4 Planetary theories with the aid of the expansions of elliptic functions

When constructing analytical solutions for the motion of planets, a difficulty arises from the magnitude of the ratio of the semimajor axes between two bodies. It is not a small quantity for two neighboring planets (0.72 for the pair Venus–Earth, and 0.54 for the pair Jupiter–Saturn, for instance). The developments of the solutions in the form of Fourier series of the mean longitudes are slowly convergent, and to keep a good accuracy in the representation of the coordinates, it is necessary to keep a great number of terms. To avoid the difficulty of the expansion of the reciprocal of the distance between two planets [29], it has been shown that, at the first order of planetary masses, the perturbations can be obtained in closed form with the use of elliptic functions. However, it is hard to extend this construction beyond the first order, keeping the explicit formulation with elliptic functions. Brumberg [8] proposes to make use of the developments of such functions in terms of their nome, which is a small quantity even for large values of the ratio of the semimajor axes ,and encourages Chapront and Simon [15] to apply his results to the study of the construction of planetary theories with the aid of elliptic functions.

4.1 Notations

We consider a planet P of mass m perturbed by a planet P' of mass m'. Notations are the same as in Section 2.1. Moreover, we note that

- ε and ε' the mean longitudes of the epoch,
- n and n' the mean motions of the planets P and P',
- $\bar{n}$ and $\bar{n}'$ the mean motions deduced from the observation,
- $\bar{\lambda} = \varepsilon_0 + \bar{n}t$ and $\bar{\lambda}' = \varepsilon_0' + \bar{n}'t$ the mean longitudes,

- x any variable of the set $a, \varepsilon, e, \varpi, \gamma, \Omega$,
- x_0 the integration constant corresponding to the variable x.

n and n' are deduced from a and a' by

$$n^2 a^3 = G(1+m) \qquad n'^2 a'^3 = G(1+m'), \tag{4.1}$$

where G is the gravitational constant. The equations of the motion are the Lagrange equations; they have the form

$$\frac{dx}{dt} = F(\bar{\lambda}, \bar{\lambda}'), \tag{4.2}$$

where λ and ε are related by

$$\frac{d\lambda}{dt} = n + \frac{d\varepsilon}{dt}. \tag{4.3}$$

4.2 Expansion of the right-hand members of the equations: A change of the time variable

A classical change of the independent variable in the equations of motion is governed by the simple form of the mutual distance Δ on coplanar circular orbits [9, 29]. With our notations, we have

$$\Delta^2 = a^2 + a'^2 - 2aa' \cos H = a'^2(1+\alpha)^2[1 - \kappa^2 \sin^2 \varphi] \tag{4.4}$$

with $\kappa^2 = 4\alpha/(1+\alpha)^2$ and $H = \pi + 2\varphi$. For circular orbits the separation angle H is the difference of the mean longitudes $\bar{\lambda} - \bar{\lambda}'$. Hence, φ is linearly related to the time with

$$\bar{\lambda} - \bar{\lambda}' = (\bar{n} - \bar{n}')t + \varepsilon_0 - \varepsilon_0' = \pi + 2\varphi. \tag{4.5}$$

Change the angular variable using the elliptic integral:

$$u = \int_0^{\varphi} \frac{d\theta}{(1 - \kappa^2 \sin^2 \theta)^{\frac{1}{2}}}. \tag{4.6}$$

The reverse form of this integral is: $\varphi = \operatorname{am} u$. Consider the nome q defined by

$$q = e^{-\pi \frac{K'}{K}}, \tag{4.7}$$

where K is the complete elliptic integral of the first kind:

$$K(\kappa^2) = \int_0^{\frac{\pi}{2}} \frac{d\theta}{(1 - \kappa^2 \sin^2 \theta)^{\frac{1}{2}}}, \tag{4.8}$$

and

$$K' = K(1 - \kappa^2)\,.$$

The nome is a small quantity even for large values of α and, as shown in [8] the developments of the elliptic fonctions expressed in terms of q are more rapidly convergent than the original expansions in α. We develop am u in the form [1]

$$\text{am}\,u = v + 2\sum_{s=1}^{\infty} \frac{q^s}{s(1+q^{2s})} \sin 2sv \tag{4.9}$$

with $v = \frac{\pi u}{2K}$. q can be easily computed for a given value of α with the process of the arithmetic- geometric mean [1]. Set $w = \pi + 2v$. From (4.5) and (4.9), we get

$$\bar{\lambda} - \bar{\lambda}' = w + 4\sum_{s=1}^{\infty}(-1)^s \frac{q^s}{s(1+q^{2s})} \sin sw\,, \tag{4.10}$$

where w is the new time variable which takes place of t in the Lagrange equation and is defined with

$$\frac{dt}{dw} = \frac{1}{\bar{n} - \bar{n}'}\left[1 + \frac{d\sigma}{dw}\right], \tag{4.11}$$

where σ and $d\sigma/dw$ are trigonometric series of the new variable w, given by

$$\sigma = 4\sum_{s=1}^{\infty}(-1)^s \frac{q^s}{s(1+q^{2s})} \sin sw\,,$$

$$\frac{d\sigma}{dw} = 4\sum_{s=1}^{\infty}(-1)^s \frac{q^s}{(1+q^{2s})} \cos sw\,. \tag{4.12}$$

Starting from (4.10) and (4.12), we find the explicit formulation of t:

$$t = \frac{w + \sigma}{\bar{n} - \bar{n}'} + \frac{\varepsilon_0' - \varepsilon_0}{\bar{n} - \bar{n}'}\,. \tag{4.13}$$

Note the new argument w':

$$w' = \frac{\bar{n}'}{\bar{n} - \bar{n}'} w + \frac{\bar{n}\varepsilon_0' - \bar{n}'\varepsilon_0}{\bar{n} - \bar{n}'}\,. \tag{4.14}$$

The mean longitudes are

$$\bar{\lambda} = w + w' + \frac{\bar{n}}{\bar{n} - \bar{n}'}\sigma\,,$$

$$\bar{\lambda}' = w' + \frac{\bar{n}'}{\bar{n} - \bar{n}'}\sigma\,. \tag{4.15}$$

The Lagrange equations (4.2) are expanded in terms of the two new arguments w and w':

$$\frac{dx}{dw} = F(w, w')\frac{dt}{dw} = \Phi[w, w')\,, \tag{4.16}$$

where dt/dw has to be replaced by (4.11), and $d\sigma/dw$ by (4.12). Equation (4.3) concerning the mean longitude becomes

$$\frac{d\lambda}{dw} = \frac{n}{\bar{n} - \bar{n}'}(1 + \frac{d\sigma}{dw}) + \frac{d\varepsilon}{dw} . \tag{4.17}$$

We build the functions $\Phi(w, w')$ using the method of double harmonic analysis, starting from relations (4.12) and (4.15).

4.3 Application to planetary problems

Chapront and Simon give several applications of their method. We describe two of them.

First-order planetary perturbations for the pairs Jupiter–Saturn and Venus–Earth

The first-order perturbations, for a given pair of planets, can be computed using Lagrange equations expanded either in the variables $\bar{\lambda}$ and $\bar{\lambda}'$ in the form (4.2) and (4.3), or in the variables w and w' in the form (4.16) and (4.17). In the first case, the developments have the form

$$\Delta^1 x = x_1 t + X^1(\bar{\lambda}, \bar{\lambda}') , \tag{4.18}$$

where x_1 is a numerical coefficient named the secular term and X^1 is a Fourier series of the arguments $\bar{\lambda}$ and $\bar{\lambda}'$.

In the second case, the first order can be written in the form

$$\Delta^1 x = \frac{x_1 w}{\bar{n} - \bar{n}'} + \frac{x_1 \sigma}{\bar{n} - \bar{n}'} + X^1(w, w') = \tilde{x}_1 w + \tilde{X}_1(w, w') . \tag{4.19}$$

Here, $\tilde{X}_1$ is a Fourier series of the arguments w and w'. Note that if it is necessary to compute such expressions explicitly in terms of the time t, it is necessary to invert the relation (4.13) which can be regarded as a generalized Kepler's equation.

Using double harmonic analysis and starting from integration constants issued from [23], Chapront and Simon have built the first-order perturbations for the two pairs of planets: Jupiter–Saturn ($\alpha = 0.544$, $q = 0.148$) and Venus–Earth ($\alpha = 0.723$, $q = 0.215$). Their results are illustrated in Table 1, which gives the number of terms with amplitude larger than $0.001''$ for the six elliptic variables, in the approximation in $\bar{\lambda}, \bar{\lambda}'$ (N) and in the approximation in w, w' ($\tilde{N}$) and the value of $N/\tilde{N}$ for the first-order perturbations of the two pairs of planets. We can see that the representation is more compact with the choice of the two fundamental arguments w and w' and that the reduction in the number of Fourier coefficients increases with α.

Table 1. First-order perturbations of the pair Jupiter–Saturn ($\alpha = 0.544$) and of the pair Venus–Earth ($\alpha = 0.723$): number of the terms with amplitude larger than $0.001''$ (or $0''.001 \times$ AU for $\Delta^1 a$) for each variable, in the approximation in $\bar{\lambda}, \bar{\lambda}'$ (N) and in the approximation in w, w' ($\tilde{N}$). (Results taken from [15]).

		$\Delta^1 a$	$\Delta^1 \lambda$	$\Delta^1 e$	$e_0 \Delta^1 \varpi$	$\Delta^1 \gamma$	$\gamma_0 \Delta^1 \Omega$	Total
Jupiter	N	170	143	143	146	64	65	731
	$\tilde{N}$	99	81	90	92	39	40	441
	$N/\tilde{N}$	1.72	1.77	1.59	1.59	1.64	1.63	1.66
Saturn	N	232	168	181	178	78	78	915
	$\tilde{N}$	146	111	125	126	58	59	625
	$N/\tilde{N}$	1.59	1.51	1.45	1.41	1.34	1.32	1.46
Venus	N	66	86	89	86	41	34	402
	$\tilde{N}$	30	41	45	46	22	18	202
	$N/\tilde{N}$	2.20	2.10	1.98	1.87	1.86	1.89	1.99
Earth	N	62	77	83	77	35	28	362
	$\tilde{N}$	35	40	46	48	21	19	209
	$N/\tilde{N}$	1.77	1.93	1.80	1.60	1.67	1.47	1.73

Computation of the mutual perturbations of Jupiter–Saturn at any order of planetary masses

The method for computing, at any order of planetary masses, the perturbations of Jupiter and Saturn, by double harmonic analysis, in the variables $(\overline{\lambda}, \overline{\lambda'})$ described in [25] is adapted to the couple of variables (w, w'). In this method, the right-hand members of Lagrange equations written in a closed form with respect to the osculating elements are computed with an iterative process, for $p + 1$ explicit values of the time. A double harmonic analysis in $(\overline{\lambda}, \overline{\lambda'})$ is performed. The resulting Poisson series are integrated. The classical formulation of the solution is written:

$$x = x_0 + x_1 t + \cdots + x_p t^p + S_0(\overline{\lambda}, \overline{\lambda'}) + t S_1(\overline{\lambda}, \overline{\lambda'}) + \cdots + t^p S_p(\overline{\lambda}, \overline{\lambda'}) \,, \qquad (4.20)$$

where $x_0, x_1, \ldots, x_p$ are numerical coefficients and $S_0, S_1, \ldots, S_p$ are Fourier series in $(\overline{\lambda}, \overline{\lambda'})$. Using the new variables w and w', the right-hand members of Lagrange equations are evaluated with the same manner for $p + 1$ values of w, by double harmonic analysis in (w, w'). After integration and iterating the process, the alternate solution has a form analogous to (4.20), i.e.:

$$x = \tilde{x}_0 + \tilde{x}_1 w + \cdots + \tilde{x}_p w^p + \tilde{S}_0(w, w') + w \tilde{S}_1(w, w') + \cdots + w^p \tilde{S}_p(w, w') \,, \quad (4.21)$$

where $\tilde{x}_0, \tilde{x}_1, \ldots, \tilde{x}_p$ are numerical coefficients and $\tilde{S}_0, \tilde{S}_1, \ldots, \tilde{S}_p$ are Fourier series in (w, w').

The process is initiated starting with a solution for the mutual perturbations of Jupiter–Saturn which has been constructed previously in [26]. Denote this solution by

$H_{\overline{\lambda},\overline{\lambda}'}$. It is represented with the formulation (4.20) where the developments in power of the time reach t^{10}. The mutual perturbations of Jupiter–Saturn are given with a precision, over 1000 years, which is approximately $0''.001$ for Jupiter and $0''.002$ for Saturn; over 6000 years it grows up $3''$ for Jupiter and $7''.3$ for Saturn. The change of variables which is defined by (4.13) for the time and by (4.15) for the mean longitudes is performed on $H_{\overline{\lambda},\overline{\lambda}'}$. The process is iterated by Fourier analysis in (w, w'). These analyses have been performed on the right-hand members of the equations of motion for 11 values of w: $w_{-5}, \ldots, w_{-1}, w_0, w_1, \ldots, w_5$ where $w_0 = 0$ and $w_p = w_0 + p\Delta w$; $\Delta w = 0.6(\overline{n} - \overline{n}')$ where the mean motions are expressed in radian per thousand years. Note the new solution $H_{w,w'}$ in the variables (w, w'). The integration constants in $H_{w,w'}$ are obtained by comparison of the two solutions $H_{w,w'}$ and $H_{\overline{\lambda},\overline{\lambda}'}$ over a time interval of 1000 years. Table 2 gives the integration constants of the solution $H_{\overline{\lambda},\overline{\lambda}'}$ taken from [23] and of the solution $H_{w,w'}$ taken from [15]. As can be verified from equations (4.14) and (4.15) the integration constants of $H_{w,w'}$ for the mean longitudes are close to $\frac{\overline{n}\varepsilon_0' - \overline{n}'\varepsilon_0}{\overline{n} - \overline{n}'}$ where $\overline{n}, \overline{n}', \varepsilon_0$, and ε_0' are the integration constants of $H_{\overline{\lambda},\overline{\lambda}'}$. For the other variables the constants in the two solutions are very similar.

The differences between the two solutions $H_{w,w'}$ and $H_{\overline{\lambda},\overline{\lambda}'}$ are smaller than $0''.001$ for Jupiter and $0''.002$ for Saturn over 1000 years and smaller than $0''.2$ for Jupiter and $0''.4$ for Saturn over 6000 years. So they are below the precision of the departure solution. This ensures to $H_{w,w'}$ a precision equivalent to $H_{\overline{\lambda},\overline{\lambda}'}$.

As has been established at the first order of the perturbations, the solutions developed in (w, w') in the form (4.21) are more compact than the solutions developed in $(\overline{\lambda}, \overline{\lambda}')$ in the form (4.20). Table 3 illustrates this fact for the mean longitudes of Jupiter and Saturn. It gives, for the two solutions, the number of terms of magnitude up to $0''.001$, over 6000 years, for the two series S_i and $\widetilde{S}_i$. We see that, for the two mean longitudes, the total number of terms in the form (4.21) is reduced by a factor 1.7 when

Table 2. Integration constants J2000 for the pair Jupiter–Saturn for the solution $H_{\overline{\lambda},\overline{\lambda}'}$ [23] and for the solution $H_{w,w'}$ [15]. Units are $''$/1000 yrs for $\overline{n}$, ua for a_0, ° for $\varepsilon_0, \varpi_0, i_0$, and Ω_0. e_0 is without dimension.

	$H_{\overline{\lambda},\overline{\lambda}'}$		$H_{w,w'}$	
	Jupiter	**Saturn**	**Jupiter**	**Saturn**
n	109 256 603.7799	43 996 098.5573	109 256 603.7205	43 996 098.7019
a_0	5.2026032092	9.5549091915	5.2027644707	9.5639969616
ε_0	34.35151874	50.07744430	60.67934784	60.67903031
e_0	0.0484979255	0.0555481426	0.0484981595	0.0555410916
ϖ_0	14.33120687	93.05723748	14.32979023	93.04909111
i_0	1.30326698	2.48887878	1.30325804	2.48861178
Ω_0	100.46440702	113.66550252	100.46605794	113.66181594

Table 3. Mean longitudes of Jupiter (λ) and Saturn (λ'): number of terms of magnitude up to $0''.001$ in the solution $H_{w,w'}$ of [15] (series $\widetilde{S}_i$) and in the solution $H_{\overline{\lambda},\overline{\lambda'}}$ of [25] (series S_i).

$H_{w,w'}$			$H_{\overline{\lambda},\overline{\lambda'}}$		
	λ	λ'		λ	λ'
$\widetilde{S}_0$	110	145	S_0	194	229
$\widetilde{S}_1$	140	181	S_1	239	284
$\widetilde{S}_2$	162	207	S_2	282	334
$\widetilde{S}_3$	173	216	S_3	314	357
$\widetilde{S}_4$	186	232	S_4	325	385
$\widetilde{S}_5$	190	230	S_5	332	405
$\widetilde{S}_6$	196	233	S_6	325	408
$\widetilde{S}_7$	193	230	S_7	314	396
$\widetilde{S}_8$	189	226	S_8	303	384
$\widetilde{S}_9$	166	205	S_9	261	339
$\widetilde{S}_{10}$	158	197	S_{10}	239	325
Total	1863	2302	Total	3128	3846

compared to the form (4.20). For the other variables, this factor is between 1.4 and 1.8, depending on the planet and the variable.

Conclusion

The method gives good results in the case of two bodies, for a large value of the ratio of the semimajor axes. Can it be extended to more planets? The time transformation given by (4.6) depends on a given pair of planets through the reciprocal of the distance. If three or more planets are concerned, the choice of the fundamental argument is not obvious. Brumberg [8] suggested to operate a transformation of the type:

$$\mathrm{am}\, u_{i,k} = \frac{\overline{n}_i - \overline{n}_k}{\overline{n}_i - \overline{n}_j}\, \mathrm{am}\, u_{i,j} + Cte\,, \tag{4.22}$$

where i, j, k are indexes of the planets. am $u_{i,j}$ and am $u_{i,k}$ are amplitudes given by relations like (4.6) for the distances between the planets i and j and the planets i and k. Depending on i, j, and k, a convenient choice of the above ratio of the semimajor axes has to be done. A complete numerical experiment has to be realized to be convinced of the success of this approach. So, the question of the interest of the use of elliptic functions for the complete problem of the eight planets of the Solar System remains open.

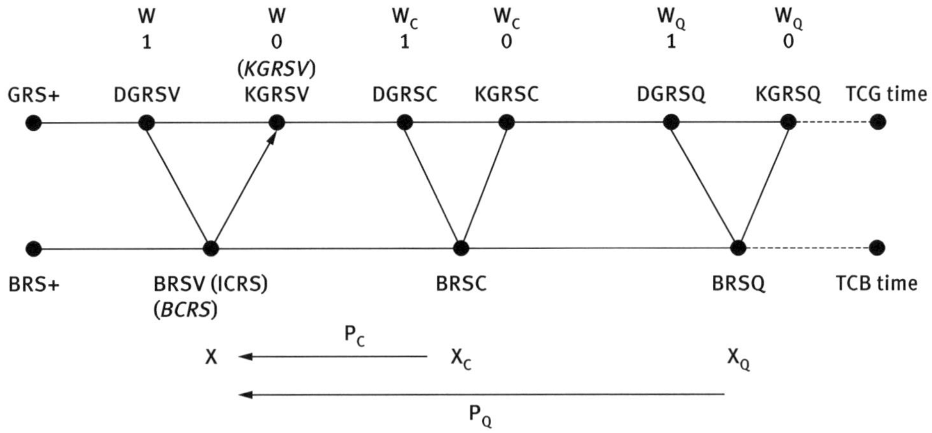

Fig. 2. Barycentric and geocentric reference systems with B stands for barycentric, G for geocentric, V for VLBI, C ecliptical, Q equatorial, D dynamical, K kinematical, and $^+$ rotating. Extracted from [11]. $\vec{X}$, $\vec{X}_C$, and $\vec{X}_Q$ are the barycentric coordinate systems

5 Reference frames, time scales, and units for planetary ephemerides

A part of Victor Brumberg works was dedicated to the definitions of concepts very common in celestial mechanics such as frames and units in a rigorous relativistic formalism. In 1991, he published with Soffel [27] a paper introducing the importance of such relativistic approaches in the perspective of (at that time) new observations such as pulsar timing, VLBI and regular tracking of spacecrafts used for the construction of planetary ephemerides. In the first subsection, we will first introduce the main concepts developed by Victor Brumberg in several of his main publications related to these questions. In the second subsection, we will put these developments in the perspective of the present accuracies achieved by the planetary ephemerides in terms of dynamical modeling and observations.

5.1 Victor Brumberg's contributions

The description of the reference frames used in classical mechanics in the frame of general relativity was one of the main contributions of Victor Brumberg as it affected in the same time our understanding of time and space. Several papers were dedicated to these questions and numerous discussions arose in the international community by Victor Brumberg about the most appropriate realizations of such complex concepts for common users. Most of these discussions lead to IAU recommendations or resolutions.

We focus here on one of them related to the transformation between the ICRS and the ITRS.

The definition of a reference frame in general relativity is an impossible question as no frame has to be preferred. One can only say that some frames are more appropriate than the others for some specific applications. However, one can always define local frames where the definition of reference frame as conserved by the special relativity can be appllied. In the Solar System, Brumberg defines a hierarchy of reference frames that can schematized in Figure 2. A frame is kinematically nonrotating relatively to another if the directions of the axis are the same in both systems. But the same frame can be dynamically rotating if centrifuge and de Coriolis accelerations appear in the equations of motions written in this system. For each reference frame defined in the barycentric reference system (BRS) but realized by different techniques (VLBI, planetary ephemerides) correspond two frames in the geocentric reference system (GRS): the dynamically nonrotating (D) appropriate for celestial mechanics and the kinematically nonrotating (K) more dedicated to astrometry. Some example could be given: planetary ephemerides are developed in the BRSC when theories of the Earth rotation are written in the DGRSC but the ICRF can be identified to the BRSV. As for the concept of reference frame, the definition of one system of fundamental constants including time does not have a sense in general relativity because such a system is not invariant through the transformation from BCRS to GCRS. Discussions about the notion of fundamental constants have then to lead. As it is described in the IAU 1992 resolution A4, the time scale of the barycentric systems is the TCB (identical for the three) when the time scale for the geocentric systems is the TCG. In practice, the TDB and the TT (as defined by the IAU 2006) are used by common users instead of TCG and TCG respectively but for high accurate modelings or observations (VLBI, LLR, Gaia) the rigorous description given by Victor Brumberg for the definitions of the barycentric and geocentric systems are applied.

We note, following the Brumberg's usual notations, $\vec{x} = (x^i)$, $\vec{x}_C = (x^i_C)$ and $\vec{x}_Q = (x^i_Q)$, (i = 1, 2, 3), the coordinate systems of the BCRS (or BCRSV in Figure 2), BCRSC and BCRSQ respectively. P_C and P_D are the constant matrices of rotation between the BCRS and the ecliptic and equatorial barycentric systems, respectively. We then have the direct relation:

$$\vec{x}_C = P_C \vec{x}\,, \qquad \vec{x}_Q = P_Q \vec{x}\,. \tag{5.1}$$

For the geocentric systems, we introduce $\hat{P}(u)$, the Earth rotation matrix giving the orientation of the ITRS axis $\vec{y} = (y^i)$ in the DGRSC (for q = 1) or in the KGRSC (for q = 0). It comes then with the index t noting the transposed matrix:

$$\vec{y} = \underset{q}{\hat{P}}(u)\, \underset{q}{\vec{w}}_C = \underset{q}{\hat{P}}(u) P_C\, \underset{q}{\vec{w}} = \underset{q}{\hat{P}}(u) P_C P_Q^t\, \underset{q}{\vec{w}}_Q, \text{pour } q = 0 \text{ et } q = 1\,. \tag{5.2}$$

The relations between the dynamically and kinematically nonrotating coordinate systems require a symmetric matrix of geodesic rotation F such as

$$\underset{0}{\vec{w}} = (E - c^{-2}F)\, \underset{1}{\vec{w}}\,, \qquad \underset{0}{\vec{w}}_C = (E - c^{-2}F_C)\, \underset{1}{\vec{w}}_C\,, \qquad \underset{0}{\vec{w}}_Q = (E - c^{-2}F_Q)\, \underset{1}{\vec{w}}_Q \tag{5.3}$$

and

$$\underset{1}{\hat{P}}(u) = \underset{0}{\hat{P}}(u)(E - c^{-2}F_C), \tag{5.4}$$

with $F_C = P_C F P_C^T$, $F_Q = P_Q F P_Q^T$. The matrix F includes the geodesic precession, the geodesic nutation and planetary terms. Victor Brumberg had worked on the computation and the analysis of this matrix and detailed description of each of its components can be found in [10]. In most of the usual applications, one only needs the direct transformation between the BRS coordinate systems x^i at the date t to the GRS w^i at the date u. Such transformation can be written as

$$w^i = r_E^i + c^{-2}[\tfrac{1}{2}\vec{v}_E\vec{r}_E v_E^i - q\varepsilon_{ijk}F^j r_E^k + \bar{U}_E(t, \vec{x}_E) r_E^i + \vec{a}_E\vec{r}_E r_E^i - \tfrac{1}{2}\vec{r}_E^{\,2} a_E^i] + \cdots \tag{5.5}$$

and

$$u = t - c^{-2}[A(t) + \vec{v}_E\vec{r}_E] + \cdots, \tag{5.6}$$

where $A(t)$ is defined by the differential equation

$$\dot{A}(t) = \tfrac{1}{2}\vec{v}_E^{\,2} + \bar{U}_E(t, \vec{x}_E) \tag{5.7}$$

$A(t)$ can then be either approximated by an analytical development such as the one given in [16] or directly integrated simultaneously with the equations of motions of the planets as it is done in INPOP [19]. $\vec{a}_E$ is the Earth acceleration in the BRS and $\bar{U}_E(t, \vec{x}_E)$ is the Newtonian potential of all the bodies of the Solar System estimated at the geocenter. The equation 5.5 is applied for the transformation from the BRS to the KCRS (with $q = 0$) as well as for the BRS to DCRS transformation (with $q = 1$). The transformation from the GRS w^i at the date u to the BCR x^i at the date t can be given by

$$x^i = w^i + z_E^i(u) + c^{-2}[\tfrac{1}{2}\vec{v}_E\vec{w} v_E^i + q\varepsilon_{ijk}F^j w^k - \bar{U}_E(t, \vec{x}_E) w^i - \vec{a}_E\vec{w} w^i + \tfrac{1}{2}\vec{w}^{\,2} a_E^i] + \cdots \tag{5.8}$$

and

$$t = u + c^{-2}[A(u) + \vec{v}_E\vec{w}] + \cdots \tag{5.9}$$

$z_E^i(u)$ is the representation of the Earth motion at the geocenter in TCG with

$$z_E^i(u) = x_E^i(t^*), \tag{5.10}$$

where t^* is an approximation of 5.6 such as

$$u = t^* - c^{-2}A(t^*) + \cdots. \tag{5.11}$$

By numerical or analytical inversion of (5.11), it is possible to deduce a relation like $t^* = t^*(u)$. In the case of an analytical development such as the one given in [3], an explicit relation between $z_E^i(u)$ and u can be given.

Thanks to his fruitful collaboration with Pierre Bretagnon, Victor Brumberg was able to provide in [3] full developments of such transformations considering the different geocentric subsystems (dynamically and kinematically nonrotating) and different time scales (TDB and TT instead of TCG and TCB). Based on these results, Brumberg was able to deduce rigorous relations between the geocentric vector of positions of any celestial bodies A, $\vec{w}_A(u)$ and its coordinates in the BRS $\vec{x}_A(t)$. In using $\vec{r}_{AE} = \vec{x}_A - \vec{x}_E$ and the functions $z^i_E(u)$ and $z^i_A(u)$ Brumberg obtained:

$$w^i_A(u) = z^i_A(u) - z^i_E(u) + c^{-2}\Big[\vec{v}_E\vec{r}_{AE}(v^i_A - \tfrac{1}{2}v^i_E) - q\varepsilon_{ijk}F^j r^k_{AE} \\ + \bar{U}_E(t, \vec{x}_E)r^i_{AE} + \vec{a}_E\vec{r}_{AE}r^i_{AE} - \tfrac{1}{2}\vec{r}^{\,2}_{AE}a^i_E\Big] + \cdots . \quad (5.12)$$

Such relation was used by Brumberg and Simon for the development of relativistic arguments of the SMART analytical Earth rotation theory in 2003.

5.2 Planetary ephemerides

The works of Victor Brumberg were intensively used by IMCCE researches for the construction of analytical planetary ephemerides such as VSOP and TOP. The VSOP solutions give an analytical description of the motion of our Solar System planets on very long intervals of time: from several thousands of years for the telluric planets to one thousand years for the big planets. The perturbations are developed with Poisson series of the mean longitudes δ such as $\delta = \delta^0 + Nt$ where N is the mean motion. The most distributed one was VSOP82 [2] providing developments in terms of elliptical variables. VSOP87 [4] was an extension of VSOP87 given in rectangular and spherical coordinates. Finally, Victor Brumberg helped for the construction of a solution developed in the frame of general relativity called VSOP2000 [21] and including the perturbations of the three biggest main belt asteroids (Ceres, Pallas and Vesta). The latest VSOP2000 shows an improvement of a factor 10 compared to VSOP82 on a n interval of several tens of years. However, as it was demonstrated in [17], these types of ephemerides did not have sufficient accuracy for being fitted directly to observations or to give accurate enough representation of the planet motions in comparison to positions of planets deduced from spacecraft tracking. This is particularly true for Mars. This planet is indeed tracked by an important number of spacecraft since the 80s and is highly perturbed by the main belt asteroids. Analytically, this problem increase greatly the size of the analytical developments of the perturbative functions as an important number of asteroids (between 60 and 300) has to be taken into account. Even if new analytical theories are still developed such as VSOP2013, since 2003, the IMCCE researchers work on a numerical planetary ephemerides called INPOP.

In INPOP, like in the other two state-of-the-art Solar System ephemerides (the American one, DE (Development Ephemeris; JPL), the Russian one, EPM (Ephemerides of Planets and the Moon; IPA, St.Petersburg)), Infeld–Hoffmann (EIH) equations of motions are numerically integrated for the whole Solar System including the Moon, Pluto, and a set of selected minor planets (about 300). Are also considered the figure effects for Earth and Moon, the solar oblateness, the rotational motion of Earth and Moon (librations) and the effect of tidal dissipation (via some lag-angle). Different versions of these ephemerides differ slightly in modeling of lunar libration, reference frames in which the ephemerides are computed, adopted value of solar oblateness, modeling of perturbation from asteroids and the set of observations to which ephemerides are adjusted.

Three versions of INPOP are presently available for users: INPOP06 [20], INPOP08 [19], INPOP10a [18]. Based on the works and discussions lead by Victor Brumberg a new definition of the TDB time scale was accepted as IAU standard in 2006. This new definition was implemented in INPOP and as it was computed previously in [16] using VSOP82, the difference TT-TDB (or TCB), important for the accurate timing procedures (pulsar timing, GAIA), is numerically estimated together with the planetary ephemerides. A perfect consistency between the timing of the observations and the motion of the Solar System bodies is then obtained. Versions of INPOP computed in TCB time scale are also provided together with a TCB-TCG (instead of TT-TDB) estimation.

Furthermore, following discussions initiated by Victor Brumberg on the appropriate definitions of units of time and length in the Solar System seen in the frame of general relativity, the IAU recommends since 2006 to fix the astronomical unit in all codes dealing with planetary motions and to estimate instead the mass of the Sun. Since INPOP08, INPOP is the first planetary ephemerides solving for the mass of the Sun for a given fixed value of astronomical unit. New estimations of the Sun mass are regularly obtained since INPOP08 together with the oblateness of the Sun.

Finally, the analysis of the planetary and Moon observations used for the construction of INPOP is based on the hierarchy of reference frames given by Brumberg (for example in 1991) and described in the previous section. His works are of special interest for the computation of the station positions given in the ITRS(GCRS) in station positions in the ICRS(BCRS). The transformation proposed by Brumberg was adopted in the IERS convention in 2003 and has an impact of about 4 mm on the Lunar Laser Ranging data over a period of 40 years [27]. This millimeter impact is small of about two order of magnitudes compared to the present differences between Moon ephemerides presented in Table 5. However, thanks to the new generation of Lunar Laser ranging facilities such as APOLLO [22] the millimeter accuracy in the lunar observations is reachable and a new interest for the Brumberg's works becomes then obvious.

More than 136 000 planetary observations (Table 4) including the latest Messenger, Mars Express, Venus Express and Cassini data points and lunar laser ranging normal points are used for the construction of INPOP. The fit of up to 200 parameters was needed for INPOP10a [18]. The latest INPOP version is INPOP13a [28] including newly analysed positions of Mercury deduced from the analysis of navigation Doppler and range data of the Messenger mission orbiting the planet since 2011. With Tables 5 and 4, one can have a better idea of the accuracies reached by the modern planetary

Table 4. Planetary ephemerides data sets given by data types (column 1), by objects (column 2) and mean accuracy (columns 3, 4, and 5) in right ascension α, declination δ and geocentric distances ρ. The numbers of each type of observations used in the INPOP10a adjustment are given in column 6.

Data type	Planets	Mean accuracy			Numbers of observations
		α	δ	ρ	
VLBI*	V, Ma, J, S	1/10 mas	1/10 mas		176
Flybys*	Me, J, S, U, N	0.1/1 mas	0.1/1 mas	1/30 m	43
Range tracking*	V, Ma, Me			2/30 m	67 768
Direct range	Me,V			1 km	951
Optical	J, S, U, N, P	300 mas	300 mas		35 437
LLR	Moon			5 cm	> 19 000

Table 5. Maximum differences between INPOP13a, INPOP10a and DE421 over 1 century in spherical geocentric coordinates and distances and barycentric Earth spherical coordinates. For the Moon, INPOP10e is used for the comparisons to INPOP10a.

Differences	INPOP10a – INPOP13a			INPOP10a–DE421		
	α mas	δ mas	ρ km	α mas	δ mas	ρ km
Mercury–Earth	2.8	1.9	1.7	5	6	1
Venus–Earth	0.9	0.6	0.1	4	2	0.3
Moon–Earth	12.2	5.0	1.5×10^{-3}	6	2.8	3.0×10^{-3}
Mars–Earth	41.6	15.9	9.5	30	15	10
Jupiter–Earth	9.1	15.1	4.7	10	30	3
Saturn–Earth	2.3	1.2	1.7	0.8	0.4	3
Uranus–Earth	217	102	1241		100	1000
Neptune–Earth	66	38	330	400	100	3000
Pluton–Earth	6200	2151	133 600	3000	1000	80 000
	long mas	lat mas	ρ km	long mas	lat mas	ρ km
Earth–SSB	2.2	0.9	1.20	4.0	1.7	0.6

ephemerides. Numerous applications are then possible in particular for testing possible violation of general relativity in consistent frames thanks to Victor Brumberg's works. Examples can be found in [18, 30] and [28].

5.3 Conclusions

The impact of Victor Brumberg's work on the development of planetary ephemerides is important for two reasons. First, he participated actively in the construction of analytical solutions for the motions of the main planets of the Solar System. By the formulims he wrote and the methods he developed, he was one of the main contributors with J. Chapront, M. Chapront, P. Bretagnon and J.-L. Simon to the modern analytical ephemerides of planets and the Moon. A second aspect of his impressive contribution was the important effort for introducing the complex concepts of the general relativity such as the tensors, the proper time etc. to the classical mechanics in such a way that common user can use and feel the importance.

References

[1] M. Abramowitz and I. A. Stegun, *Handbook of mathematical functions with formulas, graphs and mathematical tables*, 1968.

[2] P. Bretagnon, Theory for the motion of all the planets – The VSOP82 solution, *Astronomy & Astrophysics* **114** (1982), 278–288.

[3] P. Bretagnon and V. A. Brumberg, On transformation between international celestial and terrestrial reference systems, *Astronomy & Astrophysics* **408** (2003), 387–400.

[4] P. Bretagnon and G. Francou, Planetary theories in rectangular and spherical variables – VSOP 87 solutions, *Astronomy & Astrophysics* **202** (1988), 309–315.

[5] V. A. Brumberg, Application of Hill's Lunar Method in General Planetary Theory, in: *Periodic Orbits Stability and Resonances* (G. E. O. Giacaglia, ed.), p. 410, 1970.

[6] V. A. Brumberg, On a differential equation for Hansen's coefficients, *Byulleten' Instituta Teoreticheskoj Astronomii (St Petersburg)* **12** (1970), 452–457.

[7] V. A. Brumberg, *Analytical algorithms of celestial mechanics*, 1980.

[8] V. A. Brumberg, General Planetary Theory Revisited with the Aid of Elliptic Functions, in: *25th Symposium on Celestial Mechanics*, (H. Kinoshita and H. Nakai, eds.), p. 156, 1992.

[9] V. A. Brumberg, General planetary theory in elliptic functions, *Celestial Mechanics and Dynamical Astronomy* **59** (1994), 1–36.

[10] V. A. Brumberg, P. Bretagnon and G. Francou, Analytical algorithms of relativistic reduction of astronomical observations., in: *Journées 1991: Systèmes de référence spatio-temporels*, pp. 141–148, 1991.

[11] V. A. Brumberg, P. Bretagnon and B. Guinot, Astronomical Units and Constants in the General Relativity Framework, *Celestial Mechanics and Dynamical Astronomy* **64** (1996), 231–242.

[12] V. A. Brumberg and J. Chapront, Construction of a general planetary theory of the first order, *Celestial Mechanics* **8** (1973), 335–355.

[13] J. Chapront, Construction of a planetary theory to the second order of the masses, *Astronomy & Astrophysics* **7** (1970), 175.

[14] J. Chapront and J.-L. Simon, Secular variations of the first order, for the four major planets. Comparison with Le Verrier and Gaillot, *Astronomy & Astrophysics* **19** (1972), 231.

[15] J. Chapront and J.-L. Simon, Planetary theories with the aid of the expansions of elliptic functions, *Celestial Mechanics and Dynamical Astronomy* **63** (1996), 171–188.

[16] L. Fairhead and P. Bretagnon, An analytical formula for the time transformation TB-TT, *Astronomy & Astrophysics* **229** (1990), 240–247.

[17] A. Fienga, Global adjustement of analytical theories of planetary motion to observations: the first results, in: *Bulletin of the American Astronomical Society*, Bulletin of the American Astronomical Society 31, p. 1589, December 1999.

[18] A. Fienga, J. Laskar, P. Kuchynka, H. Manche, G. Desvignes, M. Gastineau, I. Cognard and G. Theureau, The INPOP10a planetary ephemeris and its applications in fundamental physics, *Celestial Mechanics and Dynamical Astronomy* **111** (2011), 363–385.

[19] A. Fienga, J. Laskar, T. Morley, H. Manche, P. Kuchynka, C. Le Poncin-Lafitte, F. Budnik, M. Gastineau and L. Somenzi, INPOP08, a 4-D planetary ephemeris: from asteroid and time-scale computations to ESA Mars Express and Venus Express contributions, *Astronomy & Astrophysics* **507** (2009), 1675–1686.

[20] A. Fienga, H. Manche, J. Laskar and M. Gastineau, INPOP06: a new numerical planetary ephemeris, *Astronomy & Astrophysics* **477** (2008), 315–327.

[21] X. Moisson and P. Bretagnon, Analytical planetary solution VSOP2000, *Celestial Mechanics and Dynamical Astronomy* **80** (2001), 205–213.

[22] T. W. Murphy, Jr., E. G. Adelberger, J. B. R. Battat, C. D. Hoyle, N. H. Johnson, R. J. McMillan, C. W. Stubbs and H. E. Swanson, APOLLO: millimeter lunar laser ranging, *Classical and Quantum Gravity* **29** (2012), 184005.

[23] J.-L. Simon, P. Bretagnon, J. Chapront, M. Chapront-Touze, G. Francou and J. Laskar, Numerical expressions for precession formulae and mean elements for the Moon and the planets, *Astronomy & Astrophysics* **282** (1994), 663–683.

[24] J.-L. Simon and J. Chapront, Second Order Perturbations for Jupiter and Saturn. Comparison with Le Verrier, *Astronomy & Astrophysics* **32** (1974), 51.

[25] J.-L. Simon and G. Francou, Improvement of the theories of Jupiter and Saturn by harmonic analysis, *Astronomy & Astrophysics* **114** (1982), 125–130.

[26] J.-L. Simon and F. Joutel, Calculation of the mutual perturbations of Jupiter and Saturn in terms of a single angular variable, *Astronomy & Astrophysics* **205** (1988), 328–334.

[27] M. H. Soffel and V. A. Brumberg, Relativistic reference frames including time scales – Questions and answers, *Celestial Mechanics and Dynamical Astronomy* **52** (1991), 355–373.

[28] A. Verma, A. Fienga, J. Laskar, H. Manche and M. Gastineau, Use of MESSENGER radioscience data to improve planetary ephemeris and to test general relativity, *ArXiv e-prints* (2013).

[29] C. A. Williams, E. A. Wright and T. van Flandern, First order planetary perturbations with elliptic functions, *Celestial Mechanics* **40** (1987), 367–391.

[30] J. G. Williams, S. G. Turyshev and D. H. Boggs, Lunar laser ranging tests of the equivalence principle, *Classical and Quantum Gravity* **29** (2012), 184004.

Gérard Petit, Peter Wolf, and Pacôme Delva

Atomic time, clocks, and clock comparisons in relativistic spacetime: a review

1 Introduction

Atomic clocks have gained about 1 order of magnitude in accuracy every decade in the past 50 years, and the present best cases claim an uncertainty budget below 1×10^{-17} in relative frequency, which accounts for all systematic effects [10, 22]. Comparing such clocks at a distance would require techniques with similar performances, for example with a time stability of the order 1 ps over one day to achieve 1×10^{-17}. Even though most clocks and comparison methods are still at the 1×10^{-15} or 1×10^{-16} level, relativistic effects are orders of magnitude larger than these numbers so that clock comparisons, and the definition of time scales, have for decades needed to be considered rigorously within the theoretical framework of general relativity. In this chapter, we examine how this relativistic treatment is successfully achieved, but we first recall the historical background that led to the present situation.

The reference systems for the Earth and for the Solar System are now accurate to the 10^{-10} level, i.e. to within several tens of microarcseconds in orientation, a few mm in the geometric shape of the Earth and 10^{-10} or below in the scale of the Solar System. The reference frames are realized through an ensemble of space geodesy techniques like very long baseline interferometry (VLBI), laser ranging to satellites (SLR) and to the Moon (LLR), ranging to satellites of global navigation satellite systems (GNSS) and to space probes. These techniques all measure the time of propagation of electromagnetic signals and were developed in the 1970s and 1980s. This was made possible by an ensemble of technical factors like the availability of new stable atomic clocks and of reliable lasers and of course by the expanding capabilities of space technology. Even though the technology has improved over the last three decades, the basic features of the present-day astro-geodetic techniques were already in place in the 1980s; the VLBI MarkIII system with 56-MHz recording at S–X bands was already operated in many observatories worldwide. The Lageos I satellite was launched in 1976 and laser reflectors were installed on the Moon in the early 1970s, providing targets to more than twenty SLR and three LLR stations, respectively. Eleven BlockI GPS satellites were launched between 1978 and 1985. The Viking lander allowed measuring the distance to Mars to

Gérard Petit: BIPM, Pavillon de Breteuil, F-92312 Sèvres Cedex, France
Peter Wolf, Pacôme Delva: SYRTE, Observatoire de Paris, CNRS, LNE, UPMC, 61 av. de l'Observatoire, F-75014 Paris, France

within meters between 1976 and 1982. See the account of Wilkins [84] and references therein, or the review by Plag et al. [67], for more details on the rapid progresses in the early years of space geodesy.

On the other hand, the reference systems for the Earth and the Solar System, as we now know them, were quite in their infancy in the early 1980s. The International Astronomical Union (IAU) celestial reference system was based on the 1976 system of astronomical constants and realized by the Fifth Fundamental Catalog (FK5) that was adopted by the IAU in 1988[1]. It was not officially replaced until 1997 by the International Celestial Reference System and Frame (ICRS-ICRF, see [2]) which, with some revisions and extensions, still provides the celestial reference. For the terrestrial system, first trials towards a global reference frame for the Earth differed from one another by several meters and were based on very few stations, e.g. 34 for the BTS84, the first realization of the series that eventually evolved to the present International Terrestrial Reference System and Frame (ITRS-ITRF, see [1]).

In the set of IAU Resolutions, relativity made its first appearance in Resolution C2 (1985)[1] which,

> recognizing ... the importance of a space-fixed reference system, independent of the mode of observation, for use in astronomy and geodesy and satisfying the requirements of relativistic theories ... invites ... to form an IAU Working Group ... which will report ... in 1988 with recommendations for (1) the definition of the Conventional Terrestrial and Conventional Celestial Reference Systems, (2) ways of specifying practical realizations of these systems, (3) methods of determining the relationships between these realizations, and 4) a revision of the definitions of dynamical and atomic time to ensure their consistency with appropriate relativistic theories.

The work of this group eventually led to the set of IAU Resolutions in 1991 and 2000 that define the present reference systems (see Section 3).

However, no relativistic formalism existed that would permit accurate definitions of reference systems for the Solar System and the Earth and of the transformations between them. It was the work of V. Brumberg and S. Kopeikin in the 1980s that first enabled such formalism. Kopeikin [47] derived the post-Newtonian equations of motion of extended bodies in an N-body system in the weak-field and slow-motion approximations and presented the matching technique that allows us to construct, for a given body, a local coordinate system in the post-Newtonian approximation with its origin moving along the world line of the center of mass of the body. This led to the Brumberg–Kopeikin (BK) formalism that develops a relativistic hierarchy of reference systems (and time scales), providing metric expressions and transformation relations for Solar System barycentric, heliocentric, Earth–Moon local, geocentric, and topocentric reference systems (Brumberg and Kopeikin, 1989 [12], see also [14]). In these years, V. Brumberg also took an active part in the IAU working group on reference systems.

1 All IAU Resolutions may be found at www.iau.org/administration/resolutions/general_assemblies/

At the same period, a similar matching technique was developed by Damour et al. [27] and resulted in the so-called DSX framework. To treat the N-body problem the DSX framework defines one global system and N local systems co-moving with each body. The formalism employs post-Newtonian mass- and spin-multipole moments (Blanchet–Damour moments, see [8]) that shape the expression of the metric potentials in the local systems. Both BK and DSX formalisms were used to define relativistic coordinate systems and metric tensors for the barycentric and geocentric systems in the IAU Resolution B1-3 adopted in 2000. During the years of discussion and after the adoption of the IAU resolutions, Brumberg also wrote or contributed to several papers explaining the issues of the new framework to a wider scientific audience [15, 76].

The chapter reviews why and how this relativistic formalism is applied in time and frequency comparisons. It follows and expands upon a previous review by two of the present authors [65]. In Section 2, we briefly present the present atomic time scales and the performances of atomic clocks. In Section 3, we recall the relativistic framework that resulted from the historical developments summarized above. In Section 4, we present how the relativistic framework is applied for time and frequency comparisons in the geocentric celestial reference system and show that the current relativistic modeling is adequate for the current and most upcoming needs of clocks and time transfer techniques, covering one-way and two-way transmissions. In Section 5, we review the main time transfer techniques, with some emphasis on two new techniques: the Time Transfer by Laser Link (T2L2) currently flying on board Jason-2 and the microwave link developed for the coming ACES mission on board the ISS. Section 6 reviews the implications of clocks in relativistic geodesy, notably the use of frequency standards to measure the gravitational potential. Finally, Section 7 summarizes the implications of the latest and upcoming developments in clock and time transfer technology.

2 Atomic time and atomic clocks

Since the first atomic clock was operated in 1955, these have gained about one order of magnitude in accuracy and stability every decade. Man-made devices therefore have taken over natural phenomena to provide the definition of time, probably forever. Because any individual device is necessarily doomed to stop or fail for some reason, atomic time scales have been developed combining several atomic clocks to provide reliability and robustness and enhance stability and accuracy. With some additional procedures for time dissemination, atomic time scales indeed provide everywhere "an ordered set of markers with an associated numbering," fullfiling the definition of the International Telecommunications Union.

Anticipating on the developments in Section 3, we can think in relativistic terms: the atomic clock, providing time to its immediate vicinity, is realizing a proper time and the atomic time scale, providing a general and unambiguous way of timing events, is realizing a coordinate time. In the following, we briefly review the present performances of atomic time scales and atomic clocks.

2.1 Atomic time scales

International Atomic Time TAI was defined in 1970 by the International Committee for Weights and Measures as "the time reference established by the BIH on the basis of the readings of atomic clocks operating in various establishments in accordance with the definition of the second." In 1980, the definition of TAI was completed by the Consultative Committee for the Definition of the Second, adding "TAI is a coordinate time scale defined in a geocentric reference frame with the SI second as realized on the rotating geoid as the scale unit." This definition explicitly refers to TAI as a coordinate time, recognizing the need of a relativistic approach (see Section 3.4). In 1988, the responsibility of establishing TAI was transferred to the International Bureau for Weights and Measures (BIPM) in Sèvres (France).

The process of calculation of TAI [3] is monthly and uses data provided by about 70 laboratories worldwide. Timing data from about 400 clocks are combined using an algorithm named ALGOS, resulting in an ensemble time scale named EAL. The stability of EAL is, by its construction, optimized for an averaging time around 1 month, and this was estimated in 2013 to be about 0.3 to 0.4×10^{-15}. Since EAL is free-running, its rate may differ from the goal set in the definition of TAI (see Section 3.4), so that a steering procedure is needed to generate TAI from EAL. In this procedure, the rate of EAL is measured by comparison with a number of primary frequency standards which aim at realizing the SI second and TAI is then derived from EAL by applying a rate correction so that the scale unit of TAI agrees with its definition. Because it has been observed for many years that EAL was affected by a systematic drift, a change has been brought to the ALGOS algorithm in 2011 to implement a quadratic prediction for the clocks frequency. Since that change, the rate of EAL with respect to the SI second has been essentially constant and the relative rate difference between EAL and TAI remains presently (2013) fixed at 6.483×10^{-13}. It may nevertheless change slightly over time, as needed to ensure the accuracy of TAI. Because this steering procedure is realized in "real time" every month and because the steps of the steering function are chosen to be small in order to not degrade the frequency stability of TAI, the scale unit of TAI does not always exactly agree with its definition. The relative difference, denoted as d in the BIPM publications[2], has been below 1×10^{-15} since mid-2012. It is however known to within $\pm 0.3 \times 10^{-15}$, an uncertainty provided by around 10 primary standards that contribute occasionally or regularly to the accuracy of TAI.

2 see, e.g. Table 7 in www.bipm.org/en/scientific/tai/time_ar2012.html

Because of the "real-time" nature of its computation and other operational constraints (e.g. no correction for a mistake discovered many days after the publication), TAI is not optimal. The BIPM therefore computes in post-processing another realization, TT(BIPMxx), where 20xx is the year of computation, which is based on EAL and on a weighted average of the primary frequency standards data [64]. Since 2003 yearly versions have been computed, the latest of which is TT(BIPM12)[3], with an estimated accuracy at 2 to 3×10^{-16} over the recent years.

Progresses in atomic time scales directly stem from progresses in atomic clocks and these are briefly summarized in the next subsection. Concurrently, improvements in clocks require improvements in the time transfer techniques used to compare them, which will be developed in Section 5.

2.2 Atomic clocks

Atomic fountain frequency standards based on laser-cooled cesium atoms have been first developed in the 1990s [23] and several units have been operated in a few metrology institutes since the 2000s. Since 2010, some 10 fountains have been more or less regularly operated in about seven laboratories, with the best uncertainty budget for all systematic effects at about 2×10^{-16} [33, 54]. Atomic fountains have improved the realization of the SI second by more than 1 order of magnitude compared to clocks using thermal atomic beams and resulted in TAI and TT(BIPM) now being accurate to the level of 2 to 3×10^{-16}.

Some of these Cs fountains, or similar devices based on the 6.8 GHz hyperfine transition in ^{87}Rb, are now operated nearly continuously so that they can be used as clocks to directly realize an accurate reference time scale [4] and some of them [61] are also operated as clocks in the EAL ensemble. Other projects for ultrastable clocks which can operate for long periods have been documented such as NASA/JPL goal to develop a frequency standard based on the hyperfine transition at 40.5 GHz in $^{199}Hg^+$ ions [18] or 29.95 GHz in $^{201}Hg^+$ ions [19] trapped in a linear ion trap. For the $^{199}Hg^+$ transition, a stability floor of less than 2×10^{-16} and a drift lower than 2.7×10^{-17}/day have been reported.

Atomic clocks with a reference transition in the optical frequency range are under development in many metrology institutes and research centers worldwide. Two different technologies have been developed. First, *ion optical clocks* [22, 38, 56, 69] based on a single trapped ion in an RF electric field, have been reported to reach record accuracies[4], as low as 9×10^{-18}. However, the stability of these clocks is limited since only a single ion is trapped. Second, a more recent technology is *optical lattice*

3 See ftp://62.161.69.5/pub/tai/scale/

4 In principle, the term accuracy is reserved to the transition which defines the unit. By extension we consider under "accuracy" the uncertainty budget for all systematic effects.

clocks [10, 36, 50, 53, 57, 60, 80]. They involve trapping of a few thousands neutral atoms in a powerful laser standing wave (or optical lattice) by the dipolar force. They have already reached an accuracy of 1×10^{-16} and are rapidly catching up with ion clocks with a recent report [10] claiming 6×10^{-18}. Furthermore, the large number of interrogated atoms allowed the demonstration of unprecedented stabilities (a few 1×10^{-16} at 1 s).

The ability to compare optical frequencies across the optical spectrum with the advent of frequency comb generators [82, 87] has been an essential step in the development and characterization of optical clocks. With progresses in frequency comparison methods using optical fibers, such clocks can now be compared at distance with an uncertainty approaching 1×10^{-19} [55, 68]. It is therefore likely that an optical transition will provide a new definition of the second in the not so distant future.

Among the numerous developments for new clocks, some specifically aim at space-qualified devices. In this respect, the most advanced and most important space project in the time and frequency domain is the Atomic Clock Ensemble in Space (ACES) project [71] whose launch date is currently estimated as 2016, to be flown on the International Space Station, attached to the Columbus module. It comprises a cold-atom clock (PHARAO) with a specified frequency accuracy of 1×10^{-16}, a space active hydrogen maser and two time transfer packages (see Section 5.2). ACES may be viewed as a technology demonstration project for future space missions proposing to fly optical clocks in space. Indeed, it is likely that in the future reference clocks need to be placed in space as clocks on Earth will be affected by uncertainties in determining the gravity potential (see Section 6).

3 Relativistic framework for time and frequency comparisons

3.1 Introduction

In relativity it turns out to be useful to distinguish locally measurable quantities from coordinates which are, by definition, dependent on conventions. One generally speaks of proper and coordinate quantities:

(i) Proper quantities are the direct results of observation without involving any information that is dependent on conventions (such as, for example, the choice of a spacetime reference frame or a convention of synchronization). In metrology, such quantities are proper time and proper length measured by a particular observer (proper to that observer). This also includes quantities that are not the real, physical results of observations, but correspond, in principle, to proper quantities (e.g. the proper time of an observer placed at the geocenter or infinitely far from the Solar System).

(ii) Coordinate quantities are dependent on conventional choices, e.g. of a spacetime coordinate system, a convention of synchronization, etc. Examples are the coordinate time difference between two events (the difference between the time coordinates of these events) or the rate of a clock with respect to the coordinate time of some spacetime reference system, which are both dependent on the chosen reference system.

For time metrology the most fundamental such quantities are the proper time of a clock (the physical, local output of an ideal clock) and the coordinate time of a conventional spacetime reference system (as defined, e.g. by the IAU, see Section 3.3). For example, a Cs primary frequency standard produces a realization of the proper time at its location, but TAI is a realization of terrestrial time (TT), a time coordinate defined by the IAU (see Section 3.4).

Due to the curvature of spacetime, the scale units of spacetime coordinate systems have, in general, no globally constant relation to proper quantities. In the framework of Newtonian mechanics (using Euclidean geometry) it is always possible to define coordinates in such a way that their scale units are equal to proper quantities everywhere, and it is therefore not necessary to explicitly distinguish between them. This is impossible in general relativity where the relation between proper quantities and coordinate scale units is dependent on the position in spacetime of the observer. For metrology this implies, for example, that the relationship between a coordinate time interval and a measured proper time interval is dependent on the position and velocity of the measuring clock.

3.2 Simultaneity and synchronization

In the context of relativity theory, there is no a priori definition of simultaneity of two distant events and therefore synchronization or, more generally, comparison of distant clocks is arbitrary. It becomes subject to a conventional choice, called a convention of simultaneity and synchronization. The first such convention was proposed by Einstein in his initial 1905 paper [29], and uses the exchange of electromagnetic signals, the so-called Einstein synchronization convention. Other conventions are, for example, "slow clock transport synchronization" or "coordinate synchronization." The latter is the natural choice when the clock comparisons are to serve the purpose of realizing coordinate time scales (e.g. the construction of TAI). It is defined as (see e.g. Klioner, 1992 [42]):

> Two events fixed in some reference system by the values of their coordinates (t_1, x_1, y_1, z_1) and (t_2, x_2, y_2, z_2) are considered to be simultaneous with respect to this reference system, if the values of time coordinate corresponding to them are equal: $t_1 = t_2$. This definition of simultaneity (and corresponding definition of synchronization) we shall call coordinate simultaneity (and coordinate synchronization).

Clearly, synchronization by this definition is entirely dependent on the chosen spacetime coordinate system, which needs to be defined consistently in a relativistic framework. Such definitions were provided by the IAU in its 1991 and 2000 resolutions (see footnote 1), the relevant parts of which are summarized in the next sections.

3.3 Relativistic definitions of spacetime coordinate systems

In order to describe the time and frequency comparison observations, one has to first choose the proper relativistic reference systems best suited to the problem at hand. The barycentric celestial reference system (BCRS) should be used for all experiments in the Solar System and not confined to the vicinity of the Earth, while the geocentric celestial reference system (GCRS) is physically adequate to describe processes occurring in the vicinity of the Earth. These systems were first defined by the IAU Resolution A4 (1991) which contains nine recommendations, the first four of which are relevant to our present discussion.

In the first recommendation, the metric tensor for spacetime coordinate systems $(t, \mathbf{x})$ centered at the barycenter of an ensemble of masses is recommended in the form

$$\begin{aligned} g_{00} &= -1 + \frac{2U(t,\mathbf{x})}{c^2} + \mathcal{O}(c^{-4}) \,, \\ g_{0i} &= \mathcal{O}(c^{-3}) \,, \\ g_{ij} &= \delta_{ij}\left(1 + \frac{2U(t,\mathbf{x})}{c^2}\right) + \mathcal{O}(c^{-4}) \,, \end{aligned} \tag{3.1}$$

where c is the speed of light in vacuum (c = 299 792 458 m/s) and U is the Newtonian gravitational potential (here a sum of the gravitational potentials of the ensemble of masses, and of a external potential generated by bodies external to the ensemble, the latter potential vanishing at the origin). The recommended form of the metric tensor can be used not only to describe the barycentric reference system of the whole Solar System, but also to define the geocentric reference system centered in the center of mass of the Earth, with U now depending upon geocentric coordinates.

In the second recommendation, the origin and orientation of the spatial coordinate grids for the barycentric and geocentric reference systems are defined.

The third recommendation defines TCB (Barycentric Coordinate Time) and TCG (geocentric coordinate time) as the time coordinates of the BCRS and GCRS, respectively, and, in the fourth recommendation, another time coordinate named TT (terrestrial time), is defined for the GCRS.

In the following years, it became obvious that this set of recommendations was not sufficient, especially with respect to planned astrometric missions with μas-accuracies (such as GAIA, now expected to be launched at the end of 2013) and with respect to the expected improvement of atomic clocks and the planned space missions involving such clocks and improved time transfer techniques (such as ACES, now expected to be launched in 2016, see Section 5.2). For that reason the IAU WG 'Relativity for astrometry and celestial mechanics' together with the BIPM-IAU Joint Committee for relativity suggested an extended set of Resolutions that was finally adopted at the IAU General Assembly in Manchester in the year 2000 as Resolutions B1.3 to B1.5 and B1.9.

Resolution B1.3 concerns the definition of the barycentric and geocentric celestial reference systems. The Resolution recommends to write the metric tensor of the BCRS in the form

$$\begin{aligned} g_{00} &= -1 + \frac{2w}{c^2} - \frac{2w^2}{c^4} + \mathcal{O}(c^{-5}) \,, \\ g_{0i} &= -\frac{4}{c^3} w^i + \mathcal{O}(c^{-5}) \,, \\ g_{ij} &= \delta_{ij}\left(1 + \frac{2w}{c^2}\right) + \mathcal{O}(c^{-4}) \,, \end{aligned} \tag{3.2}$$

where w is a scalar potential and w^i a vector potential. This extends the form of the metric tensor given in the IAU'1991 Resolutions, so that its accuracy would be sufficient for all time and frequency applications foreseen in 2000. For the GCRS, Resolution B1.3 also adds that the spatial coordinates are kinematically nonrotating with respect to the barycentric ones.

Resolution B1.4 provides the form of the expansion of the post-Newtonian potential of the Earth to be used with the metric of Resolution B1.3.

Resolution B1.5 applies the formalism of Resolutions B1.3 and B1.4 to the problems of time transformations and realization of coordinate times in the Solar System. Resolution B1.5 is based upon a mass-monopole spin-dipole model. It provides an uncertainty not larger than 5×10^{-18} in rate and, for quasi-periodic terms, not larger than 5×10^{-18} in rate amplitude and 0.2 ps in phase amplitude, for locations farther than a few Solar radii from the Sun. The same uncertainty also applies to the transformation between TCB and TCG for locations within 50 000 km of the Earth (see Section 3.5).

Resolution B1.9 concerns the definition of TT and will be developed in the next section.

3.4 Time scales in the barycentric and geocentric systems

As mentioned above, TCB and TCG are the time coordinates of the BCRS and GCRS, respectively. The IAU'1991 Recommendation 3 defined the scale unit of TCB and TCG to be consistent with the SI second. This means that if readings of proper time of an observer, expressed in SI seconds, are recomputed into TCB or TCG using the formulas from the IAU Resolutions, without any additional scaling, one gets corresponding values of TCB or TCG in the intended units. It also defines the origin of TCB and TCG by the following relation to TAI:
TCB (resp. TCG) = TAI + 32.184 s on January 1, 1977, 0 h TAI, at the geocenter.

TT is another coordinate time for the geocentric system, differing from TCG by a constant rate:

$$d(TT)/d(TCG) = 1 - L_G\,. \tag{3.3}$$

In the original definition (IAU'1991 Recommendation 4), this rate was chosen such as the scale unit of TT be consistent with the SI second on the rotating geoid, i.e. $L_G = U_G/c^2$, where U_G is the gravity (gravitational + rotational) potential on the geoid. Some shortcomings appeared in this definition of TT when considering accuracies below 10^{-17} because of uncertainties in the realization of the geoid at the level required. Therefore it was decided to dissociate the definition of TT from the geoid while maintaining continuity with the previous definition. Resolution B1.9 (2000) turned L_G into a defining constant with its value fixed at $6.969290134 \times 10^{-10}$. The origin of TT is defined, so that TCB and TCG coincide with TT in origin, by
TT = TAI + 32.184 s on January 1, 1977, 0 h TAI.

TT is a theoretical time scale and can have different realizations that are differentiated by the notation TT(realization). TT(BIPM) (see Section 2.1) is one of them, but TAI can also provide one as: TT(TAI) = TAI + 32.184 s.

Similarly, TDB is another coordinate time for the barycentric system, differing from TCB by a constant rate. In order to solve a long-standing ambiguity in the 1976 definition of TDB, Resolution B3 (2006) defined

$$TDB = TCB - L_B \times (JD_{TCB} - T_0) \times 86\,400 + TDB_0\,, \tag{3.4}$$

where JD_{TCB} is the TCB Julian date, $T_0 = 2\,443\,144.5003725$, $L_B = 1.550519768 \times 10^{-8}$, and $TDB_0 = -6.55 \times 10^{-5} s$ are defining constants.

3.5 Relativistic theory for time transformations in the Solar System (BCRS)

Following Resolutions B1.3 (2000) and B1.4 (2000), the metric tensor in the BCRS (equation (3.2)) is expressed more precisely as

$$\begin{aligned} g_{00} &= -\left(1 - \frac{2}{c^2}\left(w_0(t,\mathbf{x}) + w_L(t,\mathbf{x})\right) + \frac{2}{c^4}\left(w_0^2(t,\mathbf{x}) + \Delta(t,\mathbf{x})\right)\right) \\ g_{0i} &= -\frac{4}{c^3} w^i(t,\mathbf{x}) \\ g_{ij} &= \left(1 + \frac{2w_0(t,\mathbf{x})}{c^2}\right)\delta_{ij}\,, \end{aligned} \tag{3.5}$$

where ($t \equiv TCB, \mathbf{x}$) are the barycentric coordinates, $w_0 = G\sum_A M_A/r_A$, with the summation carried out over all Solar System bodies A, $\mathbf{r}_A = \mathbf{x} - \mathbf{x}_A$, $r_A = |\mathbf{r}_A|$, and where w_L contains the expansion in terms of multipole moments, as defined in Resolution B1.4 (2000) and references therein, required for each body. In many cases the mass-monopole approximation ($w_L = 0$) may be sufficient to reach the above mentioned uncertainties but this term should be kept to ensure the consistency in all cases. The values of masses and multipole moments to be used may be found in the IAU 2009 system of astronomical constants[5], but care must be taken that the values are in SI units (not in so-called TDB units or TT units). The vector potential $w^i(t,\mathbf{x}) = \sum_A w^i_A(t,\mathbf{x})$ and the function $\Delta(t,\mathbf{x}) = \sum_A \Delta_A(t,\mathbf{x})$ can be computed from the expressions given in the IAU Resolution B1.5 (2000).

From (3.5) the transformation between proper time and *TCB* may be derived. It reads

$$\frac{d\tau}{dTCB} = 1 - \frac{1}{c^2}\left(w_0 + w_L + \frac{v^2}{2}\right) + \frac{1}{c^4}\left(-\frac{1}{8}v^4 - \frac{3}{2}v^2 w_0 + 4v^i w^i + \frac{1}{2}w_0^2 + \Delta\right), \quad (3.6)$$

where v is the barycentric velocity.

Evaluating the Δ_A-terms for all bodies of the Solar System, we find that $|\Delta_A(t,\mathbf{x})|/c^4$ may reach at most a few parts in 10^{17} in the vicinity of Jupiter and about 1×10^{-17} close to the Earth. Presently, however, for all planets except the Earth, the magnitude of $\Delta_A(t,\mathbf{x})/c^4$ in the vicinity of the planet is smaller than the uncertainty originating from its mass or multipole moments so that it is practically not needed to account for these terms. Nevertheless, when new astrometric observations allow us to derive the mass and moments with adequate uncertainty, it will be necessary to do so. In any case, for the vicinity of a given body A, only the effect of $\Delta_A(t,\mathbf{x})$ is needed in practice for our accuracy specifications. For a clock in the vicinity of the Earth, to be compared with other clocks in the Solar System or to *TCB*, it may thus be needed to account for $\Delta_E(t,\mathbf{x})/c^4$.

Similarly, the transformation between *TCB* and *TCG* may be written as

$$\begin{aligned} TCB = TCG & \\ & + c^{-2}\left[\int_{t_0}^{t}\left(\frac{v_E^2}{2} + w_{0ext}(\mathbf{x}_E)\right)dt + v_E^i r_E^i\right] \\ & - c^{-4}\left[\int_{t_0}^{t}\left(-\frac{1}{8}v_E^4 - \frac{3}{2}v_E^2 w_{0ext}(\mathbf{x}_E)\right.\right. \\ & \left. + 4v_E^i w_{ext}^i(\mathbf{x}_E) + \frac{1}{2}w_{0ext}^2(\mathbf{x}_E)\right) dt \\ & \left. - (3w_{0ext}(\mathbf{x}_E) + v_E^2/2)v_E^i r_E^i\right], \end{aligned} \quad (3.7)$$

where t is *TCB* and the index E refers to the geocenter and "ext" refers to all bodies except the Earth. This equation is composed of terms evaluated at the geocenter

5 see http://maia.usno.navy.mil/NSFA/IAU2009_consts.html and references therein

(the two integrals) and of position-dependent terms in r_E, with position-dependent terms in higher powers of r_E having been found to be negligible. Terms in the second integral of (3.7) are secular and quasiperiodic. They amount to $\sim 1.1 \times 10^{-16}$ in rate ($dTCB/dTCG$) and primarily a yearly term of $\sim$ 30 ps in amplitude (i.e. corresponding to periodic rate variations of amplitude $\sim 6 \times 10^{-18}$). Terms in $\Delta_A(t, \mathbf{x})$ that would appear in this integral are negligible in the vicinity of the Earth where this formula is to be used. Besides the position-dependent terms in c^{-2} which may reach several microseconds, position-dependent terms in c^{-4} (the last two terms in (3.7)) are not negligible and reach, for example, an amplitude of 0.4 ps ($\sim 3 \times 10^{-17}$ in rate) in geostationary orbit. A conventional routine (HF2002_IERS.F) to compute the transformation between TCB and TCG at the geocenter, based on the work by Harada and Fukushima [34], is provided with the IERS Conventions (2010) [39] and may be found at http://tai.bipm.org/iers/conv2010/conv2010_c10.html.

4 Relativistic treatment for time and frequency comparisons in the vicinity of the Earth (GCRS)

We present here formulas or references that allow us to perform time and frequency comparisons (synchronization and syntonization) in the vicinity of the Earth, i.e. typically up to geosynchronous orbit or slightly above.

There are in principle two methods to compare distant clocks: one method is to transport a portable clock from one clock to the other, properly accounting for the coordinate time accumulated during the transport (coordinate time differs from the proper time measured by the portable clock, see Section 3.1). The clock transport technique is rather cumbersome but has been much used until the development of the GPS in the 1980s. The accuracy is limited by the instability of the transported clock, presently at a level of the order 1 ns [24]. Eventually, when a clock with a frequency stability of the order 10^{-17} can be made easily transportable, an uncertainty of about a few ps may be reached by clock transport which would make the technique again competitive. Nevertheless, we shall not cover the relativistic treatment of the clock transportation technique in more details. Interested readers can refer to, e.g. [65].

The second method, which is used in all present standard techniques for clock comparison, is to send an electromagnetic signal from one clock to the other, properly accounting for the coordinate time of propagation of the signal.

In this context, we will refer to "time comparison" as a procedure that for two distant clocks A and B determines the quantity $\tau_B(t) - \tau_A(t)$ (called desynchronization) in some well defined spacetime coordinate system with coordinate time t. This is in accordance with the convention of coordinate simultaneity (cf. Section 3.2) and will be used here in the context of clocks close to the Earth, i.e. within GCRS so that t=TCG. By a slight extension, we will also call time comparison a method that allows determining the desynchronization up to an unknown constant, that constant being unchanged from one experiment to the next, e.g. for different visibility passes of a satellite clock A above a ground clock B. In practice, an electromagnetic signal is sent from A at coordinate time t_A and received at B at coordinate time t_B with $T_{AB} = t_B - t_A$ being the coordinate time of propagation. The desynchronization is computed, e.g. at time t_B, as

$$\tau_B(t_B) - \tau_A(t_B) = \tau_B(t_B) - \tau_A(t_A) - [T_{AB}]^A , \tag{4.1}$$

where $[\cdot]^A$ represents the transformation from coordinate time to proper time at A. Therefore the procedure requires the computation of the coordinate time of propagation of the signal, which is presented in Sections 4.1 for one-way and 4.2 for two-way transmission, with the goal of 1 ps accuracy.

We will refer to "frequency comparison" as a procedure that allows determining only the time derivative of the desynchronization, i.e. the quantity $d(\tau_B(t) - \tau_A(t))/dt$ either instantaneously or averaged over some interval Δt. This quantity will be called "coordinate desyntonization" and can be determined by repeated time comparisons following equation (4.1). The "frequency comparison" procedure may also be expressed as the determination of the difference in the proper frequencies of the two clocks, as developed, e.g. by Blanchet et al. [9]:

> The frequency transfer between two clocks requires the determination of the ratio f_A/f_B between the proper frequencies f_A and f_B delivered by the clocks on the satellite (A) and on the ground (B). In practice, this is achieved using a transmission of photons from A to B and the formula
>
> $$\frac{f_A}{f_B} = \frac{f_A}{\nu_A}\frac{\nu_A}{\nu_B}\frac{\nu_B}{f_B} , \tag{4.2}$$
>
> where ν_A is the proper frequency of the photon as measured on A (instant of emission t_A), and ν_B the proper frequency of the same photon on B at t_B.

This can be interpreted by stating that, in order to coordinate syntonize two clocks, one has to make the proper frequencies of the two clocks "compensate" for the quantity ν_A/ν_B. The procedure therefore requires the computation of ν_A/ν_B, which is presented in Section 4.3.

The metric of the GCRS is provided in the IAU Resolution B1.3 (2000) as

$$\begin{aligned} G_{00} &= -1 + \frac{2W}{c^2} - \frac{2W^2}{c^4} + \mathcal{O}(c^{-5}) , \\ G_{0i} &= -\frac{4}{c^3}W^i + \mathcal{O}(c^{-5}) , \\ G_{ij} &= \delta_{ij}\left(1 + \frac{2W}{c^2}\right) + \mathcal{O}(c^{-4}) , \end{aligned} \tag{4.3}$$

where W is a scalar potential and W^i a vector potential. Evaluating the contributions of the higher order terms in the metric, it is found that the IAU'1991 framework with the metric of the form (3.1) is sufficient for time and frequency applications in the GCRS in the light of present and foreseeable future clock accuracies, i.e. at an accuracy level of 1×10^{-18} or slightly below. In this, the metric interval reads

$$ds^2 = -c^2 d\tau^2 = -(1 - 2U/c^2)c^2 dt^2 + (1 + 2U/c^2)\delta_{ij} dx_i dx_j \tag{4.4}$$

where U is the total Newtonian potential that can be split into the potential from the Earth itself and the potential from external bodies.

Note that, in applying the IAU'1991 formalism, some care needs to be taken when evaluating the Earth's potential at the location of the clock especially when an accuracy of the order 10^{-18} is required, see more developments in [42, 63, 86]. See also the discussion in Section 6.

In this framework, the proper time of a clock A located at the GCRS coordinate position $\mathbf{x}_A(t)$, and moving with the coordinate velocity $\mathbf{v}_A = d\mathbf{x}_A/dt$, is

$$\frac{d\tau_A}{dt} = 1 - \frac{1}{c^2}\left[\frac{\mathbf{v}_A^2}{2} + U_{\mathrm{E}}(\mathbf{x}_A) + V(\mathcal{X}_A) - V(\mathcal{X}_{\mathrm{E}}) - x_A^i \partial_i V(\mathcal{X}_{\mathrm{E}})\right] . \tag{4.5}$$

Here, U_{E} denotes the Newtonian potential of the Earth at the position $\mathbf{x}_A$ of the clock in the GCRS frame, and V is the sum of the Newtonian potentials of the other bodies (mainly the Sun and the Moon) computed at a location $\mathcal{X}$ in barycentric coordinates, either at the position $\mathcal{X}_{\mathrm{E}}$ of the Earth's center of mass, or at the clock location $\mathcal{X}_{\mathrm{A}}$. Only terms required for frequency transfer with uncertainty of the order 10^{-18} have been kept. For application to any experiment at a level of uncertainty greater than 5×10^{-17} on the Earth or on board a satellite in Low-Earth orbit, one can keep only the first three terms in relation (4.5) between the proper time τ_A and the coordinate time t:

$$\frac{d\tau_A}{dt} = 1 - \frac{1}{c^2}\left[\frac{\mathbf{v}_A^2}{2} + U_{\mathrm{E}}(\mathbf{x}_A)\right] . \tag{4.6}$$

The relativistic treatment of time and frequency transfer (clock synchronization and syntonization) results in the integration of a differential relation, according to the problem at hand. For example, to compute the coordinate time of propagation of a light signal, one has to first solve $ds^2 = 0$ in equation (4.4) and integrate the found solution, as done in the following subsections.

4.1 One-way time transfer

This section is based on Blanchet et al. [9], Section 3.1, see also Petit and Wolf [62].

Let A be the emitting station, with GCRS position $\mathbf{x}_A(t)$, and B the receiving station, with position $\mathbf{x}_B(t)$. We use $t = TCG$ and the calculated coordinate time intervals are in TCG. The corresponding time intervals in TT are obtained by multiplying with $(1 - L_G)$. We denote by t_A the coordinate time at the instant of emission of a light signal, and by t_B the coordinate time at the instant of reception. We put $r_A = |\mathbf{x}_A(t_A)|$,

$r_B = |\mathbf{x}_B(t_B)|$, and $\mathrm{R}_{AB} = |\mathbf{x}_B(t_B) - \mathbf{x}_A(t_A)|$. Up to the order $1/c^3$ the coordinate time of signal propagation $T_{AB} \equiv t_B - t_A$ is given by

$$T_{AB} = \frac{\mathrm{R}_{AB}}{c} + \frac{2GM_\mathrm{E}}{c^3} \ln\left(\frac{r_A + r_B + \mathrm{R}_{AB}}{r_A + r_B - \mathrm{R}_{AB}}\right), \tag{4.7}$$

where GM_E is the geocentric gravitational constant and where the logarithmic term represents the Shapiro time delay. Between a low-Earth orbit (LEO) satellite and the ground, the Shapiro time delay is a few picoseconds. For a GPS or geostationnary satellite, it is a few tens of picoseconds.

In a real experiment, the position of the receptor B is known at the time of emission t_A rather than at the time of reception t_B, i.e. we have access to $\mathbf{x}_B(t_A)$ rather than $\mathbf{x}_B(t_B)$, and formula (4.7) gets modified by Sagnac correction terms consistently to the order $1/c^3$. In this case, the formula becomes

$$\begin{aligned} T_{AB} &= \frac{\mathrm{D}_{AB}}{c} + \frac{\mathbf{D}_{AB}.\mathbf{v}_B}{c^2} \\ &+ \frac{\mathrm{D}_{AB}}{2c^3}\left(\frac{\mathbf{v}_B^2 + (\mathbf{D}_{AB}.\mathbf{v}_B)^2}{\mathrm{D}_{AB}^2 + \mathbf{D}_{AB}.\mathbf{a}_B}\right) \\ &+ \frac{2GM_\mathrm{E}}{c^3} \ln\left(\frac{r_A + r_B + \mathrm{D}_{AB}}{r_A + r_B - \mathrm{D}_{AB}}\right), \end{aligned} \tag{4.8}$$

where $\mathbf{D}_{AB} = \mathbf{x}_B(t_A) - \mathbf{x}_A(t_A)$ and $\mathrm{D}_{AB} = |\mathbf{D}_{AB}|$ and where $\mathbf{v}_B$ and $\mathbf{a}_B$ denote, respectively, the coordinate velocity and acceleration of the station B at t_A. The second term in equation (4.8) represents the Sagnac term of the order $1/c^2$ and can amount to up to 200 ns for LEOs and 133 ns for GPS. The third term, or the Sagnac term of the order $1/c^3$, is a few ps (up to 10 ps for a geostationary satellite). The fourth term is the Shapiro delay, discussed earlier.

4.2 Two-way time transfer using artificial satellites

We distinguish two types of two-way time transfer. In the first one, the signal transmission is between the two clocks A and B which are to be compared, with A on board an artificial satellite. This case is described, e.g. in ([9], Section 3.2). One signal is emitted from A at instant t_A and received at B at instant t_B, and the other signal is emitted from B at instant t'_B and received at A at instant t'_A. The intervals of time between emission and reception at A and B, $t_{AA'} = t'_A - t_A$ and $t_{B'B} = t_B - t'_B$ are measured. The quantity required for synchronization is $\Delta t = t_A - t'_B$ and derives from the application of the one-way time transfer formulas (preceding section) to the two signal transmissions involved.

This two-way time transfer technique may be performed for two different stations B and C, thereby allowing time transfer between B and C if clock A on board the satellite measures in addition the time interval between its transmission to B and to C. This type of time transfer is applied in the particular case of the T2L2 technique (see Section 5) or of the former LASSO method. In addition, the laser pulse is instantaneously retroreflected so that $t_A = t'_A$. The relativistic treatment of the LASSO clock synchronization is described in details, e.g. in ([62], Section 4).

In the second type, two clocks A and B (usually on the ground) are compared via an artificial satellite S. Two signals are transmitted from A and from B to S, where each one is immediately retransmitted towards the other station. The clocks at A and B measure the time interval between emission and reception. This is the two-way time transfer (TWTT) technique routinely used by time laboratories for clock comparisons. The relativistic treatment of TWTT is described, e.g. in ([62], Section 3) for the case of S being a geostationary satellite and by Klioner and Fukushima [44] for an arbitrary satellite. The quantity required for synchronization derives from the application of the one-way time transfer formulas (preceding section) to the four signal transmissions involved.

4.3 Frequency comparisons

The frequency comparison between clocks located at points A and B is achieved using a transmission of photons from A to B and is represented by equation (4.2). It requires the computation of the value of ν_A/ν_B, i.e. the ratio of the proper frequency of the photon at emission to the proper frequency of the photon at reception. Before we treat this, let us try to understand in simple terms what is the frequency shift of the electromagnetic signal. Indeed, it can be seen as a direct consequence of the Einstein equivalence principle (EEP), one of the pillars of modern physics [58, 85].

The Einstein equivalence principle

Let us consider a photon emitted at a point A in an accelerated reference system, toward a point B which lies in the direction of the acceleration (see Figure 1). We assume that both points are separated by a distance h_0, as measured in the accelerated frame. The photon time of flight is $\delta t = h_0/c$, and the frame velocity during this time increases by $\delta v = a\delta t = ah_0/c$, where a is the magnitude of the frame acceleration $\mathbf{a}$. The frequency at point B (reception) is then shifted because of the Doppler effect, compared to the frequency at point A (emission), by an amount

$$\frac{\nu_B}{\nu_A} = 1 - \frac{\delta v}{c} = 1 - \frac{ah_0}{c^2} . \tag{4.9}$$

Now, the EEP postulates that a gravitational field $\mathbf{g}$ is locally equivalent to an acceleration field $\mathbf{a} = -\mathbf{g}$. We deduce that in a nonaccelerated (locally inertial) frame in

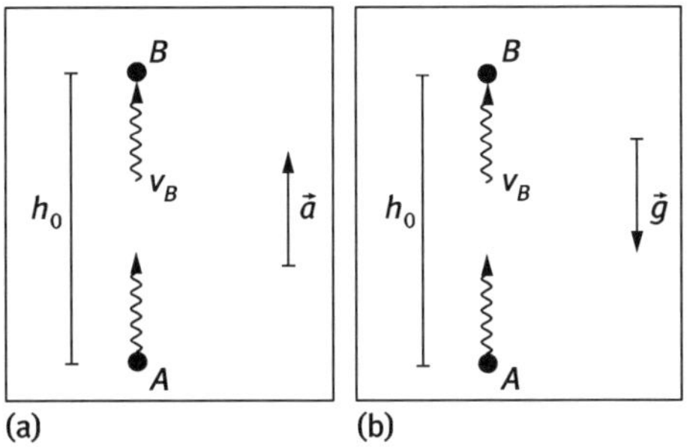

Fig. 1. A photon of frequency ν_A is emitted at point A toward point B, where the measured frequency is ν_B. (a) A and B are two points at rest in an accelerated frame, with acceleration $\vec{a}$ in the same direction as the emitted photon. (b) A and B are at rest in a nonaccelerated (locally inertial) frame in the presence of a gravitational field such that $\vec{g} = -\vec{a}$.

the presence of a gravitational field **g**:

$$\frac{\nu_B}{\nu_A} = 1 - \frac{gh_0}{c^2}, \tag{4.10}$$

where $g = |\mathbf{g}|$, ν_A is the photon frequency at emission (strong gravitational potential) and ν_B is the photon frequency at reception (weak gravitational potential). As $\nu_B < \nu_A$, it is usual to say that the frequency at the point of reception is "red-shifted." One can consider it in terms of conservation of energy. Intuitively, the photon that goes from A to B has to "work" to be able to escape the gravitational field; then it loses energy and its frequency decreases by virtue of $E = h\nu$, with h the Planck constant.

If two ideal clocks are placed in A and B and the clock at A (strong gravitational potential) is used to generate the signal ν_A, then the signal received at B (weak gravitational potential) has a lower frequency than a signal locally generated by the clock at B.

Frequency comparison in the GCRS

An electromagnetic signal locked on the clock at A is emitted with proper frequency ν_A and received at B where the received frequency ν_B is measured with respect to the proper frequency of the clock f_B. The notations are the same as those in Section 4.1 and we further note $\mathbf{N}_{AB} = (\mathbf{x}_B(t_B) - \mathbf{x}_A(t_A))/\mathrm{R}_{AB}$, the unitary vector in the direction of propagation.

The quantity $\nu_{A/B}$ is the proper frequency of the same signal at A and B so that $\nu_A/\nu_B = d\tau_B/d\tau_A$ where $d\tau_{A/B}$ is the proper period of the signal at A/B (see Figure 2).

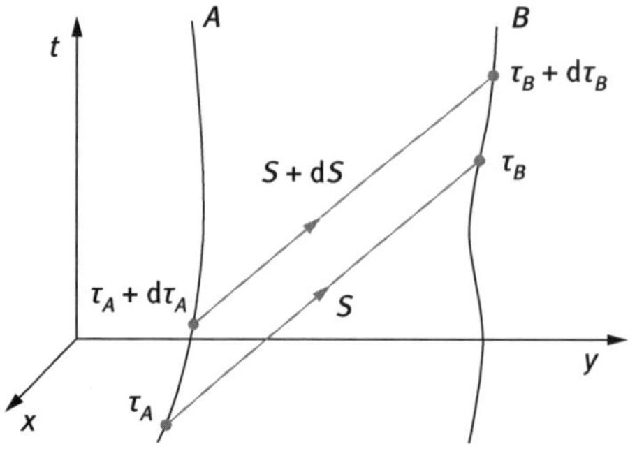

Fig. 2. Two clocks A and B are measuring proper time along their trajectory. One signal with phase S is emitted by A at proper time τ_A, and another one with phase $S + dS$ at time $\tau_A + d\tau_A$. They are received by clock B, respectively, at times τ_B and $\tau + d\tau_B$.

We have

$$\frac{\nu_A}{\nu_B} = \frac{d\tau_B}{d\tau_A} = \left(\frac{dt}{d\tau}\right)_A \frac{dt_B}{dt_A} \left(\frac{d\tau}{dt}\right)_B , \tag{4.11}$$

where $(d\tau/dt)_{A/B}$ are given by (4.5) or (4.6), depending on the desired accuracy. The quantity dt_B/dt_A in equation (4.11) can be computed with several methods. Two different approaches are presented in some detail in Appendix A of Blanchet et al. [9]: a direct integration of the null geodesic equations, and a simpler way which is the differentiation of the time transfer function. This second method is quite powerful: a general method has been developed to calculate the time transfer function as a post-Minkowskian (PM) series, up to any order in G [52, 81]. See for example [35] for the calculation of the one-way frequency shift up to the 2-PM approximation. This method does not require the integration of the null geodesic equations. The frequency shift is expressed as the integral of functions defined from the metric and its derivatives, and performed along a Minkowskian straight line.

Finally, the frequency ratio can be expressed in the GCRS as

$$\frac{\nu_A}{\nu_B} = \frac{1 - \frac{1}{c^2}\left[\frac{\mathbf{v}_B^2}{2} + U_E(\mathbf{x}_B)\right]}{1 - \frac{1}{c^2}\left[\frac{\mathbf{v}_A^2}{2} + U_E(\mathbf{x}_A)\right]} \frac{q_A}{q_B} , \tag{4.12}$$

where, if the desired accuracy is greater than 5×10^{-17},

$$q_A = 1 - \frac{\mathbf{N}_{AB}.\mathbf{v}_A}{c} - \frac{4GM_E}{c^3} \frac{\mathrm{R}_{AB}\mathbf{N}_A.\mathbf{v}_A + (r_A + r_B)\mathbf{N}_{AB}.\mathbf{v}_A}{(r_A + r_B)^2 - \mathrm{R}_{AB}^2} \tag{4.13}$$

$$q_B = 1 - \frac{\mathbf{N}_{AB}.\mathbf{v}_B}{c} - \frac{4GM_E}{c^3} \frac{\mathrm{R}_{AB}\mathbf{N}_A.\mathbf{v}_B - (r_A + r_B)\mathbf{N}_{AB}.\mathbf{v}_B}{(r_A + r_B)^2 - \mathrm{R}_{AB}^2} . \tag{4.14}$$

Note that formulas (4.13) and (4.14) have been obtained assuming that the field of the Earth is spherically symmetric. If an accuracy lower than 5×10^{-17} is required it is necessary to take into account the J_2 terms in the Newtonian potential.

The terms of the order c^{-1} correspond to the relative Doppler effect between the clocks. Terms of the order c^{-2} in equation (4.12) are the classic second-order Doppler effect and gravitational red shift[6]. Terms of the order c^{-3} amount to less than 3.6×10^{-14} for a satellite in Low-Earth orbit and 2.2×10^{-15} for the ground. Terms of the order c^{-4} omitted in equation (4.12) can reach $\sim 5 \times 10^{-19}$.

5 Time and frequency transfer techniques

Time and frequency transfer techniques may be classified in several ways. For example, one may distinguish one-way techniques where all transmitted signals originate from one end of the link and two-way techniques where the two end-points emit (or possibly reflect) at least one signal. One may also distinguish techniques that transmit signals in the radio frequency band from those that transmit in the optical domain. The latter can be further divided among transmission in free space or in a fiber.

In this section, we rather distinguish on the one hand the techniques that are presently (2013) established and have been widely used for years and, on the other hand, some novel techniques that are either already in operation or in a planning stage. The first category (Section 5.1) includes GNSS-based techniques and two-way satellite time and frequency transfer techniques (TWTT for short). In the second category, we present the T2L2 method that is already used and the ACES Microwave link that will soon fly, as well as phase-coherent optical links which offer the best perspectives in terms of performance (Section 5.2).

We will review the chosen techniques in the relativistic background discussed above. Whilst that framework is still valid for the T2L2 method and the ACES MWL, and for fiber or free space optical links for frequency comparisons, it might be necessary to re-examine and possibly expand it for time comparisons at the femtosecond level or below for future fiber or free-space methods.

5.1 Established time and frequency transfer techniques: GNSS and two-way time transfer

These techniques transmit signals in the radio frequency band (of the order 1–10 GHz) and may use the carrier phase or a modulation of the carrier providing a code, or both approaches.

6 One can notice that the separation between a gravitational red shift and a Doppler effect is specific to the chosen coordinate system. One can read the book by Synge [79] for a different interpretation in terms of relative velocity and Doppler effect only.

A first class of such techniques uses signals from navigation satellites, such as the GPS, the Russian GLONASS, the Chinese Beidou or the European GALILEO system, or from dedicated geostationary payloads, to compare the local clock with the clock on board the satellite. In such techniques, each station receives signals from a number of satellites. A pseudo-random code with a bandwidth of the order MHz and stamped by the satellite clock is compared to a locally generated code stamped by the local clock. In addition, the phase of the transmitted signals can also be measured (to within an unknown number of cycles). The comparison between the local and satellite clock is obtained by subtracting the coordinate time of propagation of the signal (see Section 4.1) from the measured time difference between the codes. Code-only techniques are widely used and provide time stability typically below 1 ns when averaging one to a few hours. Because the phase noise is about 100 times lower than the code noise, a much better short term (< 1 day) performance is obtained from phase + code techniques which allow a frequency transfer uncertainty at or slightly below 1×10^{-15} at one day averaging, decreasing as $1/T$ in principle. In all cases, the performance for time transfer is provided by the code and the time transfer accuracy is limited to a few nanoseconds by the capacity to calibrate equipment (in particular the antennas) and by effects affecting the code (such as multipath).

GNSS-based T/F transfer is a mature technique used worldwide. Although some significant improvements may be expected from using a larger number of satellites and measurements, new codes and new processing techniques (e.g. using integer phase ambiguity determination), it is not anticipated that such techniques could reach an uncertainty of the order 10^{-17} in a day, so as to allow comparing optical clocks.

A second class of techniques is two-way time transfer in which signals are sent in both directions over the transmission path and which takes advantage of the reciprocity of the path to cancel or reduce some sources of error so that the measurement uncertainty is, in principle, somewhat reduced.

In the two-way time transfer technique currently used by time laboratories [41], two stations simultaneously transmit a radio signal in the Ku band (11–14 GHz) to a geostationary communications satellite. A transponder on board the satellite retransmits the signals for reception by the stations. In current systems, a pseudorandom code stamped by the local clock is modulated at a few Mbps. The comparison between the two local clocks is obtained by a proper combination of the measured time differences between the codes and the coordinate time of propagation of the signal (see Section 4.2). Operational links show time stability at a few hundred ps or below and a frequency transfer performance at or slightly below 1×10^{-15} at one day averaging. Time transfer accuracy has been shown to be of the order 1 ns with equipment calibration (Piester et al. [66]). Limitations are mostly from constraints in the space segment, either from operational considerations (e.g. cost or set-up changes) or by technical limitations in transponders.

Techniques using phase+code are under study and could in principle improve the TWTT frequency transfer performance by some two orders of magnitude. It is however likely that successful phase+code two-way operation will need a dedicated space payload to overcome the operational constraints of using geostationary communications satellites. Such systems have been designed and hardware is under development but none is in operational use. The prototype in this category is the ACES microwave link due to fly with the ACES mission (see below).

5.2 Some novel two-way techniques

The rapid improvement in the performances of atomic clocks (see Section 2.2) is a tremendous challenge to distant clock comparison methods as it requires that the comparison method contributes noise that is below the level of the clocks within similar averaging times.

As will be shown below, the best present techniques providing time and frequency transfer for ground to space, or ground to ground through a space segment, have a measurement noise level of the order 1–10 ps. Assuming white phase noise in the comparison, a noise level of 10 ps at 1 s averaging time would provide a frequency transfer performance of 1×10^{-17} at 10 000 s averaging time, and below 1×10^{-18} at one day. However, time transfer techniques are generally not ultimately limited by white phase noise so that the actual performance of such a technique should be in the region of 1×10^{-17} at one day and subject to further improvement with longer averaging. Although not fully satisfactory to explore the potential of new atomic clocks, these techniques (T2L2, ACES microwave) will be essential for worldwide comparisons.

On the other hand, comparing the present best clocks in 1 s would require a noise level of about 50 fs at 1 s averaging time, many orders of magnitude below the best present and near future satellite methods. The only existing method satisfying such a requirement is based on two-way comparisons in fiber optic links and is presented further below.

T2L2

The most advanced satellite time transfer system currently in operation is T2L2 (time transfer by laser light) on board the Jason 2 satellite [30]. After its early predecessor LASSO [31], the T2L2 instrument, developed by CNES (Centre National d'Etudes Spatiales) and OCA (Observatoire de la Côte d'Azur), realizes the concept of time transfer based on a pulsed free-space laser link. The principle is derived from Satellite Laser Ranging and relies on the propagation of laser pulses between the clocks to be synchronized. The Jason 2 satellite, carrying the T2L2 instrument, has been successfully launched on June 20th, 2008, to a 1336 km altitude orbit. The T2L2 instrument was

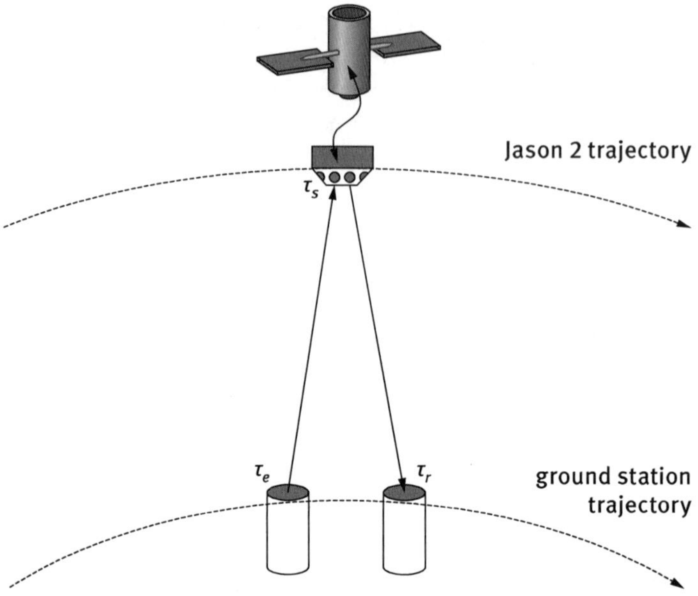

Fig. 3. Schematic of the T2L2 principle of operation (see the text for details).

turned on for the first time a few days later, on June 25th. It is connected to the onboard clock, the DORIS USO (ultra stable oscillator). Given the present good health of the instrument the operation was extended until the end of 2014. Since switch-on the instrument was operational 97% of the time. No aging or degradation of the performance has been observed to date. The operation of the system is based on the exchange of short (50 ps) laser pulses dated at emission and reception on the ground and at reflection at the satellite. The principle of operation is schematically represented in Figure 3. Laser pulses are emitted from the ground telescope, detected and reflected at the satellite, and received on their return to the ground station.

Each pulse is dated on the ground time scale at emission (τ_e) and on return (τ_r) from the satellite and on the satellite time scale at reflection (τ_s). The three events in their respective local proper times (called a triplet) can be used to extract the onboard-to-ground clock desynchronization from the simple relation

$$\tau^g(t_s) - \tau^s(t_s) = \left(\frac{\tau_e + \tau_r}{2} - \tau_s\right) + \Delta_{\mathrm{asym}} + \Delta_{\mathrm{instr}} , \tag{5.1}$$

where the left-hand-side is the desynchronization at coordinate time t_s (signal reflection), Δ_{asym} is a small correction due to the asymmetry of the optical paths on the way up and down (mainly caused by the Earth's rotation during signal propagation) that is derived from equation (4.8), and Δ_{instr} accounts for instrumental delays.

Figure 4 shows the observed performance in terms of the time stability of T2L2. In a ground to space comparison that performance is limited by the timing noise of

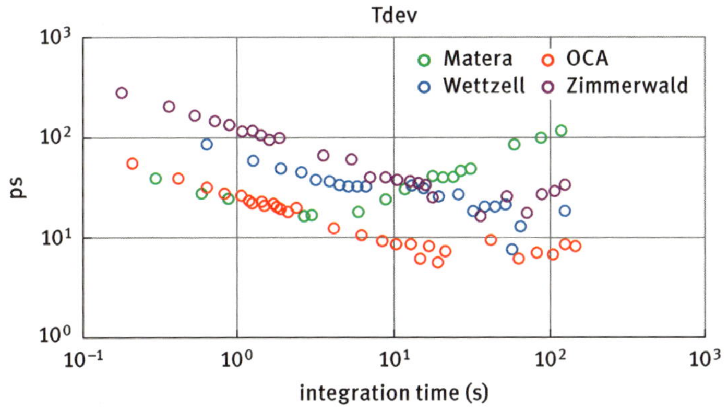

Fig. 4. Time stability of T2L2 ground-space clock comparisons for different ground stations, as a function of averaging time.

the return pulse on the ground (related to the small number of received photons) in the short term and by the on-board oscillator in the longer term ($\tau > 40$ s). At 1 s the averaging time the noise is about 30 ps for the best ground stations.

The ACES microwave link

In March 2016, the atomic clock ensemble in space (ACES) mission will be installed on board the International Space Station (ISS). It will carry two high-performance space clocks, a Swiss space hydrogen maser (SHM) and a cold atom clock (PHARAO) funded by the French space agency (CNES). It will also be equipped with two time transfer systems, the ELT (European Laser Timing) experiment which is very similar to T2L2 and the ACES MWL (MicroWave Link) that we will describe in more detail below. The principle of operation is shown in Figure 5. A signal (with frequency f_1) is generated by the ground station at coordinate time t_1^0. It is emitted by the antenna, after an internal delay, at t_1. After propagation through the atmosphere it reaches the ACES antenna phase center at t_2, and its phase difference with respect to the locally generated signal is measured at time t_2^0. Conversely, f_2 and f_3 frequency signals propagate from ACES to the ground station. The signals are radio signals at carrier frequencies of roughly 13.5 GHz, 14.7 GHz, and 2.2 GHz, respectively. The third low frequency signal is required for correction of the dispersive ionospheric delay. The signals are further modulated by a 100 Mbit/s pseudo random noise code (PRN code) which allows unambiguous determination of the carrier cycle.

The observables are the local proper time difference between a given signal generated from the local clock and the same signal arriving from the distant clock. For

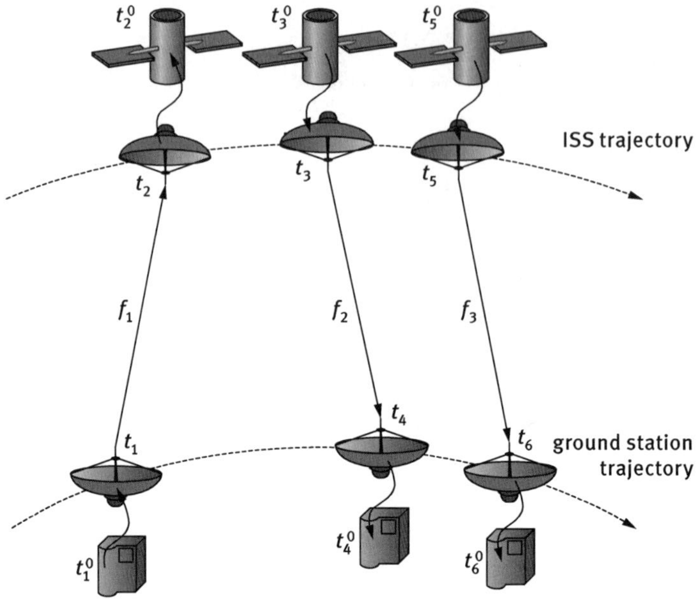

Fig. 5. Principle of operation of the ACES MWL. See the text for details.

example, for the f_1 signal

$$\Delta\tau^s(\tau^s(t_2^0)) = \tau_p^s - \tau_a^s\,, \tag{5.2}$$

where the superscript "s" denotes proper time on the space clock, the measurement is taken at $\tau^s(t_2^0)$ and is the difference in local proper time between the event when a given signal (e.g. a given bit of the PRN code, or a given zero crossing of the carrier signal) was generated by the local clock and when the corresponding signal arrived from the ground clock. Knowing that all clocks generate the same signal at the same local time, this allows the determination of the desynchronization. In principle, a single signal is sufficient to determine the desynchronization from knowledge of the ground and satellite positions and signal delays. However, using two counter-propagating signals has many advantages as numerous effects cancel, at least partially, when the two signal paths are close to each other. This is the basis of the "two-way" principle used in many modern clock comparison methods. For the ACES MWL the desynchronization is then given by

$$\begin{aligned}\tau^s(t_2) - \tau^g(t_2) = \frac{1}{2}\Bigg[&\Delta\tau^g(\tau^g(t_4^0)) + \Delta_{Tx2} + \Delta_{Rx2} - \Delta\tau^s(\tau^s(t_2^0)) - \Delta_{Tx1} - \Delta_{Rx1}\\ &+ \left(1 - \frac{GM}{|\vec{x}_g(t_2)|c^2}\right)(T_{34} - T_{12})\Bigg]\,,\end{aligned} \tag{5.3}$$

where the Δ are instrumental delays (cables, amplifiers, etc.) in the path between the clock and the antenna phase center on emission (Tx) and reception (Rx) and at fre-

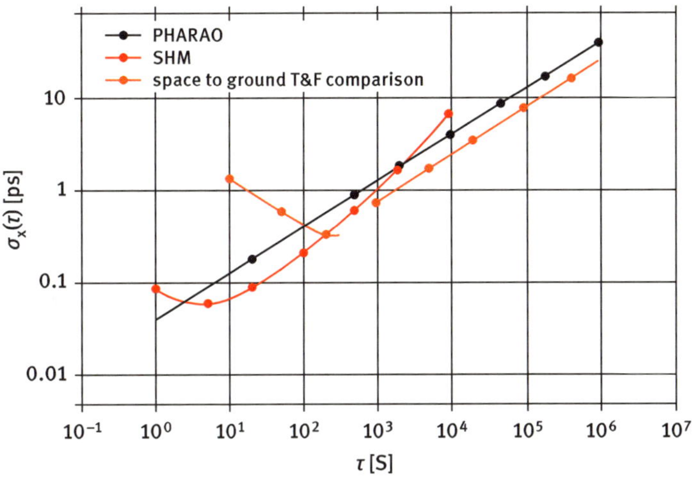

Fig. 6. Performance objective of the ACES clocks (SHM = Space Hydrogen Maser, PHARAO = Cold Atom Cs clock) and the MWL (MicroWave Link) in terms of time stability as a function of averaging time.

quencies f_1 and f_2, respectively. We define $T_{ij} = t_j - t_i$ as the coordinate times of propagation between the antenna phase centers (including atmospheric delays) that need to be calculated as described in Section 4.1. It is obvious from equation (5.3) that in this two-way configuration only the difference $T_{34} - T_{12}$ is required, which means that, e.g. precise knowledge of the satellite and ground positions is not necessary [28] nor is the precise knowledge of atmospheric delays required as long as they are the same on the two trajectories. Concerning relativistic effects, the gravitational delay (Shapiro effect) cancels in that difference, and only a small correction that reflects the transformation of the coordinate time interval $T_{34} - T_{12}$ to proper time of the ground clock is required. The expected performance of the ACES MWL and the onboard clocks is shown in Figure 6.

Note that the expected short term performance is slightly better than the observed performance of T2L2, which will make the ACES MWL the most precise satellite time transfer method in the second part of this decade. On continental scales it is expected that fiber links (see the next section) will outperform the ACES links (MWL and ELT) and T2L2. This will lead to an interesting situation where fiber links can be used to cross-check and validate the ACES links on continental scales, and to link clocks to the ACES ground terminals on the continent. The ACES links can then be used to compare clocks on intercontinental scales. It is indeed one of the core parts of the ACES program to install permanent MWL ground stations in the United States, Japan, Australia and of course in Europe.

Phase-coherent optical links

The optical methods for distant clock comparisons discussed so far all use pulsed optical signals, with the time/frequency transfer based on the timing of emission and reception of short ($\simeq$ 50 ps) laser pulses using single photon detectors. The noise of such links is thus limited at best to the order of the pulse width per pulse (cf. Figure 4). A promising method for improvement of that performance is using continuous laser links with the time/frequency transfer based on direct measurement of the optical carrier phase, or the phase of some high-frequency (several GHz) modulation of the amplitude or phase of that carrier. The ultimate noise limit is then given by photon shot noise and is many orders of magnitude below the best performance of any other comparison method.

The first implementation and by far the best developed method for such links uses optical fibers that are either dedicated, as in Predehl et al. [68], or shared with internet data traffic, as in Lopez et al. [55]. Signals are transmitted bidirectionally, with the two-way configuration used to compensate for the phase noise in the fiber to a large extent. The frequency transfer performance is about 1×10^{-19} at one day. It has so far been demonstrated at distances up to 1000 km [55, 68], but longer distances on continental scales are expected in the near future. Several time transfer techniques have also been adapted to use fiber links, providing an accuracy of around 100 ps [70, 73].

These techniques will be fundamental for frequency and time transfer within a network of sites over continental regions, and such networks are currently being implemented, in particular, in Europe. However, as they require amplifier stations about every 100 km, they are not suited for intercontinental links or, obviously, for space-to-ground or space-to-space links for clocks onboard satellites.

First attempts to extend such methods to continuous phase coherent free-space optical links were started in 2009 (see Djerroud et al. [26]) and been followed up more recently including active two-way compensation [32]. Although, and contrary to fiber links, free-space methods are limited by atmospheric turbulence, it has been demonstrated that the corresponding noise can be reduced by two-way compensation to levels below 1×10^{-18} at 10 000 s, at least over the short distances explored so far (1–2 km). The extension of those methods to ground-satellite distances is under way [20] with support from space agencies and following pioneering work on high bandwidth space to ground optical communications. Whether such long-distance free-space links will be successful is an open question, as the two-way compensation of atmospheric turbulence relies on symmetry between the turbulence induced noise on the up- and downlinks. It is not clear at present whether, and to what extent, such symmetry holds for long distance ground to satellite links.

6 Clocks in relativistic geodesy

Instead of using our knowledge of the Earth gravitational field to predict frequency shifts between distant clocks, one can revert the problem and ask if the measurement of frequency shifts between distant clocks can improve our knowledge of the gravitational field. To do simple orders of magnitude estimates it is good to have in mind some correspondences which can be deduced from equation (4.12):

$$1\ \text{meter} \leftrightarrow \frac{\Delta\nu}{\nu} \sim 10^{-16} \leftrightarrow \Delta W \sim 10\ \text{m}^2\ \text{s}^{-2}\,, \tag{6.1}$$

where 1 m is the height difference between two clocks, $\Delta\nu$ is the frequency difference in a frequency transfer between the same two clocks (see Section 4), and ΔW[7] is the gravity potential difference between the locations of the clocks.

From this correspondence, we can already recognize two direct applications of clocks in geodesy: if we are capable to compare clocks to 10^{-16} accuracy, we can determine height differences between clocks with one meter accuracy (levelling), or determine geopotential differences with $10\ \text{m}^2\ \text{s}^{-2}$ accuracy.

6.1 Review of chronometric geodesy

The first article to explore seriously this possibility was written in 1983 by Martin Vermeer [83]. The article is named "chronometric levelling." The term "chronometric" seems well suited for qualifying the method of using clocks to determine directly gravitational potential differences, as "chronometry" is the science of the measurement of time. However the term "levelling" seems too restrictive with respect to all the applications one could think of using the results of clock comparisons. Therefore, we will use the term "chronometric geodesy" to name the scientific discipline that deals with the measurement and representation of the Earth, including its gravitational field, with the help of atomic clocks. It is sometimes named "clock-based geodesy," or "relativistic geodesy." However, this last designation is improper as relativistic geodesy aims at describing all possible techniques (including, e.g. gravimetry and gradiometry) in a relativistic framework [48, 49, 59, 74, 75, 78]. The natural arena of chronometric geodesy is the four-dimensional spacetime. At the lowest order, there is proportionality between relative frequency shift measurements – corrected from the first-order Doppler effect – and (Newtonian) gravity potential differences. To calculate this relation one does not need the theory of general relativity, but only to postulate local position invariance. Therefore, if the measurement accuracy does not reach the magnitude of the higher

7 This notation should not be mixed with the notation W representing a scalar potential in equation (4.3).

order terms, it is perfectly possible to use clock comparison measurements – corrected for the first-order Doppler effect – as a direct measurement of (differences of) the gravity potential that is considered in classical geodesy. Comparisons between two clocks on the ground generally use a third clock in space. For the comparison between a clock on the ground and one in space, the model given in equations (4.12)–(4.14) must be used where the terms of the order c^{-3} in equations (4.13)–(4.14) reach a magnitude of $\sim 10^{-15}$ and $\sim 3 \times 10^{-14}$ for, respectively, the ground and the space clock, if they are separated radially by 1000 km. Terms of the order c^{-4} omitted in equation (4.12) can reach $\sim 5 \times 10^{-19}$ in relative frequency shift, which corresponds to a height difference of ~ 5 mm and a geopotential difference of $\sim 5 \times 10^{-2}\ \mathrm{m}^2\,\mathrm{s}^{-2}$. Clocks are still somewhat far from reaching this accuracy today, but it cannot be excluded for the future.

In his article, Martin Vermeer explores the "possibilities for technical realization of a system for measuring potential differences over intercontinental distances" using clock comparisons [83]. The two main ingredients are of course accurate clocks and a mean to compare them. He considers hydrogen maser clocks. For the links he considers a two-way satellite link over a geostationary satellite, or GPS receivers in interferometric mode. He has also to consider a mean to compare the proper frequencies of the different hydrogen maser clocks. Today this can be overcome by comparing primary frequency standards (PFS), which have a well-defined proper frequency based on a transition of Caesium 133, used for the definition of the second. Secondary frequency standards, i.e. standards based on a transition other than the defining one, may nevertheless be used if the uncertainty in systematic effects has been fully evaluated, in the same way as for a PFS. It often happens that this evaluation can be done more accurately than for the defining transition. This is one of the purpose of the European project "International time scales with optical clocks"[8], where optical clocks based on different atoms will be compared one each other locally, and to the PFS. It is planned also to do a proof-of-principle experiment of chronometric geodesy, by comparing two optical clocks separated by a height difference of around 1 km using an optical fiber link.

Few authors have seriously considered chronometric geodesy. Following Vermeer idea, Brumberg and Groten [16] demonstrated the possibility of using GPS observations to solve the problem of the determination of geoid heights. They consider two techniques based on frequency comparisons and direct clock readings. However they leave aside the practical feasibility of such techniques. Bondarescu et al. [11] discuss the value and future applicability of chronometric geodesy, including direct geoid mapping on continents and joint gravity-geopotential surveying to invert for subsurface density anomalies. They find that a geoid perturbation caused by a 1.5 km radius sphere with 20 per cent density anomaly buried at 2 km depth in the Earth's crust is

8 http://projects.npl.co.uk/itoc/

already detectable by atomic clocks of achievable accuracy. Finally Chou et al. demonstrated the potentiality of the new generation of atomic clocks, based on optical transitions, to measure heights with a resolution of around 30 cm [21].

6.2 The chronometric geoid

Arne Bjerhammar in 1985 gives a precise definition of the "relativistic geoid" [6, 7]:

> The relativistic geoid is the surface where precise clocks run with the same speed and the surface is nearest to mean sea level.

This is an operational definition. Soffel et al. [74] in 1988 translated this definition in the context of post-Newtonian theory. They also introduce a different operational definition of the relativistic geoid, based on gravimetric measurements: a surface orthogonal everywhere to the direction of the plumb-line and closest to mean sea level. They call the two surfaces obtained with clocks and gravimetric measurements, respectively, the "u-geoid" and the "a-geoid." They prove that these two surfaces coincide in the case of a stationary metric. In order to distinguish the operational definition of the geoid from its theoretical description, it is less ambiguous to give a name based on the particular technique to measure it. The term "relativistic geoid" is too vague as Soffel et al. have defined two different ones. The names chosen by Soffel et al. are not particularly explicit, so instead of "u-geoid" and "a-geoid" one can call them "chronometric geoid" and "gravimetric geoid", respectively. There can be no confusion with the geoid derived from satellite measurements, as this is a quasigeoid that does not coincide with the geoid on the continents [37]. Other considerations on the chronometric geoid can be found in [48, 49, 59].

Let two clocks be at rest with respect to the chosen coordinate system ($v^i = 0$) in an arbitrary spacetime, then

$$\frac{\nu_A}{\nu_B} = \frac{d\tau_B}{d\tau_A} = \frac{[g_{00}]_B^{1/2}}{[g_{00}]_A^{1/2}} . \tag{6.2}$$

In this case the chronometric geoid is defined by the condition g_{00} = const. We notice that the problem of defining a reference surface is closely related to the problem of realizing Terrestrial Time (TT). TT is defined with respect to Geocentric Coordinate Time (TCG) by the relation (3.3), where L_G is a defining constant. This constant has been chosen so that TT coincides with the time given by a clock located on the classical geoid (see Section 3.4). It could be taken as a formal definition of the chronometric geoid [86]. If so, the chronometric geoid will differ in the future from the classical geoid: a level surface of the gravity potential closest to the topographic mean sea level. Indeed, the value of the potential on the geoid, W_0, depends on the global ocean level which changes with time. In addition, there are several methods to realize that the geoid is "closest to the mean sea level" so that there is yet no adopted standard to define a reference geoid and W_0 value (see, e.g. a discussion in Sânchez 2012 [72]). Several authors have considered the time variation of W_0, see, e.g. [17, 25], but there is some uncertainty in what is accounted for in such a linear model. A recent estimate by Dayoub et al. [25] over 1993–2009 is $dW_0/dt = -2.7 \times 10^{-2}\,\mathrm{m^2 s^{-2}\,yr^{-1}}$, mostly driven by the sea level change of +2.9 mm/yr. However, the rate of change of the global ocean level could vary during the next decades, predictions are highly model dependent [40]. Nevertheless, to state an order of magnitude, considering a systematic variation in the sea level of the order 2 mm/yr, different definitions of a reference surface for the gravity potential could yield differences in frequency of the order 2×10^{-18} in a decade, following relation (6.1). Comparisons of accurate clocks could therefore help in the future to establish a worldwide vertical datum.

7 Conclusions and prospects

The relativistic theory presented in this chapter and in the references allows us to perform clock comparisons and the realization of coordinate times with an uncertainty quite below the present performances of atomic clocks. For example, it provides adequate models for frequency comparisons at a level below 1×10^{-18} in the vicinity of the Earth, while the current best reported stability of atomic clocks is $3.2 \times 10^{-16} \tau^{-1/2}$ [36] and the best reported accuracy is 6×10^{-18} [10]. Those performances are however steadily improving, which will have implications in various domains and we recall some of them below as a conclusion.

As clocks accurate to parts in 10^{17} and below now exist, they will be used (when they can be practically operated for that purpose) to provide information on the Earth's gravity potential, effectively becoming devices to measure it. Accurate clocks may then help defining a worldwide reference for the gravity potential, therefore for the height systems. This may be an essential tool in view of the present and foreseen needs of accurate measurements, e.g. of sea surface levels and sea surface topography, linked to global changes.

On the other hand, uncertainties in the knowledge of the gravity potential may create difficulties in merging the world's best clocks to form a reference time scale. To overcome this limitation, some of these ultra-accurate clocks should be placed in space to provide the reference against which the Earth clocks would be compared. First steps in this direction should happen within a few years with the ACES space clock project onboard the ISS.

At the same time, geopotential models will improve too. The recent space missions like GRACE (launched 2002) and GOCE (operated 2009–2013) have already brought significant improvements in geopotential models, particularly on their time variations. The continuous improvements in satellite laser ranging techniques, the continuation of a vigourous program of satellite altimetry missions as well as future space geodesy missions like the GRACE Follow-on will enhance this knowledge. Thus the confrontation between atomic physics and geophysics should take place in the next decade.

Techniques for clock comparisons may be a limiting factor for many years to come, because the only viable and perennial technique capable of reaching the accuracy level of 1×10^{-18} and below (phase-coherent optical link) will be for some time limited to continental links. Furthermore, many promising applications of the next generation clocks are in space (navigation, fundamental physics, geodesy, ...) for which fiber links are obviously unsuited. Even on Earth, no technology is known so far, that would allow extending fiber links over eg. intercontinental distances.

Last, but not least, it will soon be necessary to reconsider the application of the relativistic framework, in the aim of keeping a model accuracy well below the expected performance of the best clocks.

Acknowledgement: Thanks are due to Frédéric Meynadier and Etienne Samain for providing Figures 3, 4, and 5.

References

[1] Z. Altamimi, X. Collilieux and L. Métivier: ITRF2008: an improved solution of the international terrestrial reference frame, Journal of Geodesy, **85**(8), 457–473, 2011.

[2] E. F. Arias, P. Charlot, M. Feissel and J.-F. Lestrade: The extragalactic reference system of the International Earth Rotation Service, ICRS *Astronomy and Astrophysics*, **303**, 604–608, 1995.

[3] E. F. Arias, G. Panfilo and G. Petit: Timescales at the BIPM, *Metrologia*, **48**, S145, 2011.

[4] A. Bauch, S. Weyers, D. Piester, E. Staliuniene and W. Yang: Generation of UTC(PTB) as a fountain-clock based time scale *Metrologia*, **49**, 180, 2012.

[5] S. Bize, Y. Sortais, M. S. Santos, C. Mandache, A. Clairon and C. Salomon: High-accuracy measurement of the 87Rb ground-state hyperfine splitting in an atomic fountain, Europhys. Lett., **45** (5), 558, 1999.

[6] A. Bjerhammar: On a relativistic geodesy, Bulletin Géodésique, **59**(3), 207–220, 1985.

[7] A. Bjerhammar: Relativistic Geodesy, NOAA Technical Report **NON118**, NGS36, 1986.

[8] L. Blanchet and T. Damour: Post-Newtonian generation of gravitational waves, Annales de l'Institut Henri Poincaré, **50**, 377–401, 1989.

[9] L. Blanchet, C. Salomon, P. Teyssandier and P. Wolf: Relativistic theory for time and frequency transfer to order c^{-3}, *Astronomy and Astrophysics*, **370**, 320–329, 2001.

[10] B. J. Bloom, T. L. Nicholson, J. R. Williams, S. L. Campbell, M. Bishof, X. Zhang, W. Zhang, S. L. Bromley and J. Ye: A New Generation of Atomic Clocks: Accuracy and Stability at the 10^{-18} Level, arXiv:1309.1137v1, 2013.

[11] R. Bondarescu, M. Bondarescu, G. Hetényi, L. Boschi, P. Jetzer and J. Balakrishna: Geophysical applicability of atomic clocks: direct continental geoid mapping, Geophysical Journal International, **191**, 78–82, 2012.

[12] V. A. Brumberg and S. M. Kopeikin: Relativistic theory of celestial reference frames, in: *Reference Frames*, J. Kovalevsky, I. I. Mueller, B. Kołaczek (eds.), Kluwer, Dordrecht, 115–141, 1989.

[13] V. A. Brumberg and S. M. Kopeikin: Relativistic time scales in the Solar System, *Celestial Mechanics and Dynamical Astronomy*, **48**, 23–44, 1990.

[14] V. A. Brumberg: *Essential Relativistic Celestial Mechanics*, Adam Hilder, Bristol, 1991.

[15] V. A. Brumberg and E. Groten: IAU resolutions on reference systems and time scales in practice, *Astronomy and Astrophysics*, **367**(3), 1070–1077, 2002.

[16] V. A. Brumberg and E. Groten: On determination of heights by using terrestrial clocks and GPS signals, *Journal of Geodesy*, **76**, 49–54, 2002.

[17] M. Burša, S. Kenyon, J. Kouba, Z. Šíma,V. Vatrt, V. Vítek and M. Vojtíšková: The geopotential value W_0 for specifying the relativistic atomic time scale and a global vertical reference system, *Journal of Geodesy*, **81**, 103–110, 2007.

[18] E. A. Burt, W. A. Diener, R. L. Tjoelker: A compensated multi-pole linear ion trap mercury frequency standard for ultra-stable timekeeping, IEEE Trans. UFFC, **55**(12), 2586–2595, 2008.

[19] E. Burt, S. Taghavi-Larigani and R. Tjoelker: A new trapped ion atomic clock based on 201Hg+, IEEE Trans. UFFC, 57(3), 2010.

[20] N. Chiodo, K. Djerroud, O. Acef, A. Clairon and P. Wolf: Lasers for coherent optical satellite links with large dynamics, Applied Optics, **52**, 7342, 2013.

[21] C. W. Chou, D. B. Hume, T. Rosenband and D. J. Wineland: Optical clocks and relativity, Science, **329**, 1630–1633, 2010.

[22] C. W. Chou, D. B. Hume, J. C. J. Koelemeij, D. J. Wineland and T. Rosenband: Frequency comparison of two high-accuracy Al+ optical clocks, Physics Review Letters, **104**, 070802, 2010.

[23] A. Clairon, C. Salomon, S. Guellati and W. D. Phillips: Ramsey resonance in a Zacharias fountain, Europhys. Lett., **16**, 165–170, 1991.

[24] J. A. Davis and J. McA. Steele (1997): A caesium flying clock experiment between NPL and USNO, *Proc. 11th European Frequency and Time Forum*, 306, 1997.

[25] N. Dayoub, S. J. Edwards and P. Moore: The Gauss–Listing geopotential value W0 and its rate from altimetric mean sea level and GRACE, *Journal of Geodesy*, **86**(9), 681–694, 2012.

[26] K. Djerroud et al.: Coherent optical link through the turbulent atmosphere, Opt. Lett., **35**, 1479, 2010.

[27] T. Damour, M. Soffel and C. Xu: General relativistic celestial mechanics I. Method and definition of reference systems, Physical Review D, **43**, 3273–3307, 1991.

[28] L. Duchayne, F. Mercier and P. Wolf: Orbit determination for next generation space clocks, *Astronomy and Astrophysics*, **504**, 653–661, 2009.

[29] A. Einstein: Zur Elektrodynamik bewegter Körper, Annalen der Physik, **17**, 891–921, 1905.

[30] P. Exertier, E. Samain, P. Bonnefond and P. H. Guillemot: Status of the T2L2/Jason2 Experiment, Advances in Space Research, **46**, 1559–1565, 2010.

[31] P. Fridelance and C. Veillet: Operation and data analysis in the LASSO experiment, *Metrologia*, **32**, 27–33, 1995.

[32] F. R. Giorgetta et al.: Optical two-way time and frequency transfer over free space, Nature Photonics, **7**, 434–438, 2013.

[33] J. Guéna, M. Abgrall, D. Rovera, P. Laurent, B. Chupin, M. Lours, G. Santarelli, P. Rosenbusch, M. E. Tobar, R. X. Li, K. Gibble, A. Clairon and S. Bize: Progress in atomic fountains at LNE-SYRTE, IEEE Trans. UFFC, **59**, 391–410, 2012.

[34] W. Harada and T. Fukushima: Harmonic decomposition of time ephemeris TE405, The Astronomical Journal, **126**(5), 2557–2561, 2003.

[35] A. Hees, S. Bertone, C. Le Poncin-Lafitte: Frequency shift up to the 2-PM approximation in SF2A-2012, in: *Proceedings of the Annual meeting of the French Society of Astronomy and Astrophysics*, S. Boissier, P. de Laverny, N. Nardetto, R. Samadi, D. Valls-Gabaud and H. Wozniak (eds.), 145–148, 2012.

[36] N. Hinkley, J. A. Sherman, N. B. Phillips, M. Schioppo, N. D. Lemke, K. Beloy, M. Pizzocaro, C. W. Oates, A. D. Ludlow: An atomic clock with 10^{-18} instability, Science, **341**(6151), 1215–1218, 2013.
[37] B. Hofmann-Wellenhof and H. Moritz: *Physical Geodesy*, Springer, 2006.
[38] N. Huntemann, M. Okhapkin, B. Lipphardt, S. Weyers, Ch. Tamm and E. Peik: High-accuracy optical clock based on the octupole transition in $^{171}Yb^{+}$, Physics Review Letters, **108**(9), 090801, 2012.
[39] G. Petit and B. Luzum (eds.), IERS Conventions, International Earth Rotation Service, Technical Note 36, 2010.
[40] S. Jevrejeva, J. C. Moore and A. Grinsted: Sea level projections to AD2500 with a new generation of climate change scenarios, Global and Planetary Change, **80**, 14–20, 2012.
[41] D. Kirchner: Two-way time transfer via communication satellites, Proc. IEEE, **79**(7), 983–990, 1991.
[42] S. A. Klioner: The problem of clock synchronization: a relativistic approach, *Celestial Mechanics and Dynamical Astronomy*, **53**, 81–109, 1992.
[43] S. A. Klioner and A. V. Voinov: Relativistic theory of reference systems in closed form, Physical Review D, **48**, 1451–1461, 1993.
[44] S. A. Klioner and T. Fukushima: Relativistic effects in two-way time transfer via artificial satellites using laser techniques, Manuscripta Geodaetica, **19**(5), 294–299, 1994.
[45] S. A. Klioner and M. Soffel: Nonrotating astronomical relativistic reference frames, *Astronomy and Astrophysics*, **334**, 1123–1135, 1998.
[46] S. A. Klioner and M. Soffel: Relativistic celestial mechanics with PPN parameters, Physical Review D, **62**, ID 024019, 2000.
[47] S. M. Kopeikin: Celestial Coordinate Reference Systems in Curved Space-Time, *Celestial Mechanics and Dynamical Astronomy*, **44**, 87–115, 1988.
[48] S. M. Kopeikin: Relativistic Manifestations of gravitational fields in gravimetry and geodesy, Manuscripta Geodaetica, **16**, 301–312, 1991.
[49] S. M. Kopeikin, M. Efroimsky and G. Kaplan: *Relativistic Celestial Mechanics of the Solar System*, Wiley, 2011.
[50] N. D. Lemke, A. D. Ludlow, Z. W. Barber, T. M. Fortier, S. A. Diddams, Y. Jiang, S. R. Jefferts, T. P. Heavner, T. E. Parker and C. W. Oates: Spin-1/2 Optical Lattice Clock, Physics Review Letters, **103**(6), 063001, 2009.
[51] P. Lemonde et al.: Cold-Atom Clocks on Earth and in Space, in: *Frequency Measurement and Control*, A. N. Luiten (ed.), Springer, Berlin, 2001.
[52] C. Leponcin Lafitte, B. Linet and P. Teyssandier: World function and time transfer: general post-Minkowskian expansions, Class. Quantum Grav., **21**, 4463–4483, 2004.
[53] R. Le Targat, L. Lorini, Y. Le Coq, M. Zawada, J. Guéna, M. Abgrall, M. Gurov, P. Rosenbusch, D. G. Rovera, B. Nagórny, R. Gartman, P. G. Westergaard, M. E. Tobar, M. Lours, G. Santarelli, A. Clairon, S. Bize, P. Laurent, P. Lemonde and J. Lodewyck: Experimental realization of an optical second with strontium lattice clocks, Nature Communications, **4**, 2109, 2013.
[54] R. Li, K. Gibble and K. Szymaniec: Improved accuracy of the NPL-CsF2 primary frequency standard: evaluation of distributed cavity phase and microwave lensing frequency shifts, *Metrologia*, **48**, 283–289, 2011.
[55] O. Lopez, A. Haboucha, B. Chanteau, C. Chardonnet, A. Amy-Klein and G. Santarelli: Ultrastable long distance opticalfrequency distribution using the Internet fiber network, Opt. Express, **20**, 23 518–23 526, 2012.
[56] A. A. Madej, P. Dubé, Z. Zhou, J. E. Bernard and M. Gertsvolf: $^{88}Sr^{+}$ 445-THz single-ion reference at the 10^{-17} level via control and cancellation of systematic uncertainties and its measurement against the SI second, Physics Review Letters, **109**(20), 203002, 2012.

[57] J. J. McFerran, L. Yi, S. Mejri, S. Di Manno, W. Zhang, J. Guéna, Y. Le Coq and S. Bize: Neutral atom frequency reference in the deep ultraviolet with fractional uncertainty = 5.7×10^{-15}, Physics Review Letters, **108**(18), 183004, 2012.

[58] C. Misner, K. Thorne, J. A. Wheeler: *Gravitation*, Freeman, San Francisco, 1973.

[59] J. Müller, M. Soffel, S. A. Klioner: Geodesy and relativity, *Journal of Geodesy*, **82**, 133–145, 2008.

[60] T. L. Nicholson, M. J. Martin, J. R. Williams, B. J. Bloom, M. Bishof, M. D. Swallows, S. L. Campbell and J. Ye: Comparison of two independent Sr optical clocks with 1×10^{-17} stability at 10^3 s, Physics Review Letters, **109**(23), 230801, 2012.

[61] S. Peil, S. Crane, J. Hanssen, T. Swanson and C. Ekstrom: Measurements with Multiple Operational Fountain Clocks, *Proc. 2011 Joint Conference of the IEEE FCS and EFTF*, 2011.

[62] G. Petit and P. Wolf: Relativistic theory for picosecond time transfer in the vicinity of the Earth, *Astronomy and Astrophysics*, **286**, 971–977, 1994.

[63] G. Petit and P. Wolf: Computation of the relativistic rate shift of a frequency standard, IEEE Trans. IM, **46**(2), 201–204, 1997.

[64] G. Petit: A new realization of Terrestrial Time, *Proc. 35th Precise Time and Time Interval (PTTI) Meeting (San Diego, CA)*, 307–316, 2003.

[65] G. Petit and P. Wolf: Relativistic theory for time comparisons: A review, *Metrologia*, **42**, S138–144, 2005.

[66] D. Piester, A. Bauch, J. Becker, T. Polewka, A. McKinley and D. Matsakis: Time transfer between USNO and PTB: Operation and calibration results. *Proc. 35th Precise Time and Time Interval (PTTI) Meeting (San Diego, CA)*, 93–101, 2004.

[67] H. P. Plag, Z. Altamimi, S. Bettadpur, G. Beutler, G. Beyerle, A. Cazenave, D. Crossley, A. Donnellan, R. Forsberg, R. Gross et al.: The goals, achievements, and tools of modern geodesy, in: *Global Geodetic Observing System*, 15–88, Springer-Verlag Berlin Heidelberg, 2009.

[68] K. Predehl, G. Grosche, S. M. F. Raupach, S. Droste, O. Terra, J. Alnis, Th. Legero, T. W. Hänsch, Th. Udem, R. Holzwarth and H. Schnatz: A 920-Kilometer Optical Fiber Link for Frequency Metrology at the 19th decimal place, Science, **336**, 441–444, 2012.

[69] T. Rosenband, D. B. Hume, P. O. Schmidt, C. W. Chou, A. Brusch, L. Lorini, W. H. Oskay, R. E. Drullinger, T. M. Fortier, J. E. Stalnaker, S. A. Diddams, W. C. Swann, N. R. Newbury, W. M. Itano, D. J. Wineland and J. C. Bergquist: Frequency Ratio of Al^+ and Hg^+ Single-ion optical clocks; Metrology at the 17th decimal place, Science, **319**(5871), 1808–1812, 2008.

[70] M. Rost, D. Piester, W. Yang, T. Feldmann, T. Wübbena and A. Bauch: Time transfer through optical fibres over a distance of 73 km with an uncertainty below 100 ps, *Metrologia*, **49**, 772–782, 2012.

[71] C. Salomon, L. Cacciapuoti and N. Dimarcq: Atomic Clock Ensemble in Space: An Update, Int. J. Mod. Phys., **D16**, 2511, 2007.

[72] L. Sánchez: Towards a vertical datum standardisation under the umbrella of Global Geodetic Observing System, J. Geodetic Sci., **2**(4), 325–342, 2012.

[73] L. Sliwczynski, P. Krehlik, A. Czubla, L. Buczek and M. Lipinski: Dissemination of time and RF frequency via a stabilized fibre optic link over a distance of 420 km, *Metrologia*, **50**, 133–145, 2013.

[74] M. Soffel, H. Herold, H. Ruder and M. Schneider: Relativistic theory of gravimetric measurements and definition of the geoid. Manuscr. Geod., **13**, 143–146, 1988.

[75] M. Soffel: *Relativity in Astrometry, Celestial Mechanics and Geodesy*, Springer-Verlag, Berlin, Heidelberg, New York, 1989.

[76] M. H. Soffel and V. A. Brumberg: Relativistic reference frames including time scales: Questions and answers, *Celestial Mechanics and Dynamical Astronomy*, **52**(4), 355–373, 1991.

[77] M. Soffel, S. A. Klioner, G. Petit, P. Wolf, S. M. Kopeikin, P. Bretagnon, V. A. Brumberg, N. Capitaine, T. Damour, T. Fukushima, B. Guinot, T. Huang, L. Lindegren, C. Ma, K. Nordtvedt, J. Ries, P. K. Seidelmann, D. Vokrouhlicky, C. Will and C. Xu: The IAU2000 resolutions for astrometry, celestial mechanics and metrology in the relativistic framework: explanatory supplement, The Astronomical Journal, **126**(6), 2687–2706, 2003.

[78] M. Soffel and R. Langhans: *Space-Time Reference Systems*, Springer-Verlag, Berlin, Heidelberg, 2013.

[79] J. L. Synge: *Relativity: The General Theory*, North-Holland Publishing Company, Amsterdam, the Netherlands, 1960.

[80] M. Takamoto, F. Hong, R. Higashi and H. Katori: An optical lattice clock, Nature, **435**, 321–324, 2005.

[81] P. Teyssandier, C. Le Poncin-Lafitte: General post-Minkowskian expansion of time transfer functions, *Class. Quantum Grav.*, **25**, 145020, 2008.

[82] T. Udem and F. Riehle: Frequency combs applications and optical frequency standards, in: *Proceedings of the International School of Physics "Enrico Fermi" Course CLXVI "Metrology and Fundamental Constants"*, T. W. Hänsch, S. Leschiutta, A. J. Wallard and M. L. Rastello (eds.), IOS Press, Amsterdam and SIF, Bologna, 317–365, 2007.

[83] M. Vermeer: Chronometric levelling, in: *Reports from the Finnish Geodetic Institute*, Helsinki, 1983.

[84] G. A. Wilkins: *Polar Motion: Historical and Scientific Problems*, ASP Conference Series, vol. 208, S. Dick, D. D. McCarthy and B. Luzum (eds.), 2000.

[85] C. M. Will: *Theory and Experiment in Gravitational Physics*, Cambridge University Press, Cambridge, 1993.

[86] P. Wolf and G. Petit: Relativistic theory for clock syntonization and the realization of geocentric coordinate times, *Astronomy and Astrophysics*, **304**, 654–661, 1995.

[87] J. Ye and S. T. Cundiff: *Femtosecond Optical Frequency Comb: Principle, Operation and Applications*, Springer-Verlag, Berlin, Heidelberg, New York, 2004.

Index

A

ACES 251, 254, 261, 271
active galactic nuclei 160, 162
addition theorem 189, 197, 198, 201, 202
AGN 160
albedo 176, 178, 186
amplitude 48, 52, 63, 69, 177, 188, 196, 212, 237, 260, 274
approximation to a geodesic 175
argument 188–192, 195, 197–199, 201, 202, 213–218, 220, 221
associate complete elliptic integral 189
associate incomplete elliptic integral 189, 206
asteroids (in ephemeris computation) 136
astrometry 4, 5, 41, 43, 242, 257
atmospheric drag 175, 176, 178
atomic clocks 41, 249, 251–253, 256, 269, 275, 277, 278
atomic time 250–252
Avio 169

B

BCRS 107–109, 134, 149, 241, 242, 245, 256–258
BK formalism 250, 251
Brumberg V A 2, 137, 227–231, 233, 234, 240–245, 250, 251, 276
Bulirsch 209, 210, 213, 222
Bulirsch's form of complete elliptic integral 188
Bulirsch's form of incomplete elliptic integral 189

C

canonical 65, 187, 213
Carlson 211–213, 221, 222
Carlson's symmetric form of elliptic integral 189
characteristic 189–192, 199, 203, 210
Chern–Simons constant m_{CS} 164
Chern–Simons gravity 162
Chern–Simons parameter m_{CS} 163
chronometric geodesy 275, 276
complementary characteristic 189, 210
complementary modulus 189
complementary parameter 189, 195
complete elliptic integral 188–191, 202, 207–212, 218–221
complete elliptic integral of the first kind 188–190, 210, 218
complete elliptic integral of the second kind 188, 189, 219
complete elliptic integral of the third kind 188–190, 219, 220
coordinate systems 2, 118, 241, 242–243, 251, 255, 256
coordinate transformation 2, 108, 120
cosine amplitude function 188, 197
Cotton–York tensor 162
Cube Corner Reflector (CCR) 173

D

dark energy 157, 158, 162, 163
dark matter 137, 144, 157, 158
de Sitter 39, 146, 164
deflection of light 1, 32
delta amplitude function 189, 197
direction of light propagation 1, 27
Doppler tracking 1
double argument transformation 215–217, 221
drag-free satellite 175
Dragging of inertial frames 157, 161, 165, 169, 173, 177, 179, 181
DSX formalism 251
duplication 188, 195

E

Earth orientation 108, 139, 150
Earth's lowest degree even zonal harmonics 178
Earth–Yarkovsky effect 176
Ehlers–Geroch theorem 175
EIGEN-GRACE02S 168, 170, 179, 180
Einstein 39, 44, 157–159, 255
– general relativity 6, 7, 13, 39, 54, 106, 111, 113, 139, 142, 145–147, 149, 151, 157–164, 168, 169, 174–177, 181, 241, 242, 244–247, 249, 255, 275
– summation rule 162
– tensor 162, 175
– theory of gravitation 160
Einstein–Hilbert action 162
Einstein–Infeld–Hoffmann 45, 51, 107, 111, 137
element 116, 117, 123, 130, 133, 187, 234

ellipse 187, 203
elliptic function 189, 192, 193, 198, 200
elliptic integral 187–192, 199, 200, 202, 203, 206, 208, 217, 220, 235
ephemerides 106, 109, 112, 114, 117, 227, 231, 241, 242, 244–247
EPOS-OC 168, 169, 176, 180
equations of motion 44–47, 50, 68, 136, 140, 143, 145–147, 232, 239, 250
equivalence principle 39, 61, 73, 79, 85, 89, 106, 136, 141–144, 148, 151, 152, 157, 159, 264
Ernst Mach 159, 164
Euler angles 117–120, 123, 125, 129, 137, 139
European Space Agency (ESA) 169
even zonal harmonics 166, 168, 178–180

F

$f(R)$ gravity theory 158
frame dragging 40, 159–162, 164–169, 177–181
– measurement 164
– measurement error 179
frequency comparisons 251, 254, 260, 267, 276, 278
frequency shift 1, 3, 36, 39, 264, 266, 275, 276
frequency transfer 261, 262, 267–269, 274, 275
Friedmann equation 163

G

GCRS 108, 109, 134, 242, 245, 256, 257, 260–262, 265, 266
general relativity
– analogy with electrodynamics 160
– post-Newtonian corrections 176
geodesic 2–9, 18, 20, 23, 30, 42, 170, 174, 175, 177, 242, 266
geodesic motion 170, 174, 177
geodetic precession 46, 55, 69, 70–73, 82, 136, 146–148, 159, 164
GEODYN 167, 169, 176, 181
GEODYN II 176
geoid – chronometric 277
geoid model
– EIGEN 166, 168, 170, 179
– EIGEN-6C2 166
geoid-gravimetric 277
GFZ 166, 168, 170, 176
GGM02S 168, 179
global navigation satellite systems 158, 249
Global Positioning System 170
GLONASS 268
GNSS 158, 249, 267, 268
GOCE 166, 180
GP-B 160
GPS 167, 249, 260, 263, 268, 276
GRACE 167–170, 176, 181
gravitational constant, time variation 85, 140
gravitomagnetic field 161
gravitomagnetism 149, 159, 160, 161
gravity field 105, 107, 108, 115, 120, 121, 131, 139, 150, 166
Gravity Probe B 40, 73, 160, 164, 165, 175, 181
gyroscope 40, 73, 159, 160, 161, 164

H

half-argument transformation 189, 214, 217, 218, 220
hyperbola 203, 205

I

ILRS 152, 169
impact parameter of a light ray 18, 19, 21, 27, 31
incomplete elliptic integral 188–191, 199, 200, 202, 203, 206
incomplete elliptic integral of the first kind 189, 199, 217
incomplete elliptic integral of the second kind 187, 189
incomplete elliptic integral of the third kind 188–191, 199, 200, 202, 206
inertial frame 115, 117–120, 126, 134, 139, 159, 160, 264, 265
input argument 190
Italian Space Agency (ASI) 169, 182

J

Jacobi 196, 197, 199, 202
Jacobian elliptic function 200
Jacobi's amplitude function 188, 195–197, 212, 214
Jacobi's delta function 190, 196
Jacobi's epsilon function 190, 199, 200
Jacobi's form of complete elliptic integral 190, 220
Jacobi's form of incomplete elliptic integral 188–190, 199–202, 216
Jacobi's nome 189, 219

K
Kepler 48, 188, 232, 237

L
LAGEOS 160, 162–173, 176–181
LAGEOS 2 165, 167–170, 172–181, 184
LARES 160, 162–164, 167, 169–182
– geodesic motion 170, 174, 177
LARES space experiment 170, 181
laser-ranged satellites 165
Legendre 190, 203
Legendre's form of complete elliptic integral 189, 190, 211, 218, 221
Legendre's form of elliptic integral 189, 211
Legendre's form of incomplete elliptic integral 187, 189, 190, 199, 203, 217
Lense and Thirring 164
Lense–Thirring effect 36, 40, 160, 164, 165, 167–169, 179
light deflection 159
light direction triple 5, 31
light ray 1–3, 7, 9, 11, 13, 15, 18–20, 22, 27, 30, 31, 34, 36
light travel time 1, 3, 4, 9, 15, 18, 23, 27, 33–35
LLR 62, 74, 79, 80, 85, 87, 103, 109, 114, 127, 131, 134–146, 152, 159, 242, 246, 249
LLR analysis 105, 108, 114, 127, 131, 136, 138, 142, 146, 149
local inertial frames 159
longitude of the nodal line 160
longitude of the pericentre 160
Loop Quantum Gravity 158, 162
Lunar gravity field 139
Lunar interior 140
Lunar Laser Ranging 39, 151, *see also* LLR
Lunar orbit 139
Lunar secular tidal acceleration 115, 122
Lunar tidal parameters 139

M
Mach's principle 159
Mass of the Earth–Moon system 139
modular angle 190, 191
modulus 189, 190, 195, 209
Monte Carlo simulations 82, 179–181
Monte Carlo simulations of the LARES experiment 170, 180

N
NASA 165, 167, 182, 253
NASA Goddard 169, 176
no-hair theorem 91
nongravitational perturbations 170, 173, 175–178, 180, 181
normal points 104, 105, 245
null geodesic 2, 3, 5, 7–9, 18, 20, 30, 42, 266
null geodesic equations 2, 3, 9, 266

O
optical fiber links 267, 269, 273, 274, 276, 279
orbital dynamics 44, 56, 57, 75, 90, 187, 188
orbiting test particle 160, 176
origin of inertia 159
out-of-plane residual acceleration 177

P
parameter 5, 15, 49, 57, 74, 80, 90, 109, 120, 140, 146, 162, 189, 195, 215, 219
– affine 8, 18, 23
– characteristic 203
– dissipation 139
– estimation 135, 138, 145, 151
– free 83
– impact 18, 19, 21, 27, 31
– Lense-Thirring 168
– Nordtvedt 74, 135, 142, 144
– post-Newtonian 33, 35, 36
– post-Keplerian 48, 50–52, 55, 56, 59
– preferred frame 149
– small 233
– space 66
– strong equivalence principle violation 148
– timing 81
– Whitehead 84
– Yukawa 146
Pontryagin density 162
Pontryagin pseudoinvariant 161
post-Newtonian parameters 2, 7, 159
potential 117, 121, 130, 135, 141, 256, 269
– coefficients 122, 124, 128–131, 135
– differences 275, 276
– disturbing 122, 124
– Earth's 262, 278
– energy 130
– external 256, 262
– galactic 64, 83, 84
– geoid 277

– gravitational 40, 54, 90, 115, 117, 120, 121, 130, 166, 167, 251, 256, 265, 275
– gravitomagnetic 161
– gravity 254, 258, 275–278
– metric 7
– Newtonian 243, 262, 266
– perturbing 136
– post-Newtonian 257
– scalar 257, 261, 275
– tidal 121
– total 116
– variation 127
– vector 257, 259, 261
– Yukawa 144, 146
PPN parameter β 111, 147
PPN parameter γ 34, 36, 48, 52, 148
PPN parameter α_1 149
PPN parameter α_2 149
Preferred-frame effects 149

Q
quadrupole moment 167, 168
quadrupole moment J_2 35, 168
quadrupole moment of black hole 91
quadrupole moment of the Earth 166, 167
quasar 160

R
redshift gravitational 39, 48
redshift and clock measurements 159
reference system 111, 250, 255
– accelerated 264
– barycentric 242, 256
– celestial 250, 251, 256
– geocentric 242, 256
– in LLR analysis 108
– space-time 255
– theoretically defined 112
relativistic parameters from LLR 106, 137, 139–141, 151
Ricci scalar 162, 163
Riemann curvature tensor 161
rigid body 116, 119, 124, 131, 187, 191, 205
rotating supermassive black hole 160
rotational dynamics 187, 188, 197
rotational motion 117, 139, 245
Runge–Lenz vector 160

S
Satellite Laser Ranging 103, 157, 158, 269, 278
Schwarzschild metric 2, 6
secular rate of change of the two largest even zonal harmonics 180
Shapiro time delay 2, 14, 107, 159, 263
SLR 103, 151, 158, 249
solar radiation pressure 176, 178
Solar System barycenter 111, 136
Solar System experiments 1, 3, 34, 36
spacetime 4, 6, 40, 42, 58, 91, 160
– binary pulsar 41, 91
– coordinate system 255, 256, 260
– curvature 39, 157, 158, 160, 161, 174
– deformation 109
– distortion 170
– dragging 91
– dynamical 41
– geodesic 174, 175, 177
– geometry 175
– metric 40
– Minkowski 40
– relativistic 249
– Schwarzschild 6
– singularity 158
– Solar System 40
– spherically-symmetric 18, 35
– static 4, 18
– weakly curved 40
spherically symmetric metric 4
spin axis of a test gyroscope 161
standard deviation 138, 141
standard deviation of the 100 simulated measurements of frame dragging 181
STARLETTE 176, 177
string theory 158, 162
strong equivalence principle 39, 61, 73, 79, 85, 89, 106, 142, 144, 148, 151
supernovae 158
synchronization 254–256, 260, 262–264
synchronization of clocks 160, 255, 264
syntonization 260, 262

T
T2L2 251, 269–271, 273
– ground-space clock comparisons 271
– instrument 269
– method 267

– principle of operation 270
– technique 264
TAI 107, 135, 252, 253, 255, 257, 258
temporal variations of the Earth's gravitational field 178
terrestrial radiation pressure 176
terrestrial tides 178
tesseral tide 178
test particle 49, 160, 164, 174–177, 181
tests of frame dragging 181
The LARES Satellite 170, 174
thermal thrust 170, 176, 178
tidal effects 105, 121, 122, 124, 178
tidal lag angle 122, 124, 136, 139, 141, 245
time delay of electromagnetic waves 159
time scales 41, 80, 82, 85, 107, 113, 134, 241, 244, 249–253, 255, 276
time transfer 3, 251, 253, 254, 256, 262–264, 267–269, 271, 274
time transfer function 1, 2, 5, 6, 8, 11, 19–21, 26, 35, 266
timelike geodesic 174
torque-free 159, 160, 187, 188, 205
TT(BIPM) 253, 258

U
University of Texas at Austin 157, 158, 169, 176, 181
UT/CSR 176
UTOPIA 168, 169, 176, 181

V
VEGA 170, 174, 178
VEGA launcher 169
Very Long Baseline Interferometry 158
VLBI 158, 241, 242, 246, 249

W
weak equivalence principle 73, 142, 144
Weierstrass 189, 193
world function 2

Y
Yarkovsky effect 176
Yarkovsky–Rubincam effect 176
Yukawa potential 144, 146

Z
zonal harmonic of degree four 169

De Gruyter Studies in Mathematical Physics

Volume 21
Sergei M. Kopeikin (Ed.)
Frontiers in Relativistic Celestial Mechanics, Volume 1: Theory, 2014
ISBN 978-3-11-033747-1, e-ISBN 978-3-11-033749-5, Set-ISBN 978-3-11-035932-9

Volume 20
Valery A. Slaev, Anna G. Chunovkina, Leonid A. Mironovsky
Metrology and Theory of Measurement, 2013
ISBN 978-3-11-028473-7, e-ISBN 978-3-11-028482-9, Set-ISBN 978-3-11-028483-6

Volume 19
Edward A. Bormashenko
Wetting of Real Surfaces, 2013
ISBN 978-3-11-025853-0, e-ISBN 978-3-11-025879-0, Set-ISBN 978-3-11-220375-0

Volume 18
Michael V. Sadovskii
Statistical Physics, 2012
ISBN 978-3-11-027031-0, e-ISBN 978-3-11-027037-2, Set-ISBN 978-3-11-027038-9

Volume 17
Michael V. Sadovskii
Quantum Field Theory, 2013
ISBN 978-3-11-027029-7, e-ISBN 978-3-11-027035-8, Set-ISBN 978-3-11-027036-5

Volume 16
Alexander S. Holevo
Quantum Systems, Channels, Information: A Mathematical Introduction, 2012
ISBN 978-3-11-027325-0, e-ISBN 978-3-11-027340-3, Set-ISBN 978-3-11-027341-0

Volume 15
Vladimir N. Kukudzhanov
Numerical Continuum Mechanics, 2012
ISBN 978-3-11-027322-9, e-ISBN 978-3-11-027338-0, Set-ISBN 978-3-11-027339-7